THEISM, ATHEISM, AND
BIG BANG COSMOLOGY

Theism, Atheism, and Big Bang Cosmology

*

William Lane Craig
and
Quentin Smith

CLARENDON PRESS · OXFORD
1993

Oxford University Press, Walton Street, Oxford OX2 6DP
Oxford New York Toronto
Delhi Bombay Calcutta Madras Karachi
Kuala Lumpur Singapore Hong Kong Tokyo
Nairobi Dar es Salaam Cape Town
Melbourne Auckland Madrid
and associated companies in
Berlin Ibadan

Oxford is a trade mark of Oxford University Press

Published in the United States
by Oxford University Press Inc., New York

British Library Cataloguing in Publication Data
Data available

Library of Congress Cataloging in Publication Data
Craig, William Lane
Theism, atheism, and big bang cosmology
William Lane Craig and Quentin Smith.
Includes bibliographical references.
1. God—Proof, Cosmological. 2. God—Proof, Cosmological—
Controversial literature. 3. Big bang theory. I. Smith, Quentin,
1952– . II. Title.
BT102.C73 1993 215'.2—dc20 93–18830
ISBN 0–19–826348–1

Typeset by Best-set Typesetter Ltd., Hong Kong
Printed in Great Britain
on acid-free paper by
Biddles Ltd.,
Guildford and King's Lynn

Preface

In recent years, despite the general interest in the subject, there has been very little technical philosophical discussion of the metaphysical and, specifically, theistic implications of Big Bang cosmology. Most of the published discussions of these implications are found in popular science books written by physicists, which often lack the philosophical sophistication that philosophers normally seek in their work. On the other hand, the discussions of theism and atheism in contemporary journals and books written by philosophers almost always proceed in a way that takes no cognizance of Big Bang cosmology. This constitutes a surprising gap in the philosophical literature, considering that the question 'Did God create the Big Bang?' has obvious relevance to the philosophy of religion and has even become a commonplace question among members of the general public. The present book attempts to fill this gap to some extent by presenting a philosophically educated debate between a defender of the theist viewpoint (Craig) and a defender of the atheist viewpoint (Smith) regarding the implications of classical and quantum Big Bang cosmology. Our aim is to combine a scientifically informed treatment of the cosmological theories with rigorously developed philosophical arguments and counter-arguments with a view toward assessing the bearing of these theories on the question of the existence of God.

The structure of this book consists of alternating essays by Craig and Smith. Typically, an essay by one of us will consist in a criticism of the arguments developed by the other in a preceding essay. The book divides into three separate debates, corresponding to Part 1, Part 2, and Part 3.

Part 1, 'The Theistic Cosmological Argument', concerns Craig's reformulation, in light of modern cosmology, of a traditional argument for the existence of God based on considerations about the finitude of past time. Craig presents the case that there is a sound argument for theism based on Big Bang cosmology and the impossibility of an infinite past, and Smith counters that Craig's arguments are unsound. It is important to note that despite our divergent viewpoints we begin from a common ground. We agree that the empirical evidence warrants the belief that the universe began to exist with the Big Bang about 15 billion years ago, and we both present several arguments for this thesis. Our disagreement in Part 1 concerns two further theses. First, Craig believes that considerations pertaining to Cantor's theory of the infinite show that the past is necessarily finite, but Smith denies this and argues that the past is possibly infinite (although probably finite, given the empirical evidence for Big Bang cosmology). The debate about the Cantorian infinite and whether the past is necessarily finite occurs in Essays I, II, and III. The second main area of disagreement concerns whether it is reasonable to believe the Big Bang has a cause (Craig's position) or whether it is reasonable to believe that the Big Bang occurs uncaused (Smith's position). This debate about the caused or uncaused nature of the Big Bang may be found in Essays I, IV, V, and VI.

Part 2, 'The Atheistic Cosmological Argument', concerns Smith's argument for the non-existence of God that is based on Big Bang cosmology. Smith argues that Big Bang cosmology is inconsistent with God's existence and Craig argues contrariwise. Smith presents the basics of his atheistic argument in Essay VII, Craig responds to it in Essay VIII, Smith responds to Craig's criticisms in Essay IX, and Craig ends the discussion in Essay X with a reply to Smith's Essay IX. The debate in Part 2 centres on issues involving the unpredictability of the Big Bang singularity with which the universe began. Smith argues that the Big Bang singularity is physically real, that its future cannot be predicted even by a divine mind (via counterfactuals allegedly known prior to creation), that it would require irrational acts of supernatural intervention if created, and is not something a perfect being (God) would create. Craig argues that the singularity is not physically real, but that even if it were physically real, it could be predicted by God (by God's prior knowledge of counterfactuals about it) and that the mentioned acts of supernatural intervention are not irrational.

The third main debate in the book occurs in Part 3, 'Theism, Atheism, and Hawking's Quantum Cosmology', which concerns the bearing of quantum cosmology on the philosophy of religion. We discuss the most developed version of quantum cosmology, the version developed by Stephen Hawking in the 1980s. Craig argues that Hawking's quantum cosmology is not a viable alternative to theism, whereas Smith argues that it is a more plausible theory than theism. Much of this debate concerns the interpretation of Hawking's quantum cosmology and whether his cosmology even makes physical sense. Craig argues that Hawking's theory (with its notion of imaginary time, splitting universes, infinite dimensional superspace, etc.) is physically unintelligible and therefore is not a realistic alternative to theism. Smith argues that Hawking's theory does not carry any of the above-mentioned physical implications (imaginary time etc.) and that Hawking's cosmology is both inconsistent with theism and rationally preferable to theism. Part 3 supplements the discussions in Parts 1 and 2, since Parts 1 and 2, although they take into account some considerations based on quantum cosmology (in certain sections of Essays IV and V), are mostly about classical Big Bang cosmology.

This book contains both technical and non-technical accounts of the scientific theories that are considered. Readers who lack a familiarity with classical Big Bang cosmology are advised to consult Craig's presentation in Essay I and Smith's Appendix IV.2 for an introductory and relatively non-technical explanation. Only a few sections on classical Big Bang cosmology, such as the first half of Smith's Essay VI, are technically dense, and the scientifically untutored reader may safely skip the technical passages without losing the main thread of the debate between Craig and Smith. Craig's essay on Hawking in Part 3 should be accessible to philosophers not familiar with quantum cosmology, and Smith's essay on Hawking in Part 3, although more technical, should be accessible to these philosophers once they have read Craig's essay.

We hope the reader will leave this book with an increased appreciation of the profound issues involved in supporting either a theistic or atheistic interpretation of Big Bang cosmology. It does not conclude the philosophical debate about theism, atheism, and Big Bang cosmology, but begins it.

Contents

1

The Theistic Cosmological Argument

I

The Finitude of the Past and the Existence of God

WILLIAM LANE CRAIG

Introductory Note

This essay is an abridged excerpt from my 1979 book *The* Kalām *Cosmological Argument*, in which I survey the history of that argument in Christian, Islamic, and Jewish thought, before going on to reformulate and defend the argument in light of contemporary developments in philosophy and science. Since the book is already part of the literature on the present topic and forms the focus of much of Quentin Smith's essays in this volume, I thought it best not to revise it substantially, but chiefly to add a series of annotations updating the relevant aspects of observational and theoretical cosmology. Interested readers will find in the book a much richer treatment of the issues, including discussions of Zeno's paradoxes, Kant's first antinomy concerning time, thermodynamic confirmation of the beginning of the universe, God's relationship to time, and so forth.

I. PROPOSED FORMULATION OF THE *KALĀM* COSMOLOGICAL ARGUMENT

In my opinion the version of the cosmological argument which is most likely to be a sound and persuasive proof for the existence of God is the *kalām* cosmological argument based on the impossibility of an infinite temporal regress of events. In this essay I shall attempt to formulate and defend such a cosmological argument,

First pub. in *The* Kalām *Cosmological Argument*, Library of Philosophy and Religion (London: Macmillan, 1979).
Annotations, cued by letters, can be found at the end of the chapter.

taking into account the modern developments in both philosophy and science that have a bearing on the proof's cogency.

We may present the basic argument in a variety of ways. Syllogistically, it can be displayed in this manner:

(1) Everything that begins to exist has a cause of its existence.
(2) The universe began to exist.
(3) Therefore the universe has a cause of its existence.

The point of the argument is to demonstrate the existence of a first cause which transcends and creates the entire realm of finite reality. Having reached that conclusion, one may then enquire into the nature of this first cause and assess its significance for theism.

2. SECOND PREMISS: THE UNIVERSE BEGAN TO EXIST

The key premiss in our syllogism is certainly the second; therefore, let us temporarily pass by the first premiss and attempt to support the second, that the universe began to exist. This premiss may be supported by two lines of reasoning, philosophical and empirical.

2.1. *Philosophical Argument*

Turning first to the philosophical reasoning, I shall present two arguments in support of the premiss: (1) the argument from the impossibility of the existence of an actual infinite and (2) the argument from the impossibility of the formation of an actual infinite by successive addition.

Before we examine each of these arguments in detail, it is imperative to have a proper understanding of the concept of the actual infinite. Prior to the revolutionary work of the mathematicians Bernard Bolzano (1781–1848), Richard Dedekind (1831–1916), and especially Georg Cantor (1845–1918), the only infinite considered possible by philosopher and mathematician alike was the potential infinite. Aristotle had argued at length that no actually infinite magnitude can exist.[1] The only legitimate sense in which one can speak of the infinite is in terms of potentiality: something

---

[1] Aristotle, *Physics*, 3.5.204^{b}1–206^{a}8.

may be infinitely divisible or susceptible to infinite addition, but this type of infinity is potential only and can never be fully actualized. For example, space is never actually infinite, but it is infinitely divisible in that one can continue indefinitely to divide spaces. Again, number is never actually infinite, but it may be increased without limit. And time is susceptible to both infinite division and infinite increase. But while the processes of division and addition may proceed indefinitely, they never arrive at infinity: space and time are never actually infinitely divided, and number and time are never completed wholes. Since Aristotle defines the potential as that which can become actual, some have charged him with contradiction in his doctrine of the potential infinite, since a potential infinite can never be actualized.[2] But Aristotle makes it quite plain that he is here using 'potential' in another sense: 'the infinite has a potential existence. But the phrase "potential existence" is ambiguous. When we speak of the potential existence of a statue we mean that there will be an actual statue. It is not so with the infinite. There will not be an actual infinite.'[3] When Aristotle speaks of the potential infinite, what he refers to is a magnitude that has the potency of being indefinitely divided or extended. Technically speaking, then, the potential infinite at any particular point is always finite.

This conception of the infinite prevailed all the way up to the nineteenth century. The medieval scholastics adhered to Aristotle's analysis of the impossibility of an actual infinite, and the post-Renaissance thinkers, even Newton and Leibniz with their infinitesimal calculus, believed that only a potential infinite could exist.[4]

[2] Simon Van Den Bergh, Notes to Averroës, *Tahafut al-Tahafut* (The Incoherence of the Incoherence), 2 vols. (London: Luzac, 1954), ii. 8.

[3] Aristotle, *Physics*, 3.6.206^{a}15–20. For a good discussion, consult David Bostock, 'Aristotle, Zeno, and the Potential Infinite', *Proceedings of the Aristotelian Society*, 73 (1972–3), 37–57. To argue that there can be a potential infinite but no actual infinite does not commit one to the self-contradictory position that there are possibilities that cannot be actualized. (W. D. Hart, 'The Potential Infinite', *Proceedings of the Aristotelian Society*, 76 (1976), 247–64.)

[4] Abraham Robinson, 'The Metaphysics of the Calculus', in Jaakko Hintikka (ed.), *The Philosophy of Mathematics*, Oxford Readings in Philosophy (London: Oxford University Press, 1969), 156, 159. On some interesting medieval precedents for Cantor's work, see E. J. Ashworth, 'An Early Fifteenth-Century Discussion of Infinite Sets', *Notre Dame Journal of Formal Logic*, 18 (1977), 232–4. On Leibniz's view of infinitesimal terms as useful fictions, see John Earman, 'Infinities, Infinitesimals, and Indivisibles: The Leibnizian Labyrinth', *Studia Leibnitiana*, 7 (1975), 236–51.

One of the foremost mathematicians of the nineteenth century, Georg Friedrich Gauss, in an oft-printed statement, decried any use of the actual infinite in mathematics: 'I protest . . . against the use of infinite magnitude as if it were something finished; this use is not admissible in mathematics. The infinite is only a *façon de parler*: one has in mind limits approached by certain ratios as closely as desirable while other ratios may increase indefinitely.'[5] This expressed the accepted view of the infinite as the limit of a convergent or divergent process:

$$\lim_{x \to \infty} \cdots$$

But although the majority of philosophers and mathematicians adhered to this conception of the infinite, dissenting voices could also be heard. A man ahead of his time, Bolzano argued vigorously against the then current definitions of the potential infinite.[6] He contended that infinite multitudes can be of different sizes and observed the resultant paradox that although one infinite might be larger than another, the individual elements of the two infinites could none the less be matched against each other in a one-to-one correspondence.[7] It was precisely this paradoxical notion that Dedekind seized upon in his definition of the infinite: a system is said to be infinite if a part of that system can be put into a one-to-one correspondence with the whole.[8] In other words, the Euclidean maxim that the whole is greater than a part was now going by the board. According to Dedekind, this assumption could hold only for finite systems.

But it was undoubtedly Cantor who won for the actual infinite the status of legitimacy that it holds today.[9] Cantor called the potential infinite a 'variable finite' and attached the sign ∞ to it; this signified that it was an 'improper infinite'.[10] The actual infinite

---

[5] Karl Friedrich Gauss and Heinrich Christian Schumacher, *Briefwechsel*, ed. C. A. F. Peters, 6 vols. (Altona: Esch, 1860–5), ii. 269.

[6] Bernard Bolzano, *Paradoxes of the Infinite*, trans. Fr. Prihonsky, intro. Donald A. Steele (London: Routledge & Kegan Paul, 1950), 81–4.

[7] Ibid. 95–6. Despite the one-to-one correspondence, Bolzano insisted that two infinites so matched might nevertheless be non-equivalent.

[8] Richard Dedekind, 'The Nature and Meaning of Numbers', in Richard Dedekind, *Essays on the Theory of Numbers*, trans. Wooster Woodruff Beman (New York: Dover, 1963), 63.

[9] For an exposition and defence of Cantor's system, see Robert James Bunn, 'Infinite Sets and Numbers', Ph.D. thesis, University of British Columbia, 1975.

[10] Georg Cantor, *Contributions to the Founding of the Theory of Transfinite Numbers*, trans. and intro. Philip E. B. Jourdain (New York: Dover, 1915), 55–6.

he pronounced the 'true infinite' and assigned the symbol $\aleph_0$ (aleph zero) to it. This represented the number of all the numbers in the series 1, 2, 3, . . . and was the first infinite or transfinite number, coming after all the finite numbers. According to Cantor, a collection or set is infinite when a part of it is equivalent to the whole.[11] Utilizing this notion of the actual infinite, Cantor was able to develop a whole system of transfinite arithmetic which he bequeathed to modern set theory. 'Cantor's . . . theory of *transfinite* numbers . . . is, I think, the finest product of mathematical genius and one of the supreme achievements of purely intellectual human activity,' exclaimed the great German mathematician David Hilbert. 'No one shall drive us out of the paradise which Cantor has created for us.'[12]

Modern set theory, as a legacy of Cantor, is thus exclusively concerned with the actual as opposed to the potential infinite. At this point we may clarify the distinction between a potential and an actual infinite. According to Hilbert, the chief difference lies in the fact that the actual infinite is a determinate totality, whereas the potential infinite is not:

Someone who wished to characterize briefly the new conception of the infinite which Cantor introduced might say that in analysis we deal with the infinitely large and the infinitely small only as limiting concepts, as something becoming, happening, i.e., with the *potential infinite*. But this is not the true infinite. We meet the true infinite when we regard the totality of numbers 1, 2, 3, 4, . . . itself as a completed unity, or when we regard the points of an interval as a totality of things which exists all at once. This kind of infinity is known as *actual infinity*.[13]

[11] Ibid. 108.

[12] David Hilbert, 'On the Infinite', in Paul Benacerraf and Hilary Putnam (eds.), *Philosophy of Mathematics* (Englewood Cliffs, NJ: Prentice-Hall, 1964), 139, 141.

[13] Ibid. 139. Fraenkel adds, 'In almost all branches of mathematics, especially in analysis (for instance, in the theory of series and in calculus, also called "infinitesimal calculus"), the term "infinite" occurs frequently. However, mostly this infinite is but a *façon de parler* . . . the statement

$$\lim_{n \to \infty} \frac{1}{n} = 0$$

asserts nothing about infinity (as the ominous sign ∞ seems to suggest) but is just an abbreviation for the sentence: $1/n$ can be made to approach zero as closely as desired by sufficiently increasing the positive integer n. In contrast herewith the set of all integers is infinite (infinitely comprehensive) in a sense which is "actual" (proper) and not only "potential".' (Abraham A. Fraenkel, *Abstract Set Theory*, 2nd rev. edn. (Amsterdam: North-Holland, 1961), 5–6.) See also Antonio Moreno, 'Calculus and Infinitesimals: A Philosophical Evaluation', *Angelicum*, 52 (1975), 228–45.

In set theory this notion of infinity finds its place in the theory of infinite sets. According to Cantor, a set is a collection into a whole of definite, distinct objects of our intuition or of our thought; these objects are called elements or members of the set. Fraenkel draws attention to the characteristics *definite* and *distinct* as particularly significant.[14] That the members of a set are distinct means that each is different from the other. To say that they are definite means that given a set S, it should be intrinsically settled for any possible object x whether x is a member of S or not. This does not imply actual decidability with the present or even future resources of experience; rather a definition could settle the matter sufficiently, such as the definition for 'transcendental' in the set of all transcendental numbers. Unfortunately, Cantor's notion of a set as any logical collection was soon found to spawn various contradictions or antinomies within naïve set theory that threatened to bring down the whole structure. As a result, most mathematicians have renounced a definition of the general concept of set and chosen instead an axiomatic approach to set theory by which the system is erected upon several given undefined concepts formulated into axioms. But the characteristics of definiteness and distinctness are still considered to hold of the members of any set. An infinite set in Zermelo–Fraenkel axiomatic set theory is defined as any set R that has a proper subset that is equivalent to R.[15] A proper subset is a subset that does not exhaust all the members of the original set, that is to say, at least one member of the original set is not also a member of the subset. Two sets are said to be equivalent if the members of one set can be related to the members of the other set in a one-to-one correspondence, that is, so that a single member of the one set corresponds to a member of the other set and vice versa. Thus, an infinite set is one in which the whole set is not greater than a part. In contrast to this, a finite set is a set such that if n is a positive integer, the set has n members.[16] Because set theory does not utilize the notion of potential infinity, a set containing a potentially infinite number of members is impossible. Such a collection would be one in which the members are not definite in number, but may be increased without limit. It would best be described as indefinite. The crucial difference between an infinite set and an indefinite collection would be that the former is conceived as

---

[14] Fraenkel, *Abstract Set Theory*, 10.   [15] Ibid. 29.   [16] Ibid. 28.

a determinate whole actually possessing an infinite number of members, while the latter never actually attains infinity, though it increases limitlessly. Therefore, we must in our subsequent discussion always keep conceptually distinct these three types of collection: finite, infinite, and indefinite.

2.1.1. *First Philosophical Argument*

Our first argument in support of the premiss that the universe began to exist is based upon the impossibility of the existence of an actual infinite. We may present the argument in this way.

(1) An actual infinite cannot exist.
(2) An infinite temporal regress of events is an actual infinite.
(3) Therefore an infinite temporal regress of events cannot exist.

With regard to the first premiss, it is important to understand that by 'exist' we mean 'exist in reality', 'have extra-mental existence', 'be instantiated in the real world'. We are contending, then, that an actual infinite cannot exist in the real world. It is usually alleged that this sort of argument has been invalidated by Cantor's work on the actual infinite and by subsequent developments in set theory. But this allegation seriously misconstrues the nature of both Cantor's system and modern set theory, for our argument does not contradict a single tenet of either. The reason is this: Cantor's system and set theory are concerned exclusively with the mathematical world, whereas our argument concerns the real world. What I shall argue is that while the actual infinite may be a fruitful and consistent concept in the mathematical realm, it cannot be translated from the mathematical world into the real world, for this would involve counter-intuitive absurdities. It is only the real existence of the actual infinite which I deny. Far from being remarkable, this view of the actual infinite as a mathematical entity which has no relation to the real world is prevalent among the mathematicians themselves who deal with infinite sets and transfinite arithmetic. Bolzano's primary examples of infinite sets were admittedly in the 'realm of things which do not claim actuality, and do not even claim possibility'.[17] When it came to an instance of an

---

[17] Bolzano, *Paradoxes of the Infinite*, 84. See also Dedekind, 'The Nature and Meaning of Numbers', 64. These thinkers pointed to the set of all true propositions or the set of all objects of thought as examples of infinite sets, since by self-reference an infinite series is generated. But clearly such sets do not exist in the real world, and they may be considered to be potential infinites only.

actual infinite in the real world, Bolzano was reduced to pointing to God as an infinite being.[18] But of course the infinity of God's being has nothing to do with an actually infinite collection of definite and distinct finite members. Cantor's definition of a set made it clear that he was theorizing about the abstract realm and not the real world for, it will be remembered, he held that the members of a set were objects of our intuition or of our thought. According to Robinson, 'Cantor's infinites are abstract and divorced from the physical world.'[19] This judgement is echoed by Fraenkel, who concludes that among the various branches of mathematics, set theory is 'the branch which least of all is connected with external experience and most genuinely originates from free intellectual creation'.[20] As a creation of the human mind, state Rotman and Kneebone, 'the Zermelo–Fraenkel universe of sets exists only in a realm of abstract thought . . . the "universe" of sets to which the . . . theory refers is in no way intended as an abstract model of an existing Universe, but serves merely as the postulated universe of discourse for a certain kind of abstract inquiry'.[21] Such a picture of set theory ought not to surprise us, for virtually every philosopher and mathematician understands the same thing of, for example, Euclidean and non-Euclidean geometries, namely, that these represent consistent conceptual systems that may or may not hold in reality.

This being so, the novice is apt to be confused by the frequent existential statements found in books on set theory, statements such as the Axiom of Infinity that there exists an infinite set. But such statements need not be taken to imply existence in the extramental world, but only in the mathematical realm. Alexander Abian explains,

[18] Bolzano, *Paradoxes of the Infinite*, 101. Cantor also believed that his discoveries concerning the nature of infinity might be of great service to religion in understanding the infinity of God and even carried on a fascinating correspondence with Pope Leo XIII to this effect. (See Joseph W. Dauben, 'Georg Cantor and Pope Leo XIII: Mathematics, Theology, and the Infinite', *Journal of the History of Ideas*, 38 (1977), 85–108.)

[19] Robinson, The Metaphysics of the Calculus', 163. Cantor did think the number of atoms in the universe might be denumerably infinite.

[20] Fraenkel, *Abstract Set Theory*, 240.

[21] B. Rotman and G. T. Kneebone, *The Theory of Sets and Transfinite Numbers* (London: Oldbourne, 1966), 61. Thus, when one selects from an infinite set an infinite subset, the actual possibility of such an operation is not implied. 'The conception of an infinite sequence of choices (or of any other acts) . . . is a mathematical fiction—an idealization of what is imaginable only in finite cases.' (Ibid. 60.)

whenever in the Theory of Sets we are confronted with a statement such as *there exists a set* x *whose elements are sets* b *and* c, *and there exists a set* u *whose elements are the sets* x, b *and* m, then we may take this statement as implying that a table such as the following appears as part of the illusory table which describes the Theory of Sets . . .

$$
\begin{array}{ll}
a & e \\
b \in a & a \in e \\
c \in a & b \in e \\
& m \in e
\end{array}
$$

. . . The above considerations show how we may interpret more concretely the notion of *existence* in the Theory of Sets. In short if an axiom or a theorem of the Theory of Sets asserts that *if certain sets . . . exist, then a certain set . . . also exists*, we shall interpret this as: *if certain sets . . . are listed in the above illusory table, then a certain set . . . must also be listed in the same illusory table.*[22]

Thus, the existential statements in set theory have no bearing on the extra-mental existence of the entities described. When, therefore, the existence of an infinite set is postulated, no true existential import is carried by the statement, and no verdict is pronounced on whether such a collection could really exist at all. This analysis of the actual infinite says nothing about whether an actual infinite can exist in reality.

What Cantor accomplished was the establishing of the possibility of conceiving of the infinite as a completed determinate whole, thus enabling us to speak abstractly of infinite sets. Rather than attempt to synthesize the infinite mentally by counting, Cantor stood outside the infinite series of natural numbers and grasped them conceptually as a totality, and thence developed his system of transfinite arithmetic. But the ability to conceive of an actual infinite does not imply the possibility of its real existence, as Pamela Huby explains:

It is often said that Cantor legitimized the notion of the actual infinite, and it is well to get clear what this means. What it seems to mean is that we can make statements, within a certain conventional system, about *all* the members of an infinite class, and that we can clearly identify certain classes which have an infinite number of members, and even say, using new conventions, what the cardinal number of their members is. But beyond that Cantor tells us nothing about actual infinity.[23]

[22] Alexander Abian, *The Theory of Sets and Transfinite Arithmetic* (Philadelphia: W. B. Saunders, 1965), 68.

[23] Pamela M. Huby, 'Kant or Cantor? That the Universe, if Real, Must Be Finite in Both Space and Time', *Philosophy*, 46 (1971), 130.

The use of the notion of the actual infinite in modern mathematics does not, therefore, ensure the possibility of its real existence.

Now I have no intention whatsoever of trying to drive mathematicians from their Cantorian paradise. While such a system may be perfectly consistent in the mathematical realm, given its axioms and conventions, I think that it is intuitively obvious that such a system could not possibly exist in reality. The best way to show this is by way of examples that illustrate the various absurdities that would result if an actual infinite were to be instantiated in the real world.

For instance, if an actual infinite could exist in reality, then we could have a library with an actually infinite collection of books on its shelves. Remember that we are talking not about a potentially infinite number of books, but about a completed totality of definite and distinct books that actually exist simultaneously in time and space on these library shelves. Suppose further that there were only two colours of books, black and red, and every other book was the same colour. We would probably not balk if we were told that the number of black books and the number of red books is the same. But would we believe someone who told us that the number of red books in the library is the same as the number of red books *plus* the number of black books? For in the latter collection there are all the red books—just as many as in the former collection, since they are identical—plus an infinite number of black books as well. And if one were to imagine the library to have three different colours of books, or four or five or a hundred different colours of books—can we honestly believe that there are in the total collection of books of all colours no more books than in the collection of a single colour? And if there were an infinite number of colours of books, would we not naturally surmise that there was only one book per colour in the total collection? Would we believe anyone who told us that for each of the infinite colours there is an infinite collection of books and that all these infinities taken together do not increase the total number of books by a single volume over the number contained in the collection of books of one colour?

Suppose further that each book in the library has a number printed on its spine so as to create a one-to-one correspondence with the natural numbers. Because the collection is actually infinite, this means that *every possible* natural number is printed on some book. Therefore, it would be impossible to add another book to this library. For what would be the number of the new book? Clearly there is no number available to assign to it. Every possible number already has a counterpart in reality, for corresponding to every natural number is an already existent book. Therefore, there would be no number for the new book. But this is absurd, since entities that exist in reality can be numbered. It might be suggested that we number the new book '1' and add one to the number of every book thereafter. This is perfectly successful in the mathematical realm, since we accommodate the new number by increasing all the others out to infinity. But in the real world this could not be done. For an actual infinity of objects already exists that completely exhausts the natural number system—every possible number has been instantiated in reality on the spine of a book. Therefore, book 1 could not be called book 2, and book 2 be called book 3, and so on, to infinity. Only in a potential infinite, where new numbers are created as the collection grows, could such a re-count be possible. But in an actual infinite, all the members exist in a determinate complete whole, and such a re-count would necessitate the creation of a new number. But this is absurd, since every possible natural number has been used up. (If they have not been used up, then the collection is not an actual infinite after all.) Therefore, it is of no help to add the book to the beginning of the series. If it is suggested we call the new book $\omega + 1$, or $\aleph_0 + 1$, this is easily dismissed. It could not be $\omega + 1$, for this has the same cardinal number as ω, and we need a new cardinal number for this book. It could not be $\aleph_0 + 1$, for this reduces to $\aleph_0$, and yet we do have an extra, irreducibly real book on our hands. (Besides, there is no book $\aleph_0$ in the collection, since $\aleph_0$ has no immediate predecessor as books do. The symbol $\aleph_0$ just informs us that the whole collection of books as a determinate totality has a denumerably infinite number of books in it, which we already know *ex hypothesi*.) So it would therefore be impossible to add the new book to the stacks.

The same absurdity is evident in an illustration employed by David Hilbert to exhibit the paradoxical properties of the actual infinite, appropriately dubbed Hilbert's Hotel: Let us imagine a hotel with a finite number of rooms, and let us assume that all the rooms are occupied. When a new guest arrives and requests a room, the proprietor apologizes, 'Sorry—all the rooms are full.' Now let us imagine a hotel with an infinite number of rooms, and let us assume that again all the rooms are occupied. But this time, when a new guest arrives and asks for a room, the proprietor exclaims, 'But of course!' and shifts the person in room 1 to room 2, the person in room 2 to room 3, the person in room 3 to room 4, and so on . . . The new guest then moves into room 1, which has now become vacant as a result of these transpositions. But now let us suppose an *infinite* number of new guests arrive, asking for rooms. 'Certainly, certainly!' says the proprietor, and he proceeds to move the person in room 1 into room 2, the person in room 2 into room 4, and the person in room 3 into room 6, the person in room 4 into 8, and so on . . . In this way, all the odd-numbered rooms become free, and the infinity of new guests can easily be accommodated in them.[24]

In this story the proprietor thinks that he can get away with his clever business move because he has forgotten that his hotel has an *actually infinite* number of rooms, not a potentially infinite number of rooms, and that *all the rooms are occupied*. The proprietor's action can only work if the hotel is a potential infinite, such that new rooms are created to absorb the influx of guests. For if the hotel has an actually infinite collection of determinate rooms and *all* the rooms are full, then there is no more room.

These illustrations show that if an actual infinite could exist in reality, it would be impossible to add to it. But it obviously is possible to add to, say, a collection of books: just take one page from each of the first hundred books, add a title-page, and put it on the shelf. Therefore, an actual infinite cannot exist in the real world.

But suppose we could add to the infinite collection of books. The new book would have the ordinal $\omega + 1$. And yet our collection of books has not increased by a single book. But how can this be? We put the book on the shelf: there is one more book in the collection; we take it off the shelf: there is one less book in the collection. We can see ourselves add and remove the book—are we really to believe that when we add the book there are no more books in the collection and when we remove it there are no less books in the collection? Suppose we add an *infinity* of books to the collection; the ordinal number is now $\omega + \omega$. Are we seriously to believe that there are *no more* books in the collection than before? (How could a collection of books numbered $\{1, 2, 3, \ldots, 1, 2, 3, \ldots\}$ have the

[24] This story is recorded in an entertaining work by George Gamow, *One, Two, Three, . . . Infinity* (London: Macmillan, 1946), 17.

same cardinal number as a collection numbered $\{1, 2, 3, \ldots\}$, namely, $\aleph_0$? For $\aleph_0$ is the number of elements in the natural number series, and these are completely used up in the first infinite series of books, there being a one-to-one correspondence between every book and every number.) Suppose we add an infinity of infinite collections to the library $(\omega + \omega^2)$—is there actually not one more single volume in the entire collection than before? Suppose we add ω^ω books—how can one express this in words?—to the collection. Is there not one extra book in the collection? Clearly, something must be amiss here. What is it?—we are trying to take conceptual operations guaranteed by the convention of the Principle of Correspondence and apply them to the real world of things, and the results are just not believable.

But to continue. Suppose we return to our original collection (though supposedly it never really increased) and decide to loan out some of the books. Suppose book 1 is loaned out. Is not there now one book fewer in the collection? Suppose books 1, 3, 5, . . . are loaned out. The collection has been depleted of an infinite number of books, and yet we are told that the number of books remains constant. The cumulative gap created by the missing books would be an infinite distance, yet if we push the books together to close the gaps, all of the infinite shelves will remain *full* (this is Hilbert's Hotel in reverse). If we once more remove every other book from the collection, we again have removed an infinity of books, and yet the number of books in the collection is not depleted. And if we close the gaps between the books, all the shelves will remain full. We could do this infinitely many times, and the collection would never have one book fewer and the shelves would always be completely full. But suppose we were to loan out books 4, 5, 6, . . . At a single stroke the collection would be virtually wiped out, the shelves emptied, and the infinite library reduced to finitude. And yet, we have removed *exactly the same number* of books this time as when we removed books 1, 3, 5, . . . Can anyone believe such a library can exist in reality? It may be said that inverse operations cannot be performed with the transfinite numbers—but this qualification applies to the mathematical world only, not the real world. While we may correct the mathematician who attempts inverse operations with transfinite numbers, we cannot in the real world prevent people from checking out what books they please from our library.

These examples serve to illustrate that the real existence of an actual infinite would be absurd. Again, I must underline the fact that what I have said in no way attempts to undermine the theoretical system bequeathed by Cantor to modern mathematics. Indeed, some of the most eager enthusiasts of the system of transfinite mathematics are only too ready to agree that these theories have no relation to the real world. Thus, Hilbert, who exuberantly extolled Cantor's greatness, nevertheless held that the Cantorian paradise from which he refused to be driven exists only in the ideal world invented by the mathematician; he concludes,

the infinite is nowhere to be found in reality. It neither exists in nature nor provides a legitimate basis for rational thought—a remarkable harmony between being and thought. . . . The role that remains for the infinite to play is solely that of an idea—if one means by an idea, in Kant's terminology, a concept of reason which transcends all experience and which completes the concrete as a totality—that of an idea which we may unhesitatingly trust within the framework erected by our theory.[25]

Our case against the existence of the actual infinite says nothing about the use of the idea of the infinite in conceptual mathematical systems.

To return here to a point alluded to earlier, the only mathematicians who would feel uneasy with what I have contended thus far would be those who regard mathematical entities in a Platonistic way, as somehow part of the real world. The question we are raising here is, What is the ontological status of sets?[26] The question is similar to the medieval debate over the existence of universals, and the schools of thought divide along pretty much the same lines: Platonism, nominalism, and conceptualism.

1. *Platonism*, or realism, maintains that corresponding to every well-defined condition there exists a set, or class, comprised of those entities that fulfil this condition and which is an entity in its own right, having an ontological status similar to that of its members. Mathematics is a science of the discovery, not creation, of numbers and their properties, which exist independently of

---

[25] Hilbert 'On the Infinite', 151.

[26] For a good discussion of this issue, consult Abraham A. Fraenkel, Yehoshua Bar-Hillel, and Azriel Levy, *Foundations of Set Theory*, 2nd rev. edn. (Amsterdam: North-Holland, 1973), 331–45; Stephen F. Barker, *Philosophy of Mathematics*, Foundations of Philosophy (Englewood Cliffs, NJ: Prentice-Hall, 1964), 69–91.

the activity of the mathematician's mind. This viewpoint finds expression in the school of *logicism*, as represented by Frege and Russell, which attempted to reduce the laws of the mathematics of number to logic alone. According to the Platonistic perspective, Cantor's transfinite numbers do exist as a part of reality, and the existence of the actual infinite is guaranteed by the infinity of the natural number series and other mathematical examples of infinite sets.

2. *Nominalism* holds that there are no abstract entities such as numbers or sets, but that only individuals exist. Much of the task of nominalism consists in rephrasing the language of mathematics in terms of individual entities alone instead of classes or sets. But even these individual mathematical entities are not regarded as having any real existence. When it comes to Cantorian analysis of the infinite, nominalists are only too glad to jettison the whole system as a mathematical fiction.

3. *Conceptualism* contends that abstract entities such as numbers and sets are created by and exist in the mind only, and have no independent status in the real world. A well-defined condition produces a corresponding set, but this set has mental existence only—the mathematician *creates* his mathematical entities: he does not discover them. The most important modern school of conceptualistic persuasion is *intuitionism*, as represented by Kronecker and Brouwer, which argues that only those mathematical entities claim ideal existence which can be constructed by our intuitive activity of counting. Proofs involving an infinite number of steps—such as Cantor's proof of non-denumerable infinites—are ruled out of court because the mind cannot actually construct such sets. Constructible entities have a conceptual existence; non-constructible entities cannot even claim that. When applied to Cantor's theories, conceptualism could accord purely ideal existence, but not real existence, to the actual infinite, or, if it has an intuitionistic slant, it could deny any sort of existence whatsoever to the actual infinite, since it is non-constructible.

Arising out of the debate between these three schools of thought came a fourth perspective, that of *formalism*. The adherents of this position eschew all ontological questions concerning mathematical entities and maintain that mathematical systems are nothing but formalized systems having no counterparts in reality. Mathematical calculations are merely marks on paper, symbols without content, and the only condition for these is consistency. Mathematical existence is freedom from contradiction. Mathematical calculations may have utility in the real world—but that does not give the formalized systems any literal significance. Formalism's attitude toward Cantor's work is exemplified in the attitude of one of its most famous representatives, Hilbert, who regarded Cantor's system as a consistent mathematical system that carries no ontological implications.

It is highly significant that three of the four schools of thought on the question of the ontological status of mathematical entities ascribe no real existence to these entities. For the nominalist, the conceptualist, and the formalist, the mathematical validity of the Cantorian system implies no commitment to the existence of the actual infinite in the real world. Thus, the mathematical instances of the actual infinite—such as the natural number series, the set of mathematical points on a line, or the set of all functions between zero and one—have nothing to say about the real existence of the actual infinite. Only for the Platonist-realist, who accepts the independent status of mathematical entities in the real world, do Cantor's theories have ontological implications for the real world. This means that our argument against the real existence of the actual infinite would contradict Cantor's work only if the Platonist-realist position on the ontological status of numbers and sets were proven to be the correct one, for our argument would be compatible with any of the other three.

And, in fact, the Platonist-realist view is very problematic, due to the antinomies to which naïve Cantorian set theory gives rise.[27] Just at the moment when the resistance of the mathematical world to his system seemed to be dissolving, Cantor discovered in 1895 the first of several logical antinomies within his system. Not published immediately, the antinomy was also discerned in 1897 by Burali-Forti, whose name it bears. Cantor discovered a second antinomy in 1899—though it was not published until 1932—which bears his name. But Cantor did not regard these contradictions as having much significance and never abandoned a naïve view of his set theories. In 1902 the severest blow to Cantor's system was dealt

---

[27] For a thorough discussion of these, consult Fraenkel *et al.*, *Foundations of Set Theory*, 1–14; Barker, *Philosophy of Mathematics*, 82–5.

with the publication by Russell of a third antinomy at the very roots of set theory. This antinomy, named after Russell, forced a major reworking of modern set theory and so undermined the Platonist-realist view of sets.

1. *Burali-Forti's antinomy* is also known as the antinomy of the set of all ordinals. Very simply, the antinomy states that if every set has an ordinal number, then the set of all ordinal numbers would also have to have an ordinal number. But then the ordinal number would itself have to be in the set, thus requiring a larger ordinal number. Thus, there could be no ordinal number for the set of all ordinals.

2. *Cantor's antinomy* springs out of Cantor's theorem that the set of all subsets of any given set has a cardinal number greater than the set itself has. This means that the set of all sets has a power set—a set containing all the subsets of itself—that has a greater cardinal than the set itself, which is contradictory, since the original set was declared to be the set of *all* sets.

3. *Russell's antinomy* proceeds on the assumption that it is meaningful to ask whether a set is a member of itself. Some sets are clearly not members of themselves. For example, the set of all pigs is not itself a pig, and, hence, it is not a member of itself. But some sets appear to be members of themselves; for example, the set of all things mentioned in this essay is itself mentioned in this essay and so would seem to be a member of itself. But what about the set of all sets that are not members of themselves—is it a member of itself? Denoting this set by S, we discover that if S is a member of itself, then it cannot be in S, for S includes only sets that are *not* members of themselves. But if S is not a member of itself, then, since it fulfils the condition for being in S, it is a member of itself. Thus, we reach the contradictory conclusion that S is a member of S if S is not a member of S.

In the face of these antinomies, set theory had to be either abandoned or radically revised. Not being of the quitting sort, mathematicians and philosophers pursued primarily three courses of possible revision: logicism, axiomatization, and intuitionism.[28] Logicism sought to circumvent the antinomies by use of the theory

---

[28] An in-depth discussion of each of these reforms may be found in Fraenkel *et al.*, *Foundations of Set theory*, 154–209, 15–153, 210–74; see also Barker, *Philosophy of Mathematics*, 85–91.

of types, which asserted that all the entities referred to in set theory are arranged in a hierarchy of types, with each entity belonging to a certain level. At the lowest level are individuals, that is, entities which are not sets. Above this level are sets which contain the individuals of the lower level as members. Above these are sets whose members are the sets of the second level, and so forth. A set may only have members that are from the level immediately below it, and to speak of any set not fulfilling this condition is strictly nonsensical. In this way the antinomies could not arise because no set could be a member of itself.

Axiomatization chose a different course: it sought to restrict the concept of set in such a way that the paradoxical sets could not arise. This was accomplished by abandoning Cantor's naïve definition of set as any collection fulfilling a condition and adopting instead a system based on seven or eight axioms which do not attempt a definition of 'set', but rather delimit the behaviour of sets so that sets like the set of all ordinals cannot appear in the system.

Intuitionism tended to welcome the antinomies, for they exposed the weakness of non-intuitionistic mathematics with its non-constructible sets. Only those sets can be granted mathematical existence which are constructible; therefore, sets such as the power set of the set of all cardinals cannot exist, since it is not constructible. The antinomies are actually helpful in that they aid in defining the scope of human mathematical creativity and thus the realm of mathematical entities.

All this strikes at the heart of the Platonist-realist thesis that numbers and sets are component parts of independently existing reality. The logical antinomies in naïve set theory are damaging to this thesis because if numbers and sets do exist extra-mentally, then such sets as are encountered in the antinomies seem inevitable. There is no reason for denying that the set of all ordinals or the power set of all cardinals should exist. On this basis, Stephen Barker scores the logicist theory of types as without foundation: 'Russell's avowed philosophy was that of realism, and realism offers no philosophical rationale for rejecting impredicative definitions [definitions which, in defining a thing, refer to some totality to which the thing being defined belongs]. If a set has independent reality, then why may not members of the set be defined by reference to the set itself?'[29] Logicism, therefore, was not a very convincing revision of the Platonist-realist position and failed to generate much support, as contrasted with the axiomatic method. Therefore, the Platonist-realist view of numbers and sets as independently existing entities awaiting discovery is exposed as inadequate by the antinomies of naïve set theory. If it could be proved that the Platonist-realist view of the ontological status of mathematical entities is correct and that such a view could escape the logical antinomies implicit in naïve set theory, then our argument that an actual infinite cannot exist in reality would stand opposed to Cantor's analysis. But as it is, either axiomatization or intuitionistic reform seem much more plausible alternatives. And under either of these two views, the world which Cantor created is clearly a purely theoretical one: in the one case deduced from presumed axioms just as a Euclidean or a non-Euclidean geometry might be deduced without reference to the real world, and in the other case mentally constructed within the bounds of the mind's finite operations. Thus, out of the four schools of thought concerning the ontological status of sets—Platonism, nominalism, conceptualism, and formalism—only the first is rendered untenable by the antinomies. All the others could acceptably escape these contradictions of naïve set theory, and any of them are compatible with our case against the real existence of the actual infinite. Therefore, the Cantorian theory of the actual infinite does not imply that an actual infinite can really exist; indeed, we have seen that this theory itself makes it intuitively obvious that such a conceptual system cannot be instantiated in the real world.

Up to this point we have assumed that the Cantorian analysis of the actual infinite is correct, but it should now be noted that one important school of mathematicians, the intuitionists, do not regard it as so. Their contention is significant, for although the school of intuitionists is not large, it numbers among its members some of the most brilliant mathematicians of the past few generations from several nations. We have seen that the root presupposition of intui-

---

[29] Barker, *Philosophy of Mathematics*, 87. Barker himself adopts a syncretistic view, arguing that because transfinite arithmetic and the set theory from which it is deduced cannot be regarded as statements about counting, these parts of mathematics ought to be regarded as 'games with marks', albeit games 'of great intellectual interest'. (Ibid. 103.)

tionism is that the basis of mathematics is found in the pure intuitition of counting. Thus, constructibility by actual operations becomes the prerequisite of any legitimate mathematical operation. Since an actual infinite cannot be constructed by the human mind, it follows that the infinite is not a well-defined totality. And not being well defined, the actual infinite cannot be said to exist in the mathematical realm. Thus, both Kronecker and Brouwer deny that the natural number series is a complete and determinate ideal totality. The natural number series is only an indefinite series, surpassing each limit it reaches. In other words, only the existence of the potential infinite is granted, and intuitionism is thus the heir of the Aristotelian tradition of basing mathematics on the potential infinite. This sort of infinite causes no problems for mathematics because any statement about a potential infinite can be translated into a statement about a finite but extendable entity.

So the Cantorian analysis of the actual infinite is far from unchallenged among mathematicians. If the intuitionists are correct, then not only the real existence but even the conceptual existence of the actual infinite is inadmissible. And why not?—for it is consistent to conceive of the natural number series as a potential infinite, never arriving at infinity, but increasing according to a rule, that of adding one, so that new numbers are created, not discovered, by the mind. One cannot help but wonder if the resistance to intuitionism among many mathematicians is not due more to a stubborn refusal to abandon the Cantorian paradise than to the inadequacy of intuitionistic theories. Of course, intuitionists have no doubt generated opposition to their theories by arguing, for example, that the Law of Excluded Middle does not hold for certain operations (because they concern entities that are not well defined and are thus insusceptible to determination of their truth value), but it would seem that their view of the infinite as potential and not actual is of itself significant, apart from what might be considered 'objectionable' tenets. If the Platonist-realist view of the ontological status of numbers is discarded, then there does not appear to be any *proof* that the number series is a mathematical instance of an actual infinite—usually this seems to be just taken for granted.[30] But there appears to be no necessity for regarding it

[30] If the class of natural numbers is a potential infinite, then it is indefinite and cannot be said to possess an actually infinite number of elements. See Hermann

as such. If, then, the number series is conceived of as a potential infinite, the case against the real existence of the actual infinite would be even stronger, since there would not only be no real instances of such an entity but no mathematical instances either.

Finally, we may venture an opinion as to *why* it is that an actual infinite cannot exist in reality without entailing the various absurdities described. It seems to me that the surd problem in instantiating an actual infinite in the real world lies in Cantor's Principle of Correspondence. The principle asserts that if a one-to-one correspondence between the elements of two sets can be established, the sets are equivalent. This principle is simply adopted in set theory as a convention; for how could it be proved? One may cite empirical examples of the successful use of the principle for comparing finite real collections such as eggs and apples, beads and coins, persons and seats; but it would be impossible to conduct such an empirical proof for infinite collections. Therefore, in the mathematical realm equivalent sets are simply *defined* as sets having a one-to-one correspondence. The principle is simply a convention adopted for use in the mathematical system created by the mathematician. This is why, given this principle, Cantor can consistently assert, for example, that $\omega + \omega$ has a cardinal number of $\aleph_0$, which, we have argued, is not a condition realizable in the real world. For given the Principle of Correspondence, the set $\{1, 2, 3, \ldots\}$ *is* equivalent to the set $\{1, 2, 3, \ldots, 1, 2, 3, \ldots\}$, odd as this appears. This is also why Cantor can consistently maintain that a proper subset of an infinite set is equivalent to the whole set. For given the Principle of Correspondence, the set $\{\ldots, -3, -2, -1, 0, 1, 2, 3, \ldots\}$ *is* equivalent to the set $\{1, 2, 3, \ldots\}$, strange as this may seem.

Should someone naïvely object to Cantor's system on the basis that in it the whole is not greater than a part, mathematicians will remind him that Euclid's maxim holds only for finite magnitudes, not infinite ones. But surely the question that then needs to be asked is, How does one know that the Principle of Correspondence does not also hold only for finite collections, but not for infinite ones? Here the mathematician can only say that it is simply defined

---

Weyl, 'Mathematics and Logic', *American Mathematical Monthly*, 53 (1946), 2–13. That a potential infinite need not imply an actual infinite, as Cantor contended, is argued by Hart, 'The Potential Infinite', 254–64.

as doing so. For all the finite examples in the world cannot justify the extrapolation of this principle to the infinite; its provability is precisely the same as Euclid's maxim. One can show that both of these principles hold true for finite collections, but neither can be proved to be true for infinite collections. Moreover, it is clear that they cannot *both* be true for infinite collections, since they are, in this case, contradictory principles: one asserts that the whole is greater than a part, while the other maintains that the whole is not greater than a part. But which principle is to be sacrificed? Both seem to be intuitively obvious principles in themselves, and both result in counter-intuitive situations when either is applied to the actual infinite. The most reasonable approach to the matter seems to be to regard both principles as valid in reality and the existence of an actual infinite as impossible.

In summary, we have argued in support of the first premiss of our syllogism: (1) that the existence of an actual infinite would entail various absurdities; (2) that the Cantorian analysis of the actual infinite may represent a consistent mathematical system, but that this carries with it no ontological import for the existence of an actual infinite in the real world; and (3) that even the mathematical existence of the actual infinite has not gone unchallenged and therefore cannot be taken for granted, which would then apply doubly to the real existence of the actual infinite. Therefore, we conclude that an actual infinite cannot exist.

The second premiss states that *an infinite temporal regress of events is an actual infinite.* By 'event' we mean 'that which happens'. Thus, the second premiss is concerned with change, and it asserts that if the series or sequence of changes in time is infinite, then these events considered collectively constitute an actual infinite. The point seems obvious enough, for if there has been a sequence composed of an infinite number of events stretching back into the past, then the set of all events would be an actually infinite set.

But manifest as this may be to us, it was not always considered so. The point somehow eluded Aristotle himself, as well as his scholastic progeny, who regarded the past sequence of events as a potential infinite. Aristotle contended that since things in time come to exist sequentially, an actual infinite never exists at any one moment; only the present thing actually exists.[31] Similarly, Aquinas,

---

[31] Aristotle, *Physics*, 3.6.206ª25–206ᵇ1.

after confessing the impossibility of the existence of an actual infinite, nevertheless proceeded to assert that the existence of an infinite regress of past events is possible.[32] This is because the series of past events does not exist in actuality. Past events do not now exist and hence do not constitute an infinite number of actually existing things. The series is only potentially infinite, not actually infinite, in that it is constantly increasing by addition of new events. But surely this analysis is inadequate. The fact that the events do not exist simultaneously is wholly irrelevant to the issue at hand; the fact remains that since past events, as determinate parts of reality, are definite and distinct and can be numbered, they can be conceptually collected into a totality. Therefore, if the temporal sequence of events is infinite, the set of all past events will be an actual infinite. It is interesting that at least one prominent Thomist agrees that Aquinas and Aristotle fail to carry their case; thus Fernand Van Steenberghen states, 'For him [Aristotle] an infinity in act is impossible; now a universe eternal in the past implies an infinite series in act, since the past is *acquired*, is *realized*; that this realization has been successive does not suppress the fact that the infinite series is *accomplished* and constitutes quite definitely an infinite series in act.'[33] Accordingly, Van Steenberghen maintains that Aquinas clearly contradicts himself by adhering both to the possibility of an infinite temporal regress of events and to the impossibility of an infinite multitude.[34] Aquinas's own example of the blacksmith working from eternity who uses one hammer after another as each one breaks furnishes a good example of an actual infinite. For the set of all hammers employed by the smith is an actual infinite. In the same way, the set of all events in an infinite temporal regress of events is an actual infinite.

The point raised by Aristotle and Aquinas serves to bring out an important feature of past events that is not shared by future events, namely their actuality. For past events have really existed; they have taken place in the real world, while future events have not, since they have not occurred. In no sense does the future actually

[32] Thomas Aquinas, *Summa Theologiae*, 1a.7.4.

[33] F. Van Steenberghen, 'Le "Processus in infinitum" dans les trois premières "voies" de saint Thomas', *Revista Portuguesa de Filosofia*, 30 (1974), 128. Another Thomist of the same judgement is Lucien Roy, 'Note philosophique sur l'idée de commencement dans la création', *Sciences Ecclesiastiques*, 1 (1949), 223.

[34] Van Steenberghen, 'Le "Processus in Infinitum"', 129.

exist—we must not be fooled by Minkowski diagrams of four-dimensional spacetime depicting the world line of some entity into thinking that future events somehow subsist further down the line, waiting for us to arrive at them. As P. J. Zwart rightly urges, Minkowski spacetime is only a diagrammatical method of displaying the relations of an object to time and space; the fact that we can mark out the future world line of an object in no way implies that these future events actually exist.[35] Only the sequence of past events can count as an actual infinity.

The importance of this difference between future and past events becomes evident when we turn to questions concerning the actual infinite. For clearly, past events are actual in a way in which future events are not. In the real sense, the set of all events later than any point in the past is not an actual infinite at all, but a potential infinite. It is an indefinite collection of events, always finite and always increasing. But the series of past events is an actual infinite, for at any point in the past the series of prior events remains infinite and actual.

Because the series of past events is an actual infinite, all the absurdities attending the real existence of an actual infinite apply to it. In fact, far from alleviating these absurdities, as Aristotle and Aquinas would have us believe, the sequential nature of the temporal series of events actually intensifies them. For example, we argued that it would be impossible to add to a really existent actual infinite, but the series of past events is being increased daily. Or so it appears. For if Cantor's system were descriptive of reality, the number of events that have occurred up to the present is no greater than the number that have occurred *at any point in the past*. Here the reader may be reminded of the argument of al-Ghazālī concerning the concentric spheres which revolved such that the innermost sphere completed one rotation in a year while the outermost sphere required thousands of years to complete a single rotation. If the sequence of past events is infinite, then which sphere has completed the most rotations? According to Cantor, if his system were descriptive of reality, the number of revolutions would be equal, for they could be placed in a one-to-one correspondence. But this is simply unbelievable, since every revolution of the great sphere generated thousands of revolutions in the little

---

[35] P. J. Zwart, *About Time* (Amsterdam: North-Holland, 1976), 179.

sphere, and the longer they revolved the greater the disparity grew. Therefore, in demonstrating that an infinite temporal regress of events is an actual infinite, we not only find that the absurdities pertaining to the existence of an actual infinite apply to it, but also that these absurdities are actually heightened because of the sequential character of the series.

Since an actual infinite cannot exist and an infinite temporal regress of events is an actual infinite, we may conclude that an infinite temporal regress of events cannot exist. But, it might be objected, do we not have decisive counter-examples to this conclusion? For any event can be divided up into phases, each of which constitutes an event in its own right, and each of these can be divided up again, and each again, *ad infinitum*. Thus, the transpiring of any event, however brief, involves the transpiring of an infinite number of events. Imagine, for example, two otherwise immobile rocks passing each other in outer space. That might seem to count as one event. But it can be broken down into a number of subevents: first, the frontal end-points of the rocks passed each other, then their mid-points, and finally their posterior end-points passed. But further subdivision is possible. Before the front point of one could pass the mid-point of the other, it had to pass the point midway between the frontal end-point and the mid-point, and so on. The reader will readily recognize that Zeno's paradoxes of motion are upon us. If the temporal regress of past events cannot be actually infinite, then no event can elapse, since the transpiring of any event involves an actually infinite number of subevents.

Now one way to respond to this objection would be to insist that the events one is talking about, whatever they might be, must be of the same duration. Since an event is a change, there are no instantaneous events. Neither could there be infinitely slow events. Therefore, any event one picks as one's standard will be of finite, non-zero duration, whether it be the orbits of the earth, the swing of a pendulum, or the periods of an atomic clock. By stipulating that the events must all be of the same duration, one precludes counting as events the phases of the event chosen as standard. Of course, we hasten to add, one is at liberty to drop the standard event and choose one of its phases as standard instead; but then one must apply that standard in talking about the set of events elapsed—one cannot count as an event the longer event, and then the subevent, and then a subsubevent, and so on. Our conclusion may be more properly stated as 'An infinite temporal regress of events of equal duration cannot exist'.

This solution might appear to be threatened by *metric conventionalism*, the view that what counts as isochronous intervals is wholly arbitrary, set by stipulation.[36] There is no factual truth of the matter concerning whether two temporal intervals are equal in length; we adopt an arbitrary standard clock and define isochronous intervals to be those corresponding to equal units of the conventional standard clock. Hence, the notion of events as equal in duration is not factual, but merely stipulative. Therefore, an interval reckoned to be of infinite duration according to one metric, that is, reckoned to contain an actually infinite number of events of equal duration, might be calculated to be finite according to another metric, that is, to contain only a finite number of events of equal duration. Since there literally is no matter of fact concerning the isochrony of temporal intervals, any temporal series can be regarded as finite or infinite.

But metric conventionalism, except in the trivial semantic sense,[37] is groundless and counter-intuitive. It is generally recognized, I think, that arguments for metric conventionalism, such as Grünbaum's, are unsound.[38] And any proposed metric of time will have to make peace with one's pre-philosophical intuitions in order to be acceptable.[39] The measure of a property which determines that that property is shared to an equal degree, for example, by the era of galaxy formation and by my lunch-break just is not a measure of temporal congruence. The same may be said of any metric according to which my lunch-break constitutes an infinite stretch of time, that

---

[36] For conventionalist viewpoints, see Henri Poincaré, *Science and Hypothesis*, in *The Foundations of Science* (Science Press, 1913; repr. Washington, DC: University Press of America, 1982), 57–66, 92–3; Hans Reichenbach, *The Philosophy of Space and Time*, trans. Maria Reichenbach and John Freund, Intro. Rudolf Carnap (New York: Dover, 1958), 11–36; Adolf Grünbaum, *Philosophical Problems of Space and Time*, 2nd edn., Boston Studies for the Philosophy of Science, xii (Dordrecht: D. Reidel, 1973), chs. 1–2, 16.

[37] Trivial Semantic Conventionalism states that the truth of a sentence like 'Five seconds elapsed between *A* and *B*' depends upon the definition given to the word 'second'. For a good discussion see Paul Gordon Horwich, 'On the Metric and Topology of Time', Ph.D. thesis, Cornell University, 1975.

[38] See the incisive critiques of Philip L. Quinn, 'Intrinsic Metrics on Continuous Spatial Manifolds', *Philosophy of Science*, 43 (1976), 396–414; Graham Nerlich, *The Shape of Space* (Cambridge: Cambridge University Press, 1976).

[39] A point made with special effectiveness by Michael Friedman, 'Grünbaum on the Conventionality of Geometry', Patrick Suppes (ed.), in *Space, Time and Geometry*, Synthese Library (Dordrecht: D. Reidel, 1973), 217–33.

is, an interval during which an actually infinite number of events of equal, non-zero duration occur. Given its fantastic and groundless character, metric conventionalism poses no threat to our requiring that the events in question be of equal duration.

But, for my part, I should like to go one step further than this and maintain that it makes no difference whether one stipulates that the events must be equal in duration or not: an actually infinite number of events cannot exist in either case. This is the thorough-going Aristotelian position on the infinite: only the potential infinite exists. I agree with G. J. Whitrow when he writes, 'although the hypothesis that time is truly continuous has definite *mathematical* advantages, it is an idealization, and not an actual characteristic of physical time'.[40] This position does not imply that minimal time atoms, or chronons, exist; time, like space, is infinitely divisible, in the sense that division can proceed indefinitely, but time is never actually infinitely divided, neither does one arrive at an instantaneous point. As Grünbaum explains, it is not infinite divisibility as such which gives rise to Zeno's paradoxes; the paradoxes presuppose the postulation of an actual infinity of points *ab initio*. '. . . any attribution of (infinite) "divisibility" to a Cantorian line must be based on the fact that *ab initio* that line and the intervals are already "divided" into an actual dense infinity of point-elements of which the line (interval) is the aggregate. Accordingly, the Cantorian line can be said to be already actually *infinitely divided*.'[41] By contrast, if we think of the line as a whole as logically prior to any points designated on it, then it is not an ordered aggregate of points nor actually infinitely divided. Time as a whole duration is thus logically prior to the (potentially infinite) divisions we make of it. Specified instants are not temporal intervals, but merely the boundary points of intervals, which are always non-zero in duration. Treating instants as degenerate temporal intervals of zero duration does seem to land one in Zeno's clutches, since temporal becoming would require the actualization of consecutive instants, which is incoherent.[42] But if every interval of time is of non-zero duration

[40] G. J. Whitrow, *The Natural Philosophy of Time*, 2nd edn. (Oxford: Clarendon Press, 1980), 200.

[41] Grünbaum, *Philosophical Problems of Space and Time*, 169.

[42] For a good discussion see Adolf Grünbaum, 'Relativity and the Atomicity of Becoming', *Review of Metaphysics*, 4 (1950–1), 143–86. Grünbaum succeeds in defending the continuity of time only at the expense of sacrificing temporal becoming, which his interlocutors James and Whitehead held to. I take up this issue in my forthcoming *God, Time, and Eternity*.

and there is only a potentially infinite number of such intervals in any finite interval, it follows that if an actually infinite number of temporal intervals existed, then time would be infinite. If, as we have argued, an actual infinite cannot exist, then (metric) time must be finite. So if we are correct, it makes no difference whether we stipulate that the events at issue be equal in duration or not; even if we take events of progressively shorter durations, we shall never come up with an actually infinite number of them in any finite interval. Thus, our conclusion that an actually infinite temporal regress of events cannot exist rules out an infinite, but not a finite, past.

Therefore, we conclude: the universe began to exist. And this is the second premiss of our original syllogism which we set out to prove. To recapitulate: since an actual infinite cannot exist and an infinite temporal regress of events is an actual infinite, we can be sure that an infinite temporal regress of events cannot exist, that is to say, the temporal regress of events is finite. Therefore, since the temporal regress of events is finite, the universe began to exist.

2.1.2. *Second Philosophical Argument*

We may now turn to our second philosophical argument in support of the premiss that the universe began to exist, the argument from the impossibility of the formation of an actual infinite by successive addition. The argument may be exhibited in this way:

(1) The temporal series of events is a collection formed by successive addition.

(2) A collection formed by successive addition cannot be an actual infinite.

(3) Therefore the temporal series of events cannot be an actual infinite.

Here we do not assume that an actual infinite cannot exist. Even if an actual infinite can exist, the temporal series of events cannot be such, since an actual infinite cannot be formed by successive addition, as the temporal series of events is.

The first premiss seems obvious enough. First, the collection of all past events prior to any given point is not a collection whose members all coexist. Rather it is a collection that is instantiated sequentially or *successively* in time, one event following upon the heels of another. Secondly, nor is the series formed by subtraction or division but by *addition* of one element after another. This elementary point merits underscoring since neglect of it can lead to confusion.[43] For although we may *think* of the past by subtracting events from the present, as when we say an event occurred ten years ago, it is none the less clear that the series of events is *formed* by addition of one event after another. We must be careful not to confuse the realms of thought and reality. Even the expression 'temporal regress' can be misleading, for the events themselves are not regressing in time; our thoughts regress in time as we mentally survey past events. But the series of events is itself progressing in time, that is to say, the collection of all past events grows progressively larger with each passing day. Nor is the series of events a tenselessly existing continuum from which events are formed by division. Neither subtraction nor division accounts for the sequential formation of the collection of past events. Therefore, the temporal series of events is a collection formed by successive addition.

The second premiss asserts that a collection formed by successive addition cannot be an actual infinite. Sometimes this is described as the impossibility of counting to infinity. For each new element added to the collection can be counted as it is added. It is important to understand exactly *why* it is impossible to form an actual infinite by successive addition. The reason is that for every element one adds, one can always add one more. Therefore, one can never arrive at infinity. What one constructs is a potential infinite only, an indefinite collection that grows and grows as each new element is added. Another way of seeing the point is by recalling that $\aleph_0$ has no immediate predecessor. Therefore, one can never reach $\aleph_0$ by successive addition or counting, since this would involve passing through an immediate predecessor to $\aleph_0$. *Notice that the argument has nothing to do with any time factor.* Sometimes it is wrongly alleged that the reason an actual infinite cannot be formed by

---

[43] For example, William James distinguishes between a 'standing' infinity and a 'growing' infinity, by which he means an infinity that is simply given and an infinity formed by successive addition. He acknowledges that a growing infinity can never be completed and that the infinite cannot be traversed, but he holds that past time is nevertheless infinite because it is a standing infinity. (William James, *Some Problems of Philosophy* (London: Longman, Green, 1911), 167–70, 182.) But this is clearly wrong-headed, for the past was formed precisely by successive addition and was never simply given. Thus, if it is infinite, it is an example of a growing infinite that has been completed, an actual infinite which has been traversed.

successive addition is because there is not enough time.[44] But this
is wholly beside the point. Regardless of the time involved an
actual infinite cannot be completed by successive addition due to
the very nature of the actual infinite itself. No matter how many
elements one has added, one can always add one more. A potential
infinite cannot be turned into an actual infinite by any amount of
successive addition; they are conceptually distinct. To illustrate:
suppose we imagine a man running through empty space on a path
of stone slabs, a path constructed such that when the man's foot
strikes the last slab, another appears immediately in front of him. It
is clear that even if the man runs for eternity, he will never run
across all the slabs. For every time his foot strikes the last slab, a
new one appears in front of him, *ad infinitum*. The traditional
cognomen for this is the impossibility of traversing the infinite. The
impossibility of such a traversal has nothing at all to do with the
amount of time available: it is of the essence of the infinite that it
cannot be completed by successive addition. As Russell himself
states, 'classes which are infinite are given all at once by the
defining properties of their members, so that there is no question of
"completion" or of "successive synthesis"'.[45] The only way in
which an actual infinite could come to exist in the real world would
be to be instantiated in reality all at once, simply given in a
moment. To try to instantiate an actual infinite progressively in the
real world would be hopeless, for one could always add one more
element. Thus, for example, if our library of infinite books were to
exist in the real world, it would have to be instantaneously created

[44] For example, Russell declares, 'when Kant says that an infinite series can
"never" be completed by successive synthesis, all that he has even conceivably a
right to say is that it cannot be completed *in a finite time*'. (Bertrand Russell, *Our
Knowledge of the External World*, 2nd edn. (New York: W. W. Norton, 1929),
171.)

[45] Ibid. 170. But suppose someone objects that the man in question has been
running from eternity past: if his foot strikes a stone every second and there are in
eternity past an infinite number of seconds, will he not have completed his course
successfully? In one sense, yes; *if* an infinite number of seconds could elapse, then
an infinite number of stones could be traversed. But this only pushes the issue one
step backwards: how can an infinite number of seconds elapse? One does not
eliminate the problem of forming an infinite collection by successive addition by
superimposing another collection on top of the first; for if one is possible, both are
possible, and if one is absurd, both are absurd. Since *any* collection formed by
successive addition cannot be infinite, an infinite number of seconds cannot have
elapsed. This means that time either had a beginning or that measured time was
preceded by an undifferentiated time.

ex nihilo by the divine fiat, 'Let there be . . . !' But even God could not instantiate the infinite library volume by volume, one at a time. No reflection on His omnipotence, for such a successive completion of an actual infinite is absurd.

This brings to mind Russell's account of Tristram Shandy, who, in the novel by Sterne, writes his autobiography so slowly that it takes him a whole year to record the events of a single day. Were he mortal, he would never finish, asserts Russell, but if he were immortal, then the entire book could be completed, since by the method of correspondence each day would correspond to each year, and both are infinite.[46] Such an assertion is wholly untenable, since the future is in reality a potential infinite only. Though he write for ever, Tristram Shandy would only get farther and farther behind so that instead of finishing his autobiography, he will progressively approach a state in which he would be *infinitely* far behind. But he would never reach such a state because the years and hence the days of his life would always be finite in number though indefinitely increasing.[47] Russell has confounded the actual infinite status which past events possess (if an infinite temporal regress exists) with the merely potential infinite status belonging to future events. But let us turn the story about: suppose Tristram Shandy has been writing from eternity past at the rate of one day per year. Would he now be penning his final page? Here we discern the bankruptcy of the Principle of Correspondence in the world of the real. For according to that principle, Russell's conclusion would be correct: a one-to-one correspondence between days and years could be established so that given an actual infinite number of years, the book will be completed. But such a conclusion is clearly ridiculous, for Tristram Shandy could not yet have written *today's* events down. In reality he could never finish, for every day of writing generates another year of work. But if the Principle of Corres-

---

[46] Bertrand Russell, *The Principles of Mathematics*, 2nd edn. (London: Allen & Unwin, 1937), 358–9. Remarkably, even Fraenkel appears to agree with Russell on this score, though a mathematician of his status ought to be acquainted with the difference between a potential and an actual infinite (Fraenkel, *Abstract Set Theory*, 30.)

[47] Russell's fallacy is also discerned by Whitrow, *The Natural Philosophy of Time*, 149. Whitrow argues that Russell presupposes the incompletable series of events in question may be regarded as a whole, when in fact it is not legitimate to consider the events of Tristram Shandy's life as a completed infinite set, since the author could never catch up with himself.

pondence were descriptive of the real world, he should have finished—which is impossible.

But now a deeper absurdity bursts into view.[48] For if the series of past events is an actual infinite, then we may ask, why did Tristram Shandy not finish his autobiography yesterday or the day before, since by then an infinite series of events had already elapsed? No matter how far along the series of past events one regresses, Tristram Shandy would have already completed his autobiography. Therefore, at no point in the infinite series of past events could he be finishing the book. We could never look over Tristram Shandy's shoulder to see if he were now writing the last page. For at any point an actual infinite sequence of events would have transpired and the book would have already been completed. Thus, at no time in eternity will we find Tristram Shandy writing, which is absurd, since we supposed him to be writing from eternity. And at no point will he finish the book, which is equally absurd, because for the book to be completed he must at some point have finished. What the Tristram Shandy story really tells us is that an actually infinite temporal regress is absurd.

The only way a collection to which members are being successively added could be actually infinite would be for it to have an infinite 'core' to which additions are being made. But then it would not be a collection *formed* by successive addition, for there would always exist a surd infinite, itself not formed successively but simply given, to which a finite number of successive additions have been made. But clearly the temporal series of events cannot be so characterized, for it is by nature successively formed throughout. Thus, prior to any arbitrarily designated point in the temporal series, one has a collection of past events up to that point which is successively formed and completed and cannot, therefore, be infinite.

Contemporary critics have faltered in the face of this reasoning. John Hospers, himself no friend of philosophical theism, acknowledges that it is insufficient simply to assert that an infinite series of events is possible because an infinite series of integers is possible. For, he asks, 'If an infinite series of events has preceded the present moment, how did we get to the present moment? How could we get to the present moment—where we obviously are

---

[48] See a similar argument in David A. Conway, 'Possibility and Infinite Time: A Logical Paradox in St Thomas' Third Way', *International Philosophical Quarterly*, 14 (1974), 201–8.

now—if the present moment was preceded by an infinite series of events?'[49] Concluding that this difficulty has not yet been overcome and that the issue is still in dispute, Hospers passes on to other forms of the cosmological argument, leaving this one unrefuted. Similarly William L. Rowe, after expositing Bonaventure's argument for creation, comments rather weakly, 'It is difficult to show exactly what is wrong with this argument,' and with that remark moves on without further ado to discuss the Leibnizian argument.[50] But contemporary thinkers have discussed more fully two other philosophical puzzles that embody issues closely related to the present argument: Zeno's paradoxes of motion and the thesis of Kant's first antinomy of pure reason. In order to keep this essay to a reasonable length, I have chosen to bypass those issues here and simply to refer the reader to my discussion of them in appendices 1 and 2 of *The* Kalām *Cosmological Argument*.

Hence, we may conclude that a collection formed by successive addition cannot be an actual infinite. Since the temporal series of events is a collection formed by successive addition, we conclude: therefore, the temporal series of events cannot be an actual infinite. This means, of course, that the temporal series of events is finite and had a beginning. Therefore, the universe began to exist. This is, again, the second premiss in our cosmological argument.

This completes the philosophical case in support of the second premiss. We have argued that the impossibility of the existence of an actual infinite implies that the universe began to exist and that even if an actual infinite could exist, the inability of this infinite to be formed by successive addition implies that the universe began to exist. We may now turn to the empirical confirmation of this argument.

2.2. *Empirical Confirmation*

Some persons may be sceptical about philosophical arguments concerning the universe—what has been characterized as 'armchair

---

[49] John Hospers, *An Introduction to Philosophical Analysis*, 2nd edn. (London: Routledge & Kegan Paul, 1967), 434. Hosper's own statement of the argument is defective, for he argues that it would take infinite time to get through an infinite series, and this is the same as never getting through. It is not the same, of course, but the argument has nothing to do with the amount of time allowed: it is *inherently* impossible to form an actual infinite by successive addition.

[50] William L. Rowe, *The Cosmological Argument* (Princeton, NJ: Princeton University Press, 1975), 122.

cosmology'. They distrust metaphysical arguments, considering them to be misguided attempts to legislate for reality what can and cannot be. They are liable to be more impressed by empirical facts than by abstract arguments and are apt to ask for scientific evidence that the universe began to exist. I shall now present such evidence, which is drawn from remarkable discoveries made within the last twenty years in what is undoubtedly one of the most exciting and rapidly developing areas of scientific research: astronomy and astrophysics. With astounding rapidity, one breakthrough has come upon the heels of another so that now the prevailing cosmological view among scientists is that the universe did have a beginning. Our empirical argument is divided into two parts: (1) the argument from the expansion of the universe and (2) the argument from thermodynamics. To put the empirical evidence into a proper framework, I shall argue that a model of the universe in which the universe has an absolute beginning is not only logically consistent but also 'fits the facts' of experience.

Again, in the interest of space, we shall in this essay consider only the argument from the expansion of the universe leaving the reader to consult *The* Kalām *Cosmological Argument* for the argument from thermodynamics.[51] When Einstein formulated his relativity theories, he assumed that (1) the universe is homogeneous and isotropic, so that it appears the same in any direction from any place and (2) the universe is in a steady state, with a constant mean mass density and a constant curvature of space. But finding that his original general relativity theory would not permit a model consistent with these two conditions, he was forced to add to his gravitational field equations the cosmological constant Λ in order to counterbalance the gravitational effect of matter and so ensure a static model of the universe. Another solution to Einstein's difficulty was noted by de Sitter, who observed that in an empty universe the conditions and field equations would be satisfied. The model is static because there is no matter. But should particles of matter be introduced, the Λ factor would cause them to be repulsed from each other, thus leading to an expanding universe. Such a model of

[51] For an introduction to scientific cosmology, see P. J. E. Peebles, *Physical Cosmology*, Princeton Series in Physics (Princeton, NJ: Princeton University Press, 1971); D. W. Sciama, *Modern Cosmology* (Cambridge: Cambridge University Press, 1971); J. D. North, *Measure of the Universe* (Oxford: Clarendon Press, 1965); and S. Weinberg, *Gravitation and Cosmology* (New York: Wiley, 1972).

an expanding universe became known as an Einstein–de Sitter model, and it serves as a special limiting case for a luxuriant jungle of expanding universe models that have grown up since then. Such models were derived independently by Friedman and Lemaître; they posited an expanding universe which began in a state of high density. At the same time that this purely theoretical work was going on, observational cosmology spearheaded by Hubble was progressing toward the same sort of picture of the real universe. Slipher had noted in 1926 that the optical spectra of the light from distant galaxies were shifted toward the red end of the spectrum. This was thought to be the result of the Doppler effect, which displaces the spectral lines in radiation received from a source because of the relative motion of the source in the line of sight.[52]

---

[52] There has been some controversy over the value of red-shifts as distance indicators. (For a synopsis of the debate, consult George B. Field, Halton Arp, and John N. Bahcall, *The Redshift Controversy*, Frontiers in Physics (Reading, Mass.: W. A. Benjamin, 1973). A lucid history of the dispute may also be found in Daniel Weedman, 'Seyfert Galaxies, Quasars and Redshifts', *Quarterly Journal of the Royal Astronomical Society*, 17 (1976), 227–62.) Pointing to various discrepancies in the red-shift data, especially those from quasi-stellar objects (QSOs), some have argued that some other factor may account for the observed red-shifts. But the weight of the evidence supports the expansion hypothesis. Bahcall enumerates six observational tests which the theory of red-shifts as distance indicators has passed, as well as three successful predictions of the theory. (Field *et al.*, *The Redshift Controversy*, 77–9, 108.) According to Lang and his colleagues, the uncertainty concerning QSO red-shifts is due to small sampling, and the QSO slope, when corrected, is comparable to that of galaxies. (Kenneth R. Lang, Steven D. Lord, James M. Johanson, and Paul D. Savage, 'The Composite Hubble Diagram', *Astrophysical Journal*, 202 (1975), 583–90.) The discrepancies of red-shifts among closely related stars, moreover, do not suffice to overturn the Doppler effect theory, according to P. C. Joss, D. A. Smith, and A. B. Solinger, 'On Apparent Associations among Astronomical Objects', *Astronomy and Astrophysics*, 47 (1976), 461–2. See also D. Wills and R. L. Ricklefs, 'On the Redshift Distribution of Quasi-stellar Objects', *Monthly Notices of the Royal Astronomical Society*, 175 (1976), 65p–70p; M. Rowan-Robinson, 'Quasars and the Cosmological Distance Scale', *Nature*, 262 (1976), 97–101; Richard F. Green and Douglas O. Richstone, 'On the Reality of Periodicities in the Redshift Distribution of Emission-Line Objects', *Astrophysical Journal*, 208 (1976), 639–45. In the early autumn of 1976 an international astronomical conference held in Paris devoted itself largely to a debate of the red-shift controversy. The results of this important conference are published in the *International Astronomical Union Colloquium*, 37, *Déclages vers le rouge et expansion de l'univers* (Paris: Edition de Centre National de la Recherche Scientifique, 1977). According to Peebles, there is no serious competitor to the expansion hypothesis, and he lists seven points in favour of an expanding universe: (1) the red-shifts from galaxies, (2) the frequency shifts of radio lines, (3) the fact that Hubble's law fits a homogeneous and isotropic universe, (4) the harmony of theory and observation, (5) the fact that the Hubble time coincides with the age of the stars and the elements, (6) the presence of the

When a source is moving toward an observer, there is a blue-shift in the spectral lines; when the source is receding, a red-shift occurs. Hubble demonstrated in 1929 that not only are all measured galaxies receding, but that their velocity of recession is proportional to their distance from us.[53] On what has become known as a Hubble diagram, he plotted the apparent magnitude, or measure of luminosity, of galaxies against their red-shifts. Taking magnitude as a measure of distance, Hubble concluded that red-shift and distance are linearly related: the greater the distance the greater the source's red-shift. This is known as Hubble's law. Since red-shift is indicative of velocity, the Hubble diagram showed that the further sources are from us the faster they are receding, so that a source twice as far away from us as another source is receding twice as fast. The constant expressing the ratio between the recession velocities of galaxies and their distance from us is known as the Hubble constant and is abbreviated H_0.

But not only was it found that the universe is expanding, but also that it is expanding isotropically; it is the same in all directions. No matter where in the sky a galaxy is measured, the ratio of its velocity to its distance is the same. Hence, the relational aspects of the universe do not vary (see Fig. I.1). If A, B, and C are three galaxies, then as the universe expands, they will recede from each other, but their relations remain constant. The work of Hubble and his colleagues appeared to establish conclusively that the universe is expanding, just as the theoretical models had predicted. This dramatic harmony of theory and observation was hailed by many as one of the greatest successful predictions in the history of science, and Einstein promptly urged that his original static model of the universe be discarded, along with the cosmological constant which he had introduced.

If the universe is expanding, then the obvious question next to be asked was, How long? The simplest model of the universe would be one in which the recessional velocity of the galaxies would

black body background radiation, and (7) the fact that relativistic corrections to stellar magnitudes brings these into harmony with the theory. (Peebles, *Physical Cosmology*, 25–7.) Most of these points will be explained later, but this seems the best place to list them all.

[53] Edwin Hubble, 'A Relation between Distance and Radial Velocity among Extra-galactic Nebulae', *Proceedings of the National Academy of Sciences*, 15 (1929), 168–73.

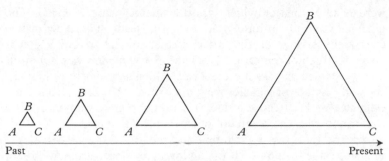

FIG. I.I. The configuration of galaxies *A*, *B*, and *C* remains constant over time, but as space expands, the distances separating them increase.

remain unchanged through time. In this case the expansion would have been going on for the time it would take any given galaxy at its present velocity to reach its present position, or, in other words, by the inverse of the Hubble constant. This is called the Hubble time and is the time elapsed from the beginning of the expansion until the present. The staggering implication of this is that by thus extrapolating back into the past, we come to a point in time at which *the entire known universe was contracted into an arbitrarily great density*; if one extrapolates the motion of the galaxies into the past as far as possible, one reaches a state of contraction of *infinite density*. If the velocity of the galaxies has remained unchanged, then one Hubble time ago, the universe began to expand from a state of infinite density in what has come to be called the 'Big Bang'.

Hubble's original estimate of the Hubble time was 2 billion years. This, however, conflicted sharply with the age of stellar objects and particularly of the earth itself as given by radioactive decay dating methods. Clearly, either the simplest models of the expanding universe were wrong or the value of H_0 was incorrect. New Big Bang models were devised, and an alternative model to the Big Bang theory was broached: the steady state model, expounded in 1948 by Bondi, Gold, and Hoyle.[54] The steady state model also involved an expanding universe, but not a Big Bang. Instead, it proposed that the gaps left between receding galaxies

[54] H. Bondi and T. Gold, 'The Steady-State Theory of the Expanding Universe', *Monthly Notices of the Royal Astronomical Society*, 108 (1948), 252–75; F. Hoyle, 'A New Model for the Expanding Universe', *Monthly Notices of the Royal Astronomical Society*, 108 (1948), 372–82.

40   *William Lane Craig*

were filled by matter which was drawn into being *ex nihilo*. The
universe was thus infinitely old and in a steady state, new matter
continually arising in the space vacated by the receding stellar
systems. Much of the rationale for these new models was removed,
however, when various inaccuracies in the measurement of luminosity and distance of galaxies were discovered. All these affected the
estimates of H_0, whose value was brought more and more into
conformity with the results of other dating procedures.

Perhaps the greatest triumph for the Big Bang model of the
universe came in 1965 with the discovery by Penzias and Wilson of
a microwave background radiation that permeates the entire
universe.[55] In 1946 Gamow had predicted the existence of such a
background radiation from purely theoretical considerations of the
physics of the early phases of a Big Bang universe. The earliest
phase of the universe[a] has been characterized as a 'primeval fireball',
with a temperature in excess of $10^{12\circ}$ K.[56] During this phase, called
the hadron era, which represents the first one-hundred-thousandth
(10^{-5}) of a second after the Big Bang, the temperatures were so
high that matter could not exist in a structured form—the universe
was filled with nuclear particles that were in equilibrium with the
field of radiation, that is to say, particles interacted to form photons
and vice versa. When the temperature had dropped to about $10^{12\circ}$ K,
the universe entered a phase called the lepton era, which lasted
from about one-ten-thousandth (10^{-4}) of a second after the Big
Bang to 10 seconds after the Big Bang. Before the temperature
reached $10^{11\circ}$ K, the density was low enough to allow neutrinos and
antineutrinos to cease interacting with photons and to maintain
independent existence.[57] At this point the universe consisted mostly
of photons, electrons (e^-), positrons (e^+), neutrinos, antineutrinos,
and a trace of protons and neutrons. As the temperature dropped to
about $10^{9\circ}$ K, all the positrons were annihilated by combination

[55] A. A. Penzias and R. W. Wilson, 'A Measurement of Excess Antenna
Temperature at 4080 Mc/s', *Astrophysical Journal*, 142 (1965), 419–21; see also
R. H. Dicke, P. J. E. Peebles, P. G. Roll, and D. T. Wilkinson, 'Cosmic Black-
Body Radiation', *Astrophysical Journal*, 142 (1965), 414–19.
[56] For a good synopsis of phases of the early universe, see E. R. Harrison,
'Standard Model of the Early Universe', *Annual Review of Astronomy and Astrophysics*,
11 (1973), 155–86.
[57] A neutrino is a stable subatomic particle that has no charge and zero mass
when at rest (which it never is, since it is travelling at the speed of light as long as it
exists).

with electrons, whose number exceeded that of the positrons. This marks the beginning of the radiation era, which lasts from 10 seconds until 10^{12} seconds after the Big Bang. Protons and neutrons combined to form helium; most of the present helium in the universe was synthesized at that time. All the neutrons were used up in this process, so that the universe consisted of protons and helium nuclei together with the electrons that made up an ionized gas; photons and neutrinos were also present. As the universe expanded, the temperature continued to drop until at about 3000°K the nuclei and the electrons of the ionized gas recombined to form an un-ionized gas. This marked the beginning of the matter era, around 10^{13} seconds after the Big Bang. The photons emitted during this recombination are what we detect as the microwave background today. From far beyond the faintest galaxies perceptible comes this low-frequency radiation, bathing the whole universe. This radiation field has the spectral characteristics of the thermal radiation emitted by a black body, that is, an ideal body which is a perfect emitter of radiation. Its amazing isotropy—it varies only about one part in a thousand[b]—supports the conclusion that this cosmic radiation is indeed the relic of an early era of the universe in which the universe was very hot and very dense. Due to the expansion of the universe, this radiation is Doppler shifted into the microwave region of the spectrum and cooled to about 2.76°K. Gamow, nearly twenty years before its discovery, had predicted 5°K. The agreement between the predictions based on Gamow's theoretical model and the observations of Penzias and Wilson constitutes powerful evidence in favour of the Big Bang cosmology. It was the second time that this model of the universe was successfully corroborated by empirical confirmation of theoretical predictions.

But when did the Big Bang occur? Estimates of the date of this event are dependent upon two factors: the value of H_0 and the effect of gravity in slowing down the galaxies' recessional velocity. During the 1970s scientists began to close in upon accurate estimations of these two values in what Allan Sandage, who has pioneered the drive, described as a thirty-year 'search for two numbers'.[58] During the past several years, in a very important series of articles,[c] Sandage and Tammann have described their steady progress toward a

[58] Allan R. Sandage, 'Cosmology: A Search for Two Numbers', *Physics Today* (Feb. 1970), 34.

determination of these figures.[59] In order accurately to determine H_0, more precise distance estimates had to be first obtained for galaxies sufficiently remote to have significant expansion velocities. This was accomplished in three progressive steps, moving from regions of ionized hydrogen in the interstellar space of nearby galaxies to intermediate isolated galaxies and finally to remote galaxies that are distant enough to permit reliable determination of the Hubble constant. They found the distances calculated to be greater than Hubble's estimates by a factor of ten. The first indications of an accurate Hubble constant were given by estimates obtained for the Virgo cluster, a cluster of about 2500 galaxies at a distance of 19.5 ± 0.8 Mpc.[60] The value of H_0 for this cluster is 57.0 ± 6 km s^{-1} Mpc^{-1}. (Or, in other words, the cluster is receding at about 57 kilometres per second for every megaparsec it is distant.) When Sandage and Tammann combined these results with individual red-shifts, they obtained a local Hubble constant of 57.0 ± 3 km s^{-1} Mpc^{-1}, which was the same as that for the Virgo cluster alone. The final results obtained from the remotest galaxies yields a global value for the Hubble constant very close to the local value: $H_0 = 55.0 \pm 5$ km s^{-1} Mpc^{-1}. This result is the most precise determination to date.[d] The inverse of the Hubble constant gives the Hubble time (H_0^{-1}): 17.7×10^9. Thus, assuming that acceleration is constant, the universe began about 18 billion years ago in a primordial cataclysmic event.

But the acceleration has undoubtedly not been constant. For the mutual gravitational attraction of the galaxies acts as a restraint on their acceleration from each other. As the universe expands the gravitational effect of matter will cause deceleration of the receding galaxies. Thus, in the past, galaxies must have been moving faster

[59] Allan Sandage and G. A. Tammann, 'Steps toward the Hubble Constant', i: 'Calibration of the Linear Sizes of Extragalactic H_{II} Regions', *Astrophysical Journal*, 190 (1974), 525–38; ii: 'The Brightest Stars in the Late-Type Spiral Galaxies', *Astrophysical Journal*, 191 (1974), 603–21; iii: 'The Distance and Stellar Content of the MIOI Group of Galaxies', *Astrophysical Journal*, 194 (1974), 223–43; iv: 'Distances to Thirty-Nine Galaxies in the General Field Leading to a Calibration of the Galaxy Luminosity Classes and a First Hint of the Value of H_0', *Astrophysical Journal*, 194 (1974), 559–68; v: 'The Hubble Constant from Nearby Galaxies and the Regularity of the Local Velocity Field', *Astrophysical Journal*, 196 (1975), 313–28; vi: 'The Hubble Constant Determined from Redshifts and Magnitudes of Remote Sc I Galaxies: the Value of q_0', *Astrophysical Journal*, 197 (1975), 265–80.

[60] Mpc is the abbreviation for megaparsec. A parsec is equal to 3.26 light years. Mega is a prefix meaning 10^6.

than they are now. Neglecting the decelerating effect of gravity leads to an overestimate of the age of the universe. Only if the acceleration of the galaxies has been constant would the Big Bang be exactly one Hubble time ago; if there has been a deceleration, then the universe actually began less than a Hubble time ago. Thus, in order to determine more accurately the date of the Big Bang we need to know the value of the deceleration of the expansion; this is called the deceleration parameter and is symbolized q_0. Since q_0 is a measure of the density of the universe (since it is the density that determines the deceleration), I shall reserve a discussion of the value of q_0 until later. For now I shall simply report that when the value of q_0 is taken into account, the age of the universe is reduced to about 15 billion years.[61][e]

Thus, according to the Big Bang model, the universe began with a great explosion from a state of infinite density about 15 billion years ago. Four prominent astrophysicists describe that event in these words:

the universe began from a state of infinite density about one Hubble time ago. Space and time were created in that event and so was all the matter in the universe. It is not meaningful to ask what happened before the big bang; it is somewhat like asking what is north of the North Pole. Similarly, it is not sensible to ask where the big bang took place. The point-universe was not an object isolated in space; it was the entire universe, and so the only answer can be that the big bang happened everywhere.[62]

This event that marked the inception of the universe becomes all the more remarkable when one reflects that a condition of 'infinite density' is precisely equivalent to 'nothing'. There can be no object in the real world that possesses infinite density, for if it had any size at all, it would not be *infinitely* dense.[63] As Hoyle points out, the steady state model requires the creation of matter from nothing, but so does the Big Bang; this is because as one follows the

---

[61] Sandage and Tammann, 'Steps toward the Hubble Constant', vi, 277.

[62] J. Richard Gott III, James E. Gunn, David N. Schramm, and Beatrice M. Tinsley, 'Will the Universe Expand Forever?' *Scientific American* (Mar. 1976), 65.

[63] It is sometimes erroneously asserted that black holes are infinitely dense; I have even seen them described in one popular magazine as infinitely dense and hence literally nothing! In reality a black hole (so called because light cannot escape its gravitational field, thus causing it to appear only as a dark area) is simply a region whose radius becomes smaller than $2GM/c^2$ (G = gravitational constant; M = mass; c = speed of light), and its density can be quite small if the mass is large.

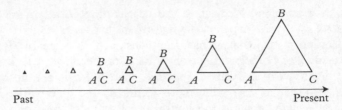

FIG. I.2. As one extrapolates the cosmic expansion into the past, the universe becomes denser and denser until one arrives at a point in the finite past at which the universe was infinitely dense and before which it did not exist.

expansion back in time, one reaches a time at which the universe was 'shrunk down to nothing at all'.[64]

This is the state of 'infinite density'. What a literal application of the Big Bang model really requires, therefore, is *creatio ex nihilo*.[65] A literal application of the Big Bang model in which the universe originates in an explosion from a state of infinite density, that is, from nothing, provides a simple, consistent, and empirically sound construction of how the universe began.

Some are unhappy about a theory of the origin of the universe which implies a beginning *ex nihilo*. But if one denies such an origin, then one is left with two alternatives: a steady state model or an oscillating model.[f]

Hoyle, the perennial champion of the steady state model, recoils at the notion of the origin of the universe from nothing (we shall see why later):

According to our observations and calculations, this was the situation some 15,000 million years ago. This most peculiar situation is taken by many astronomers to represent the *origin of the universe*. The universe is supposed

---

[64] Fred Hoyle, *From Stonehenge to Modern Cosmology* (San Francisco: W. H. Freeman, 1972), 36; Fred Hoyle, *Astronomy and Cosmology: A Modern Course* (San Francisco: W. H. Freeman, 1975), 658.

[65] In the words of Zel'dovich and Novikov: 'Quite recently the question has been discussed as to whether singularity, infinite density of matter at the onset of the expansion, actually exists. Perhaps singularity is characteristic only of an ideal, homogeneous, isotropic model. . . . *Today it has been rigorously shown that singularity did exist in the real universe*, even if early stages of expansion differed sharply from homogeneous isotropic expansion. (I. B. Zel'dovich and I. D. Novikov, 'Contemporary Trends in Cosmology', *Soviet Studies in Philosophy*, 14 (1976), 46–7 (my italics).)

to have begun at this particular time. From where? The usual answer, surely an unsatisfactory one, is: from nothing! The elucidation of this puzzle forms the most important problem of present day astronomy, indeed, one of the most important problems of all science.[66]

He elsewhere admits that the steady state model sought to bypass the conceptual problems involved in the origin of the universe.[67] But the steady state theory has taken a pretty heavy hammering since it was proposed in 1948, and even Hoyle seems ready to throw in the towel.[g]

Why, then, has he persisted in so doggedly opposing the Big Bang model of the universe? We have seen that, according to Hoyle's own admission, the steady state model sought to bypass the conceptual difficulties of the origin of the universe. But why would a man who proposes a theory requiring the continuous generation of matter *ex nihilo* have conceptual difficulties with the beginning of the universe from nothing? I get the strong impression that it is because Hoyle, unlike the vast majority of scientists, realizes the metaphysical and theological implications of such a beginning, and he recoils from these implications. Hoyle appears to have a strong philosophical, almost religious, streak, and he looks to the universe itself as a sort of God surrogate:

If you ask, 'Why investigate the structure of matter, of galaxies, of the universe?' the answer is no different in principle from the motives of the builders of Stonehenge. The motive is religious, ironically, more truly religious than the crude ritualistic survivals from the Stone Age that pass for religion in our modern communities. . . . It is to the structurally elegant and beautiful laws which govern the world that the modern scientist looks in his religious impulses. Discovering deeper levels of significance in these laws has the same religious meaning to the modern scientist that eclipses of the Sun and Moon had to Stone-Age man.[68]

The Big Bang model tends to undermine such religious veneration of the universe itself. Hoyle realizes that an absolute beginning of

---

[66] Fred Hoyle, *Astronomy Today* (London: Heinemann, 1975), 165.

[67] Fred Hoyle, 'The Origin of the Universe', *Quarterly Journal of the Royal Astronomical Society*, 14 (1973), 278.

[68] Hoyle, *From Stonehenge to Modern Cosmology*, 2. Hugo Meynell reports, 'In his contribution to a symposium, [*Religion and the Scientists*, ed. Mervyn Stockwood, p. 56] Hoyle says that he believes in a God in the sense of an intelligence at work in nature, but then he qualifies this by saying that he means by "God" the universe itself rather than some being who transcends the universe.' (Hugo Meynell, *God and the World: The Coherence of Christian Theism* (London: SPCK, 1971), 18–19.)

the universe points beyond the universe to a reality more ultimate than itself, since to say it simply sprang into being for no reason out of nothing is 'unsatisfactory'.[69] But this opens the door to theism and threatens the sort of pantheistic sentiments expressed above. This Hoyle simply will not have; in his text on astronomy and cosmology, he writes with regard to the absolute beginning of the universe:

Many people are happy to accept this position. . . . The abrupt beginning is regarded as *meta*physical—i.e., *outside* physics. The physical laws are therefore considered to break down at $\tau = 0$, *and to do so inherently*. To many people this thought process seems highly satisfactory because a 'something' outside physics can then be introduced at $\tau = 0$. By a semantic manœuvre, the word 'something' is then replaced by 'god', except that the first letter becomes a capital, God, in order to warn us that we must not carry the enquiry any further.[70]

Hence, Hoyle inveighs against those who import 'metaphysical intrusions' into the world.[71] It is this desire to avoid the intrusion of theism, in my opinion, that largely accounts for Hoyle's adherence to steady state models beyond the limits of plausibility.[72] In a sense, Hoyle's own philosophical streak is partly to blame for the intrusion of metaphysical considerations into cosmology, for the majority of scientists who adhere to a Big Bang model of the universe probably see no theistic implications in it whatsoever. Thus, when I asked Dr Tinsley of Yale what relevance the model had to the question of the existence of God, she replied, 'I don't see that all this has any bearing on the question. . . . I asked your . . . question to a group of my colleagues, and their initial reactions were the same as mine—no relevance.'[73] It is Hoyle the philosopher, not Hoyle the scientist, who cannot adhere to the Big Bang model of the universe. The scientific evidence points to a beginning of the universe, not a steady state universe. The only motivation for adhering to the steady state model today is philosophical. As

[69] Hoyle, *Astronomy Today*, 165.

[70] Hoyle, *Astronomy and Cosmology*, 684–5.

[71] Ibid. 685.

[72] Noting that the steady state theory failed to secure 'a single piece of experimental verification', Jaki points out that the theory's exponents often had 'overtly anti-theological, or rather anti-Christian motivations' for propounding this cosmological model. (Stanley L. Jaki, *Science and Creation* (Edinburgh and London: Scottish Academic Press, 1974), 347.)

[73] Beatrice M. Tinsley, personal letter.

Webster remarks, the steady state model seeks to side-step the issues of a beginning of time and the universe.[74] Hoyle is, I think, more astute than his colleagues in realizing that there *are* philosophical questions involved in the origin of the universe, but the desire to avoid these questions cannot change the drift of empirical fact, which has increasingly discredited the steady state model. According to Ivan King, 'The steady state theory has now been laid to rest, as a result of clearcut observations of how things have changed systematically with time.'[75] It serves now as 'an example of the lengths to which a philosophical system can stretch itself, in the absence of a sufficiently clear factual picture'.[76]

The other model of the universe which attempts to escape the necessity of an absolute beginning is the oscillating model. John Gribbin comments,

The biggest problem with the Big Bang theory of the origin of the Universe is philosophical—perhaps even theological—what was there before the bang? This problem alone was sufficient to give a great initial impetus to the Steady State theory; but with that theory now sadly in conflict with the observations, the best way round this initial difficulty is provided by a model in which the universe expands from a singularity, collapses back again, and repeats the cycle indefinitely.[77]

It is only within recent years that such a model has been rendered untenable by further advances in observational cosmology. In 1974 four of the world's leading astronomers concluded on the combined weight of several different lines of evidence that the universe is open, that is, will continue to expand for ever,[h] contrary to the oscillating hypothesis.[78] The key consideration is whether the density of the universe is great enough to overcome by gravitational attraction the recessional velocity of the galaxies and so pull the universe back together again. As I indicated earlier, the velocity of the galaxies is not constant, but is decelerating due to the force of

---

[74] Webster, 'The Cosmic Background Radiation', *Scientific American* (Aug. 1974), 26–8.

[75] Ivan R. King, *The Universe Unfolding* (San Francisco: W. H. Freeman, 1976), 462.

[76] Ibid.

[77] John Gribbin, 'Oscillating Universe Bounces Back', *Nature*, 259 (1976), 15.

[78] J. Richard Gott III, James E. Gunn, David N. Schramm, and Beatrice M. Tinsley, 'An Unbound Universe?' *Astrophysical Journal*, 194 (1974), 543–53. The previously cited article by the same authors in the *Scientific American* is a rewrite and revision of this article.

gravity. But if their velocity is great enough, they will continue to recede for ever, though at progressively slower speeds, just as a rocket reaching escape velocity will leave the earth's atmosphere and continue into space, though its speed will be slowed by the pull of the earth's gravity. In the latter case, the determining factor in whether the speed is great enough to escape the tug of the earth's gravity is the mass of the earth. In the same way the crucial factor in whether the universe will expand for ever is the average density of the universe. If the amount of matter per cubic volume is more than some critical value, then the gravitational force of matter will eventually overcome the velocity of the receding galaxies, and the universe is closed. The galaxies would reach the limit of their expansion, and then in an accelerating contraction they would rush together again. The critical density of the universe is estimated to be around 5×10^{-30} grams per cubic centimetre or about three hydrogen atoms per cubic metre. The ratio of the actual density to the critical density is called the density parameter and is designated by Ω. If $\Omega > 1$, then the universe is closed; if $\Omega < 1$, then the universe is open; and if $\Omega = 1$, then the universe is expanding precisely at escape velocity, and the universe is open.

There are several possible methods of determining whether the universe is open or closed. First, *one may seek to establish directly a value for q_0.* Since the deceleration parameter is a measure of the density of the universe, a determination of q_0 will yield the value of Ω. If $q_0 \leqslant \frac{1}{2}$, then $\Omega \leqslant 1$, and the universe is open. If $q_0 > \frac{1}{2}$, then $\Omega > 1$, and the universe is closed. Unfortunately, the value of q_0 cannot be determined by direct observation because the deceleration of any galaxy's velocity would be negligible within the span of a human lifetime. But it is possible to measure the velocities of very remote galaxies and compare them with the velocities of nearer galaxies. Because the light from the remote galaxies has taken billions of years to reach us, as we look at these galaxies, it is just as though we were looking back in time to the state of these systems as they were in the remote past. Galaxies nearer to us are more representative of the present, since the light from them is 'younger'. By comparing the Doppler shifts of these galaxies, astronomers can determine whether galaxies are today moving significantly slower than they were in the past. If the velocity of expansion was greater in the past, then at extreme distances the velocity of the galaxies should be greater than that predicted by Hubble's law. Distance estimates are, however, difficult to ascertain with precision. The distance of galaxies must be estimated on the basis of their apparent magnitude. Presumably the faintest galaxies are furthest away. But because galaxies are not all of the same intrinsic luminosity, it is possible for a fainter source to be actually closer than a brighter source. Fortunately, such individual anomalies tend to be ironed out when a sufficiently large sampling is taken because the discrepancies cancel each other out. Another factor that needs to be taken into account is the change of luminosity of stars in the course of their evolution. The light from the stars of an isolated galaxy probably declines a few percentage points every billion years, which means that galaxies were brighter in the past. If one were to neglect this decline in luminosity, one would conclude that the galaxies are closer than they really are, since the light received from them originates from a time when they were brighter. Thus, the distances would be underestimated, and consequently the value of q_0 would be overestimated—the galaxies would not have decelerated as much as it appears. According to Gott, Gunn, Schramm, and Tinsley, the best observable results, taking into account the decline in luminosity, suggest a value for q_0 that is closer to o than to $\frac{1}{2}$, and, therefore, the universe is open. They hasten to add, however, that these results are very uncertain[i] and that therefore one cannot conclude from this test alone that $q_0 < \frac{1}{2}$. But very large values for q_0, such as $q_0 = 2$, do appear to be excluded.

Secondly, *one may determine the age of the universe.* If there were no deceleration, then the age of the universe would equal one Hubble time. But if the expansion is decelerating, then a comparison of the universe's true age with the Hubble time should enable one to calculate the amount of deceleration and thus determine q_0. There are two possible methods in estimating the universe's age, both yielding lower limits only since they measure objects in the universe which were formed later than the Big Bang. (1) One may determine the age of the oldest stars. These are found in the globular clusters within our own galaxy. These formations are tightly packed, symmetrical groups of thousands of very old stars, whose ages have been estimated between 8 and 16 billion years old.[79] (2) One may measure the abundance of certain heavy

[79] See Icko Iben, 'Post Main Sequence Evolution of Single Stars', *Annual Review of Astronomy and Astrophysics*, 12 (1974), 215–56. See also Icko Iben, 'Globular Cluster Stars', *Scientific American* (July 1970), 26–39.

elements. Radioactive elements such as ^{232}Th and ^{187}Re are thought to have been formed in supernovas, which have probably been exploding since the formation of the galaxy. Dating of these elements by radioactive decay methods provides an estimate of their age and hence the age of the galaxy. These indicate an age between 6 and 20 billion years.[80] If the two ages are to be consistent, they suggest a date for the Big Bang of between 8 and 18 billion years ago. Unfortunately, these estimates are so generous that it is not possible to determine from them whether the universe is open or closed.

Thirdly, *one may measure the average density of matter in the universe in order to determine directly* Ω. This may be accomplished in two ways. (1) One may count the galaxies in a given volume of space, multiply by the masses of the galaxies, and divide by the volume. The masses of galaxies may be estimated by their observed gravitational effects upon one another. Estimates of the masses of a great many galaxies in different counts of large volumes of space indicate that if all the mass in the universe is associated with galaxies then Ω can be only about 0.04, and the universe is definitely open. Even if the uncertainty is of a factor of three such that $\Omega = 0.12$, this is still far below the density needed to halt the expansion. (2) One may compare the behaviour of distant galaxies with the behaviour of those in the local supercluster of galaxies of which our own galaxy is a member. The average density of galaxies in the local supercluster is two and a half times greater than that in the universe as a whole. If all mass is associated with galaxies, then the average density of the matter in the supercluster must be two and a half times greater than that outside it. Because of the greater density of the supercluster, nearby galaxies should be decelerated. If Ω is small, the greater deceleration would be imperceptible; but if Ω is large, there ought to be a significant difference. In fact, the difference is undetectable. Therefore, Ω must be very small, no larger than 0.1. These estimates indicate that the galaxies themselves cannot close the universe.

It might be objected that substantial amounts of matter are not, in fact, associated with the galaxies, but exist in the form of intercluster gas, dust, and other material. However, there appears

[80] See David N. Schramm, 'Nucleo-Cosmochronology', *Annual Review of Astronomy and Astrophysics*, 12 (1974), 383–406. See also David N. Schramm, 'The Age of the Elements', *Scientific American* (Jan. 1974), 69–77.

to be no evidence to substantiate this. If the galaxies condensed from a smoothly distributed gas, then they would eventually have pulled in the extraneous material as well. It might be contended that the missing mass needed to close the universe consists of some uniformly distributed medium with enough internal pressure not to be affected by the galaxies' gravitation. But this is untenable, according to Gott and his collaborators, because then the galaxies would have been prevented from forming in the first place.

Fourthly, *one may measure the amount of deuterium in the universe in order to discover* Ω. Within a few minutes after the Big Bang, subatomic particles begin to interact to form the lighter elements. The simplest of these is deuterium, which is formed by the synthesis of one proton and one neutron.[81] Deuterium nuclei would then combine to form helium, which is composed of two protons and two neutrons. The proportion of deuterium to helium is closely related to the density of the universe at that time. If the universe possessed a great density, most or all of the deuterium would have been converted into helium. But, in fact, this was not the case, for in the present universe deuterium is abundant, 4×10^{-31} grams per cubic centimetre in nearby interstellar space. For any value of the Hubble time between 13 and 19 billion years, a value of Ω as great as 1 is inconsistent with the deuterium abundance.

This argument assumes that no deuterium has been synthesized since the Big Bang. It has been suggested by Hoyle and Fowler that deuterium could possibly be synthesized in shocks in the envelopes of supernovae.[82] But according to Gott and his colleagues, this possible production of 'deuterium *ex machina*' 'cannot be taken seriously' because the amounts of boron and beryllium also produced would be greater than their observed abundance ratios to deuterium.[83] According to Audouze and Tinsley, even if all observed

---

[81] For a good general review of the deuterium question, see Jay M. Pasachoff and William A. Fowler, 'Deuterium in the Universe', *Scientific American* (May 1974), 108–18.

[82] Fred Hoyle and William A. Fowler, 'On the Origin of Deuterium', *Nature*, 241 (1973), 384–6.

[83] Gott *et al.*, 'An Unbound Universe?' 548. If deuterium were produced in supernovae, then we would have too much ^{7}Li, ^{9}Be, and ^{11}B, according to Richard I. Epstein, W. David Arnett, and David N. Schramm, 'Can Supernovae Produce Deuterium?' *Astrophysical Journal*, 190 (1974), L13–16. See also Harrison, 'Standard Model of the Early Universe', 166–9; Hubert Reeves, Jean Audouze, William A. Fowler, and David N. Schramm, 'On the Origin of Light Elements', *Astrophysical Journal*, 179 (1973), 909–30.

B and Be were produced by supernovae shocks—which is unlikely—
the amount of D produced is still much less than that observed.[84]
Therefore, the present abundance of deuterium is probably due to
primordial factors.

These four constraints—the deceleration parameter, the age of
the universe, the density of galaxies, and the abundance of
deuterium—all combine to point to a value of Ω between 0.04 and
0.09, a value well below that required to close the universe. And it
is noteworthy that two other observations are consistent with this
range of values. (1) The age of globular cluster stars is extremely
sensitive to the primordial helium abundance, which is, in turn,
dependent upon the density of the universe. The age of the stars
determined by nucleochronology is consistent with the age required
by the helium abundance in a universe of $\Omega < 1$. (2) These
constraints require a Hubble time between 13 and 20 billion years.
This is consistent with the estimates both of Sandage and Tammann
and of Kirschner and Kwan, who employed a method completely
independent of all steps in the classic approach of the former
team.[85] The evidence straightforwardly interpreted therefore
suggests a universe which will expand for ever.

Since the time Gott and his collaborators first broached their case
for an open universe, developments in this fast-moving field of
science have confirmed their conclusions. Perhaps most striking is
the work of Sandage and Tammann to home in on accurate values
for H_0 and q_0, which served as only the most general of constraints
for Gott and his colleagues. We have seen that Sandage and
Tammann determined $H_0 = 55 \pm 5\,\mathrm{km\,s^{-1}\,Mpc^{-1}}$. The best way
to determine q_0, they contend, is to derive the age of the universe
from other sources.[86] The oldest known objects in the universe, the
halo globular clusters, are estimated to have an age of $14 \pm 1 \times 10^9$

[84] Jean Audouze and Beatrice M. Tinsley, 'Galactic Evolution and the Formation
of the Light Elements', *Astrophysical Journal*, 192 (1974), 487–500.

[85] R. P. Kirschner and J. Kwan, cited in Gott *et al.*, 'An Unbound Universe?'
552.

[86] Sandage and Tammann, 'Steps toward the Hubble Constant', vi. 276. Davis
and May have reported that they have determined as a result of observations of the
839.4 MHz absorption line in the spectrum of the quasar 3C 286 a red-shift accuracy
of one part in 10^6; observations over several decades could place 'useful limits on q_0'.
(Michael M. Davis and Linda S. May, 'New Observations of the Radio Absorption
Line in 3C 286, with Potential Application to the Direct Measurement of Cosmological Deceleration', *Astrophysical Journal*, 219 (1978), 3.) Thus, indirect estimates
of q_0 may not be the exclusive possibility in the future.

years, which serves as an upper limit for q_0. But one must allow additional time for the collapse of the galaxy in question and for the formation of the protogalaxy out of the pre-galactic gas; this pushes the origin of the universe back a billion years to 15 billion years ago. From this data it may be determined what the value of q_0 is:

q_0	T_0 ($H_0 = 55$)
0.000	17.73
0.025	16.58
0.100	15.00
0.200	13.81
0.300	12.98
0.500	11.83
0.600	10.12

If the universe is at least 15 billion years old, then q_0 is no greater than 0.1, which is far below the $q_0 < \frac{1}{2}$ needed to guarantee an open universe. This means, Sandage and Tammann conclude, that *the universe has happened only once and that the expansion will never stop.*[87] In a later article, Sandage reviews six methods for determining q_0: (1) direct measurement of the deceleration by comparison of nearby to remote galaxies' red-shifts; (2) comparison of the Hubble time with the age of events in the early universe to give an accurate estimate of T_0; (3) attempts to determine the average density of the universe by dynamical arguments concerning the infall of matter into clusters yielding the observed X-ray flux; (4) comparison of the observed deuterium abundance with the calculated production of D in the Big Bang as a function of density; (5) summing up the density of galaxies, and (6) use of local velocity perturbations to measure the effect of gravity on the expansion.[88] Utilizing the last method, Sandage derives a value for $q_0 < 0.28 \pm 0.09$. This accords closely with the results obtained by all the other five methods and suggests, according to Sandage, (1) that the deceleration is almost negligible, (2) the universe is open, and (3) the expansion will not reverse.[89] It has even been suggested by

---

[87] Ibid. 278.

[88] Allan Sandage, 'The Redshift–Distance Relation', viii: 'Magnitudes and Redshifts of Southern Galaxies in Groups: A Further Mapping of the Local Velocity Field and an Estimate of q_0', *Astrophysical Journal*, 202 (1975), 563–82.

Gunn and Tinsley that far from decelerating, the expansion may actually be accelerating.[90] Thus, the case for an open universe has been considerably strengthened by more precise calculations of q_0.[j]

A value of $\Omega < 1$ is confirmed by Gott and Rees in their analysis of galaxy formation.[91] If galaxies evolved from primordial density fluctuations, then the data fit much better with a model in which $\Omega = 0.1$ than a model in which $\Omega = 1.0$. Elsewhere Turner and Gott also argue that it is much easier to maintain a stable population of galaxies in an open cosmology.[92] If the universe were closed, the population would be unstable, tending to the formation of more galactic clusters. In another study Gott and Turner estimate the mean luminosity density of galaxies and their characteristic mass to light ratio and from these attempt to discover the contribution Ω_G of galaxies to the critical density.[93] They calculate $\Omega_G = 0.08$ and report, 'our conclusion is an old one: if the $\Lambda = 0$ Friedmann world model applies and if galaxies and their environs contain more than a small fraction of all matter, the Universe is open by a wide margin'.[94] An independent confirmation of Gott and his colleagues' estimates of Ω is reported by Fall, who calculates $0.01 \leqslant \Omega \leqslant$

[89] Ibid. 579.

[90] James E. Gunn and Beatrice M. Tinsley, 'An Accelerating Universe', *Nature*, 257 (1975), 454–7. There are complications with this theory, however, and it has been suggested by Ostriker and Tremaine that a counter-evolutionary effect may restore deceleration, though the universe would still be open. (J. P. Ostriker and Scott D. Tremain, 'Another Evolutionary Correction to the Luminosity of Giant Galaxies', *Astrophysical Journal*, 202 (1975), L113–17.) According to Tinsley, 'all these effects are so uncertain that we should stick to the view given in the earlier papers, that estimates of the density constitute the strongest tests for the type of cosmological model, and they point to monotonic expansion from the big bang. The arguments in our paper based on density and age estimates have been strengthened, but not substantially altered, by subsequent developments.' (Tinsley, personal letter.) Cf. Beatrice M. Tinsley, 'The Cosmological Constant and Cosmological Change', *Physics Today*, 30 (1977), 32–8.

[91] J. Richard Gott III and Martin J. Rees, 'A Theory of Galaxy Formation and Clustering', *Astronomy and Astrophysics*, 45 (1975), 365–76.

[92] Edwin L. Turner and J. Richard Gott III, 'Evidence for a Spatially Homogeneous Component of the Universe: Single Galaxies', *Astrophysical Journal*, 197 (1975), L89–93.

[93] J. Richard Gott III and Edwin L. Turner, 'The Mean Luminosity and Mass Densities in the Universe', *Astrophysical Journal*, 209 (1976), 1–5.

[94] Ibid. 4, 5. By weighting each galaxy by its luminosity rather than weighting giant and dwarf galaxies equally, Turner and Ostriker determine a mass to light ratio that also yields $\Omega \sim 0.08$. (Edwin L. Turner and Jeremiah P. Ostriker, 'The Mass to Light Ratio of Late-Type Binary Galaxies: Luminosity versus Number-Weighted Averages', *Astrophysical Journal*, 217 (1977), 24–36.)

0.05.[95] He supports the view that the density of matter must be low in order that inhomogeneities do not disrupt the recessional expansion of galaxies more than is observed. The universe is therefore open by a large margin, he concludes. The calculations of Eichler and Solinger as to the amount of unseen matter in the universe in the form of burned out stars and black holes confirm the previous estimates that it is not sufficient to close the universe.[96 k]

The researches of Epstein and Petrosian confirm that the abundance of deuterium points to an open universe.[97] They note that the deuterium abundance would have to be less than one-fifth of the observed lower limit in order for the universe to be closed. If the universe is to be closed then either (1) our galaxy was formed out of especially deuterium-rich matter or (2) much deuterium-deficient matter (including black holes) is preferentially excluded from the interstellar gas. The former possibility is unlikely because in the more dense regions, which would presumably be the sites for galaxy formation, deuterium is underabundant. As to the latter possibility, either the black holes formed prior to or subsequent to nucleosynthesis of the elements. If prior, they could not affect deuterium production, and the abundance of deuterium is characteristic of diffuse matter. If subsequent, the higher-density regions are deficient in deuterium and so would enhance its abundance in the remaining gas. These considerations weigh against the possibility of a closed universe. According to Epstein and Petrosian, the mass density of the universe is at least a factor of three below the critical density needed to close it. Audouze asserts that arguments from the

---

[95] S. Michael Fall, 'The Scale of Galaxy Clustering and the Mean Matter Density of the Universe', *Monthly Notices of the Royal Astronomical Society*, 172 (1975), 23p–26p.

[96] David Eichler and Alan Solinger, 'The Electromagnetic Background: Limitations on Models of Unseen Matter', *Astrophysical Journal*, 203 (1976), 1–5. Field and Perrenod argue further that both observational and theoretical constraints (including limitations of the energy source needed to heat the gas, the observed deuterium abundance, and the lack of evidence for any clumping of gas clouds in intergalactic space, without which they could not persist) combine to make a cosmologically significant amount of hot intergalactic gas uncertain. (George B. Field and Stephen C. Perrenod, 'Constraints on a Dense Hot Intergalactic Medium', *Astrophysical Journal*, 215 (1977), 717–22.)

[97] Richard I. Epstein and Vahe Petrosian, 'Effects of Primordial Fluctuations on the Abundances of Light Elements', *Astrophysical Journal*, 197 (1975), 281–4. See also Donald G. York and John B. Rogerson, Jr., 'The Abundance of Deuterium Relative to Hydrogen in Interstellar Space', *Astrophysical Journal*, 203 (1976), 378–85.

nucleosynthesis of elements like helium and deuterium actually constitute the best indicators in favour of an open universe.[98] [l]

The evidence therefore appears to preclude an oscillating model of the universe, since such a model requires a universe of closure density.[m] At any rate, this model would seem to be only a theoretical, not a real, possibility; as Tinsley comments with regard to oscillatory models: 'even though the mathematics says that the universe oscillates, there is no known physics to reverse the collapse and bounce back to a new expansion. The physics seems to say that those models start from the big bang, expand, collapse, then end.'[99] In such a case one does not escape the necessity of an absolute beginning of the universe.

In summary, we have seen that (1) the scientific evidence related to the expansion of the universe points to an absolute beginning of the universe about 15 billion years ago; (2) the steady state model of the universe cannot account for certain features of observational cosmology, and (3) the oscillating model of the universe violates several constraints of observational cosmology which indicate that the universe is open. Therefore, we conclude that the universe began to exist.

This concludes our empirical support for our second premiss. I have argued that scientific evidence concerning the expansion of the universe indicates that the universe is finite in duration, beginning to exist about 15 billion years ago. This is a truly remarkable confirmation of the conclusion to which philosophical argument alone led us. In support of our second premiss, then, we have maintained four distinct arguments, two philosophical and two

---

[98] Jean Audouze, 'L' Univers est-il ouvert ou fermé?' *Recherche*, 6 (1975), 462–5. For a good summary of possible mechanisms for deuterium production, see Richard I. Epstein, James M. Lattimer, and David N. Schramm, 'The Origin of Deuterium', *Nature*, 263 (1976), 198–202. After exploring the possible production of deuterium by spallation reactions, pre-galactic cosmic rays, shock waves in a low density medium, hot explosive events, and disrupted neutron stars, they conclude, 'very severe restrictions can be placed on mechanisms for producing deuterium. In fact it seems unlikely that objects or events which are currently known or inferred to have existed in our Galaxy or in well-observed extra-galactic objects could be capable of producing the observed deuterium abundance. . . . Big-bang nucleosynthesis . . . requires only the simplest cosmological assumptions and is consistent with all well-established cosmological data. Contrasted to this, post-big-bang deuterium production requires extremely violent and exotic events, the existence of which is certainly doubtful. . . . Big-bang nucleosynthesis is by far the most reasonable site for the origin of deuterium.' (Ibid. 199, 202.)

[99] Tinsley, personal letter.

empirical: (1) the argument from the impossibility of the existence of an actual infinite; (2) the argument from the impossibility of the formation of an actual infinite by successive addition; (3) the argument from the expansion of the universe; and (4) the argument from thermodynamics. In the light of these considerations, I think we are amply justified in concluding our second premiss: the universe began to exist.

3. FIRST PREMISS: EVERYTHING THAT BEGINS TO EXIST HAS A CAUSE OF ITS EXISTENCE

We may now return to a consideration of our first premiss, that everything that begins to exist has a cause of its existence. The phrase 'cause of its existence' needs clarification. Here I do not mean sustaining or conserving cause, but creating cause. We are not looking here for any continual ground of being, but for something that brings about the inception of existence of another thing. Applied to the universe, we are asking, Was the beginning of the universe caused or uncaused? In this book I do not propose to construct an elaborate defence of this first premiss. Not only do considerations of time and space (in their practical, not philosophical, sense!) preclude such, but I think it to be somewhat unnecessary as well. For the first premiss is so intuitively obvious, especially when applied to the universe, that probably no one in his right mind *really* believes it to be false. Even Hume himself confessed that his academic denial of the principle's demonstrability could not eradicate his belief that it was none the less true.[100] Indeed the idea that anything, especially the whole universe, could pop into existence uncaused is so repugnant that most thinkers intuitively recognize that the universe's beginning to exist entirely uncaused out of nothing is incapable of sincere affirmation. For, as Anthony Kenny emphasizes, 'According to the big bang theory, the whole matter of the universe began to exist at a particular time

---

[100] Hume wrote, 'But allow me to tell you that I *never* asserted so absurd a Proposition as *that anything might arise without a cause*: I only maintain'd, that our Certainty of the Falshood of that Proposition proceeded neither from Intuition nor Demonstration; but from another Source.' (David Hume to John Stewart, Feb. 1754, in *The Letters of David Hume*, ed. J. Y. T. Greig, 2 vols. (Oxford: Clarendon Press, 1932), 1. 187.)

in the remote past. A proponent of such a theory, at least if he is an atheist, must believe that the matter of the universe came from nothing and by nothing.'[101] Now this is a pretty hard pill to swallow. Thus Broad, after asserting that the universe either (1) had no beginning and is temporally infinite, (2) had no first event but is temporally finite, being analogous to the series . . . , $\frac{1}{8}, \frac{1}{4}, \frac{1}{2}$, or (3) had a first event and is temporally finite, makes this very interesting comment:

> Now . . . I find no difficulty in supposing that the world's history had no beginning and that its duration backwards from its present phase is infinite. Nor do I find any insuperable difficulty in supposing that the world's history had no beginning, but that its duration backwards from its present phase does not exceed a certain finite limiting value. But I must confess that I have a very great difficulty in supposing that there was a first phase in the world's history, i.e., a phase immediately before which there existed neither matter, nor minds, nor anything else. . . . I suspect that my difficulty about a first event or phase in the world's history is due to the fact that, whatever I may *say* when I am trying to give Hume a run for his money, I can not really *believe in* anything beginning to exist without being *caused* (in the old-fashioned sense of *produced* or *generated*) by something else which existed before and up to the moment when the entity in question began to exist. . . . I . . . find it impossible to give up the principle; and with that confession of the intellectual impotence of old age I must leave this topic.[102]

It seems that old age had actually brought a measure of insight to Broad, for while his first two alternatives are untenable in the light of our arguments in favour of the second premiss, his reasoning concerning the third alternative is entirely correct. That the universe began to exist is true enough, but that it should begin to exist utterly uncaused out of nothing is too incredible to be believed. And Broad is not alone in this; for example, in the Maimonidean–Thomist argument that a necessary being must exist or else given infinite time nothing would exist,[103] virtually no one ever questions the premiss that if in the past nothing existed then nothing would exist now. That something should spring into existence out of

---

[101] Anthony Kenny, *The Five Ways: St Thomas Aquinas' Proofs of God's Existence* (New York: Schocken Books, 1969), 66.

[102] C. D. Broad, 'Kant's Mathematical Antinomies', *Proceedings of the Aristotelian Society*, 40 (1955), 9–10.

[103] Moses Maimonides, *Guide for the Perplexed*, 2.1; Thomas Aquinas, *Summa Theologiae*, 1a.2.3.

nothing is so counter-intuitive that to attack Maimonides and Aquinas at this point seems to colour one's intellectual integrity. The old principle *ex nihilo nihil fit* appears to be so manifestly true that a sincere denial of this axiom is well-nigh impossible.[104] Reluctant to fly in the face of this principle, many philosophers find it easier to believe that the universe must be temporally infinite. But we have already shown that both philosophical and empirical reasoning preclude a temporally infinite universe. The alternatives, then, are two: either the universe was caused to exist or else it sprang into existence wholly uncaused out of nothing a finite number of years ago. The first alternative appears eminently more reasonable.

It has been contended, however, that the second alternative is equally plausible. Thus, Hume writes,

as all distinct ideas are separable from each other, and as the ideas of cause and effect are evidently distinct, 'twill be easy for us to conceive any object to be non-existent this moment, and existent the next, without conjoining to it the distinct idea of a cause or productive principle. The separation, therefore, of the idea of a cause from that of a beginning of existence, is plainly possible for the imagination; and consequently the actual separation of these objects is so far possible, that it implies no contradiction nor absurdity; and is therefore incapable of being refuted by any reasoning from mere ideas; without which 'tis impossible to demonstrate the necessity of a cause.[105]

What this argument amounts to, as Anscombe points out, is that because we can imagine something's coming into existence without a cause, it is possible that something really can come into existence without a cause.[106] But, she continues,

The trouble about it is that it is very unconvincing. For if I say I can imagine a rabbit coming into being without a parent rabbit, well and good: I imagine a rabbit coming into being, and our observing that there is no parent rabbit about. But what am I to imagine if I imagine a rabbit coming into being without a cause? Well, I just imagine a rabbit coming into

---

[104] It has been alleged that *creatio ex nihilo* also violates this principle. (Leroy T. Howe, 'God and the Being of the World', *Journal of Religion*, 53 (1973), 411.) But *creatio ex nihilo* asserts only that creation lacks a *material* cause, not an *efficient* cause. In this premiss we are maintaining the necessity of an efficient cause of the universe.
[105] David Hume, *A Treatise of Human Nature*, I. iii. 3.
[106] G. E. M. Anscombe, '"Whatever Has a Beginning of Existence Must Have a Cause": Hume's Argument Exposed', *Analysis*, 34 (1974), 150.

being. That this *is* the imagination of a rabbit coming into being without a cause is nothing but, as it were, the *title* of the picture. Indeed I can form an image and give my picture that title. But from my being about to do *that*, nothing whatever follows about what is possible to suppose 'without contradiction or absurdity' as holding in reality.[107]

What is true of rabbits is equally true of the universe. We can in our mind's eye picture the universe springing into existence uncaused, but the fact that we can construct and label such a mental picture does not mean the origin of the universe could have really come about in this way. As F. C. Copleston puts it, 'even if one can imagine first a blank, as it were, and then X existing, it by no means follows necessarily that X can begin to exist without an extrinsic cause'.[108] All Hume has really shown is that the principle 'Everything that begins to exist has a cause of its existence' is not analytic and that its denial, therefore, does not involve a contradiction or a *logical* absurdity. But just because we can imagine something's beginning to exist without a cause it does not mean this could ever occur in reality. There are other absurdities than logical ones. And for the universe to spring into being uncaused out of nothing seems intuitively to be really, if not logically, absurd. Therefore, of the two alternatives presenting themselves, namely that the universe has a cause of its existence or the universe came into existence uncaused, the first is inherently more plausible.

Although we have declined an elaborate defence of the principle that everything that begins to exist has a cause of its existence, two possible lines of support might be elucidated.

1. *The argument from empirical facts*: The causal proposition could be defended as an empirical generalization based on the widest sampling of experience. The empirical evidence in support of the proposition is absolutely overwhelming, so much so that Humean empiricists could demand no stronger evidence in support of any synthetic statement. To reject the causal proposition is therefore completely arbitrary. Although this argument from empirical facts is not apt to impress philosophers, it is nevertheless undoubtedly true that the reason we—and they—accept the prin-

[107] Ibid. See also Kenny, *The Five Ways*, 66–8; Aziz Ahmad, 'Causality', *Pakistan Philosophical Journal*, 12 (1973), 17–24.
[108] Frederick C. Copleston, *A History of Philosophy*, v: *Hobbes to Hume* (London: Burns, Oates, & Washbourne, 1959), 287.

ciple in our everyday lives is precisely for this very reason, because it is repeatedly confirmed in our experience. Constantly verified and never falsified, the causal proposition may be taken as an empirical generalization enjoying the strongest support experience affords.

2. *The argument from the a priori category of causality*: Hackett formulates a neo-Kantian epistemology and defends the validity of the causal principle as the expression of the operation of a mental a priori category of causality which the mind brings to experience.[109] Kant had argued that knowledge is a synthesis of two factors: the sense data of experience and the a priori categorical structure of the mind. The categories are primitive forms of thought which the mind must possess in order to make logical judgements without which intelligible experience would be impossible. Kant attempted to compile a list of these categories by correlating a category with each of the types of logical judgement; the category associated with the hypothetical judgement type is the category of causality. Kant argued that these categories are not simply psychological dispositions in which we think, but that they are objectively valid mental structures which the mind brings a priori to experience. For without them no object of knowledge could be thought; if the mind does not come to experience with the a priori forms of thought, thought could never arise. Therefore, the categories must be objectively real. Kant made two crucial limitations on the operation of the categories: (1) the categories have no application beyond the realm of sense data, and (2) the categories furnish knowledge of appearances only, but not of things in themselves.

Hackett makes three critical alterations in Kant's formulation of a categorical epistemology.[110] First, the number of the categories must be reduced. It is universally recognized that Kant's tables of categories and logical judgement types are highly artificial; accordingly, Hackett eliminates the categories of totality and limitation and equates the category of existence with that of substance. The remaining categories he regards as validly derived. Secondly, the categories have application beyond the realm of sense data. Kant's position is self-refuting: for if the categories are restricted in operation to the realm of sense data alone, then no knowledge of

---

[109] Stuart C. Hackett, *The Resurrection of Theism* (Chicago: Moody Press, 1957), 37–113.
[110] Ibid. 46–55.

the categories themselves would be possible, since they are characterized by the very absence of sense data. Yet we do possess speculative knowledge of the categories; therefore, they cannot be restricted to the realm of sense experience. Thirdly, the categories do furnish knowledge of things in themselves. For either things in themselves exist or they do not. If they do not, we are reduced to solipsism. Besides the obvious problems of solipsism, the crucial point is this: since the categories do apply to the phenomena and these are all that exist, it follows that the categories do give a knowledge of reality in itself. But suppose things in themselves do exist (as Kant undoubtedly believed); then it becomes impossible to deny that the categories provide knowledge of things in themselves. For at least the categories of reality and causality must apply to them (since they cause the phenomena which we apprehend), unless one is willing to relapse into solipsism. Thus, to assert 'No knowledge of the noumena is possible' is self-refuting, since it itself purports to be an item of knowledge about the noumena. This means, concludes Hackett, that the categories are both forms of thought and forms of things—thought and reality are structured homogeneously.

How can it be shown that the a priori categories are objective structural features of the mind? Hackett sums up his positive defence:

either the categories are thus *a priori* or they are derived from experience. But an experiential derivation of the categories is impossible because only by their means can an object be thought in the first place. Since the categories are preconditions of all possible knowledge, they cannot have been derived from an experience of particular objects: the very first experience would be unintelligible without a structure of the mind to analyze it. . . . After all either thought starts with some general principles with which the mind is initially equipped, or it cannot start at all. Thought consists of ideas and judgments, as we have seen: and the very first act of judging presupposes that the thinker has a structure of thought in terms of which subject and predicate may be united according to certain relations.[111]

The argument, which is basically Kant's, is not that without the categories we could not experience sensations much as an animal does, but that self-conscious thought could not arise unless the human mind were structured so that it could.

[111] Ibid. 57.

Since the categories are objective features of both thought and reality and since causality is one of these categories, the causal relation must hold in the real world, and the causal principle would be a synthetic a priori proposition. It is a priori because it is universal and necessary, being a pre-condition of thought itself. But it is synthetic because the concept of an event does not entail the concept of being caused. Hackett's attempt to thus found the causal principle on an a priori mental category merits further investigation outside the scope of this book; for, as Bella Milmed observes, although much of Kant's work is obsolete,

surely most if not all of his categories are still recognizable as relevant to the interpretation of the empirical world; and the increased flexibility of logic means that it should be easier to find logical foundations for such categories, avoiding those of Kant's derivation that appear strained. Moreover, some of the most important of his derivations, e.g., those of substance and causality, do not appear strained at all.[112]

These two arguments suggest possible ways of defending the principle that everything that begins to exist has a cause of its existence. But probably most people do not really need convincing. In summary, we have contended that (1) it is intuitively obvious that anything that begins to exist, especially the entire universe, must have a cause of its existence; (2) Hume's attempt to show that the universe could have sprung uncaused out of nothing fails to show this to be a *real* possibility; and (3) the causal principle could be more elaborately defended in two ways. Therefore, we conclude our first premiss: everything that begins to exist has a cause of its existence.

4. CONCLUSION: THE UNIVERSE HAS A CAUSE OF ITS EXISTENCE

Since everything that begins to exist has a cause of its existence, and since the universe began to exist, we conclude, therefore, that the universe has a cause of its existence. We ought to ponder long and hard over this truly remarkable conclusion, for it means that transcending the entire universe there exists a cause which brought

---

[112] Bella Milmed, *Kant and Current Philosophical Issues* (New York: New York University Press, 1961), 45.

the universe into being *ex nihilo*. If our discussion has been more than a mere academic exercise, this conclusion ought to stagger us, ought to fill us with a sense of awe and wonder at the knowledge that our whole universe was caused to exist by *something* beyond it and greater than it.[113] For it is no secret that one of the most important conceptions of what theists mean by 'God' is Creator of heaven and earth.

But even more: we may plausibly argue that the cause of the universe is a personal being. Here we may pick up again the threads of the argument discussed in relation to Kant's first antinomy. There Kant argued in the antithesis that the universe could not have had a beginning in time. He asserted,

[113] Objections to a First Cause of the universe hardly merit refutation. Laird argues that the universe is the 'theatre' of causes but does not itself need a cause. (John Laird, *Theism and Cosmology* (London: George Allen & Unwin, 1940), 95.) This is manifestly untrue, since the 'theatre' itself had a beginning, and no doubt did not spring into being uncaused. Hospers and Matson object that the First Cause would also have to have a cause, which is impossible. (Hospers, *An Introduction to Philosophical Analysis*, 431; Wallace I. Matson, *The Existence of God* (Ithaca, NY: Cornell University Press, 1965), 61.) This is incorrect because the causal principle concerns only what *begins* to exist, and God never began to exist, but is eternal. MacIntyre contends that a causal relationship involves two observable events; since God is not observable, the relationship between God and the world cannot be causal. (Alasdair C. MacIntyre, *Difficulties in Christian Belief* (London: SCM Press, 1959), 60.) But MacIntyre's stipulation is entirely arbitrary and unwarranted; for unobservable entities such as cosmic rays cause observable effects. And could not an unobservable spirit being like an angel or demon, if there be such, cause observable effects, such as the levitation of an object? Why then could not God cause the world? The significance of this conclusion may be seen in the fact that it thoroughly undermines the foundation of modern process theology. According to process theologians, God and the universe are coeternal poles of the one divine being, which develops through the process of interrelation between these poles. But if our analysis is correct, such an understanding of the universe (to say nothing of God) is both philosophically unsound and empirically unscientific. We have argued that the temporal series of events cannot be infinite; therefore, the universe cannot have been developing coeternally with God. (And even if it could, then the process of God's development would have already been completed by now.) To retreat to the position that God and the world lay dormant from eternity and began a process of mutual development a finite number of years ago completely removes any rationale for process theology, since according to this school, process and development are essential to God's very nature, and He cannot exist without development. Moreover, such a theology is anti-empirical, for it appears to presuppose some sort of continuous steady state universe. To try to wed process theology with modern cosmology can only make us smile at the incongruity: a pitiable God, indeed, whose poor body explodes from a point of infinite density and perishes in the cold reaches of outer space! It is ironic that theologians should have developed such a view of God and the universe at precisely the same time that scientists were accumulating evidence that tended to confirm *creatio ex nihilo*.

For let us assume that it has a beginning. Since the beginning is an existence which is preceded by a time in which the thing is not, there must have been a preceding time in which the world was not, i.e. an empty time. Now no coming to be of a thing is possible in an empty time, because no part of such a time possesses, as compared with any other, a distinguishing condition of existence rather than of non-existence; and this applies whether the thing is supposed to arise of itself or through some other cause. In the world many series of things can, indeed, begin; but the world itself cannot have a beginning, and is therefore infinite in respect of past time.[114]

Now Kant's argument is to a large degree cogent. He is not arguing that there is any inherent absurdity in the notion of a void time, as is sometimes thought. Rather he is contending that prior to the existence of the universe no moment is distinguishable from another and, therefore, no condition exists at one moment rather than another which would account for the universe's beginning to exist at that moment rather than earlier or later.[115] Whitrow attempts to refute Kant's antithesis by arguing that on a relational view of time, time begins with the first event, and therefore the question of why the universe did not begin earlier or later cannot arise.[116] But this only partly solves the problem, for it works only if the beginning of the universe is *uncaused*. If the beginning of the universe is uncaused, then whenever it springs into being is irrelevant, since its beginning is wholly unrelated to determinate conditions. Thus, if it springs into being in Newtonian absolute time, no distinguishing condition of one moment from another is necessary. And if it comes into existence with relational time, no problem arises because no prior conditions are necessary for the beginning of time. But what if the beginning of the universe is *caused*? Here Whitrow's response is of no avail. For the necessary and sufficient conditions for the produc-

---

[114] Kant, *Critique of Pure Reason*, trans. Norman Kemp Smith (London: Macmillan, 1978), A 427/B 455.

[115] See Gottfried Martin, *Kant's Metaphysics and Theory of Science*, trans. P. G. Lucas (Manchester: Manchester University Press, 1955; repr. Westport, Conn.: Greenwood Press, 1974), 48. According to Martin, 'In this empty time before the beginning of the world there was the passage of time, but no events.' (Ibid.) Accordingly, the antithesis asks why the world began at a certain point in time, when all moments are alike. (Sadik J. al-Azm, *The Origins of Kant's Arguments in the Antinomies* (Oxford: Clarendon Press, 1972), 44–5.)

[116] G. J. Whitrow, *What is Time?* (London: Thames & Hudson, 1972), 566; Whitrow, *The Natural Philosophy of Time*, 32; G. J. Whitrow, 'The Age of the Universe', *British Journal for the Philosophy of Science*, 5 (1954), 217.

tion of the first event are either present from eternity or not. If they are, then the effect will exist from eternity, that is to say, the universe will be eternal. But if they are not, then the first event could not possibly occur, since the necessary and sufficient conditions for the production of the first event could never arise. Kant may be assuming that the first event must have been caused, since he speaks of conditions for its existence. The antithesis of the first antinomy, which like the thesis echoes so clearly the arguments of the medieval Islamic *mutakallimūn*, really asks, Why did the universe begin to exist when it did instead of existing from eternity? The answer to Kant's conundrum was carefully explained by al-Ghazālī and enshrined in the Islamic principle of determination. According to that principle, when two different states of affairs are equally possible and one results, this realization of one rather than the other must be the result of the action of a personal agent who freely chooses one rather than the other. Thus, al-Ghazālī argues that while it is true that no mechanical cause existing from eternity could create the universe in time, such a production of a temporal effect from an eternal cause is possible if and only if the cause is a personal agent who wills from eternity to create a temporally finite effect. For while a mechanically operating set of necessary and sufficient conditions would produce the effect either from eternity or not at all, a personal being may freely choose to create at any time wholly apart from any distinguishing conditions of one moment from another. For it is the very function of will to distinguish like from like.[117] Thus, on a Newtonian view of time, a personal being could choose from eternity to create the universe at any moment he pleased.[118] On a relational view of time, he could will timelessly to create and that creation would mark the inception of time.[119] Thus, Kant's antithesis, far from disproving the

[117] Al-Ghazālī, *Tahafut al-Falasifah*, trans. S. A. Kamali (Lahore: Pakistan Philosophical Congress, 1963), 163–7.

[118] Hospers objects that it makes no sense to speak of God's choosing from eternity to create because choice is always a temporal decision. (Hospers, *An Introduction to Philosophical Analysis*, 433.) The objection is incorrect because choice does not imply changing one's mind, but is simply a determination of the mind to execute freely a certain course of action. Hence, it is perfectly intelligible to speak of God's choosing from eternity.

[119] As that wise old man C. D. Broad remarks, 'On this relational view of time the question: "Why did the world begin when it did, and not at some earlier or later moment?" would reduce to the question: "Why did the particular event, which in fact had no predecessors, not have predecessors?" ... But I cannot help doubting

beginning of the universe, actually provides a dramatic illumination of the nature of the cause of the universe; for if the universe began to exist, and if the universe is caused, then the cause of the universe must be a personal being who freely chooses to create the world.

Further than this we shall not go. The *kalām* cosmological argument leads us to a personal Creator of the universe, but as to whether this Creator is omniscient, good, perfect, and so forth, we shall not enquire. These questions are logically posterior to the question of His existence. But if our argument is sound and a personal Creator of the universe really does exist, then surely it is incumbent upon us to enquire whether He has specially revealed Himself to man in some way that we might know Him more fully or whether, like Aristotle's unmoved mover, He remains aloof and detached from the world that He has made.

Annotations to Essay I

a. During the 1980s, through the marriage of particle physics and cosmology, scientists have attempted to push back the frontiers of our knowledge of the early universe ever closer to the Big Bang. Weinberg, Salam, and Glashow had predicted that at energies of 100 GeV (i.e. 10^2 billion electron volts), which would have obtained at about 10^{-12} sec after the Big Bang, the electromagnetic and weak atomic force would unite into a single force, and this prediction was verified in 1983 by the discovery of the intermediate vector bosons $W\pm$ and Z^0, particles which carry the weak force. Prior to 10^{-12} sec, however, the physics becomes speculative. The object of Grand Unified Theories (GUTs) is to provide an account of the unification of the strong atomic force with the electroweak force into a single force carried by hypothetical X, Y particles. It is thought that at $\sim 10^{-14}$ GeV, conditions obtaining at 10^{-35} sec, the strong force and the electroweak force separated out from the grand unified force, which had existed from the Planck time

whether it is a significant question, except in a rather special theistic context; and in that context the only answer is: "God knows!" (Broad, 'Antinomies', 7.) More seriously, Sturch argues that God may eternally wish that a temporal world exist; since He is omnipotent, God's wish is done, and a temporal world exists. (R. L. Sturch, 'The Problem of Divine Eternity', *Religious Studies*, 10 (1974), 488–9.) On the question of whether God could have created the universe sooner, see Bas Van Fraassen, *An Introduction to the Philosophy of Time and Space* (New York: Random House, 1970), 24–30.

10^{-43}. Unfortunately none of the predictions of GUTs have been experimentally verified. Indeed, some of the predictions run decidedly contrary to observational evidence. For example, GUTs necessarily predict the existence of magnetic monopoles, enormously heavy, stable particles 10^{16} times the proton mass, each of which is a single magnetic pole. In fact, there should be as many monopoles as particles of ordinary baryonic matter! But observationally, there are none. This problem is solved by the further speculation of an inflationary epoch between 10^{-35} sec and 10^{-33} sec, during which the universe expanded at an exponential rate, increasing in size by a factor of 10^{28} or more. Inflation is said to occur because the energy density of the vacuum state of the primordial universe has a non-zero value, or in the language of classical cosmology, there is a positive cosmological constant. In such a false vacuum, the energy density of the universe does not decrease as it expands and so continues to drive the expansion faster until a phase transition to the present zero energy density vacuum. The inflationary expansion blasts almost all the monopoles out beyond our event horizon (indeed, it is hypothesized that only one monopole is left lurking somewhere in our observable universe). Perversely, the observed absence of monopoles is sometimes actually taken as *evidence for* the inflationary scenario, when in fact the latter is an attempt to explain away the former. Inflation also serves to solve two other embarrassments related to the fine-tuning of the initial conditions of the universe: (1) the horizon problem (the microwave background radiation is astonishingly homogeneous, despite the fact that it was emitted from regions of the universe which were at that time causally unconnected, so that 'the accuracy of the Creator's aim', to borrow Penrose's phrase, in selecting our universe must have been at least one part in $10^{10(123)}$); (2) the flatness problem (in order for the universe to be expanding at its present rate, the density parameter Ω must at 10^{-35} sec have differed from 1.0 by less than one part in 10^{49}, the chances of which Lawrence Krauss has compared to someone's guessing the precise number of atoms in the sun). Inflation solves these by pushing inhomogeneities out beyond our event horizon and suppressing the cosmological curvature term so that spacetime is virtually flat. The solution to (1), however, tends by its very nature to be unverifiable, since inflation serves as 'the magic trick by which the universe was able to erase all . . . memory of any gross defects or blemishes'. (Joseph Silk, *The Big Bang*, 2nd edn. (San Francisco: W. H. Freeman, 1989), 120.) As for the solution to (2), this has the consequence that $\Omega = 1$ even today, which contradicts observational data. None the less most theorists hold to an inflationary scenario largely due to the fine-tuning arguments. But those factors are just part of a much wider series of amazing coincidences which have served to fuel discussions of the Anthropic Principle and the teleological argument for a Divine Designer (John D. Barrow and Frank J. Tipler, *The Anthropic Cosmological Principle* (Oxford: Clarendon Press, 1986); John Leslie, *Universes* (London: Routledge, 1989)). As a theist, such fine-tuning does not surprise me and so furnishes no independent grounds to adopt an inflationary scenario in contradiction to the evidence. Perhaps there was such an inflationary epoch in the early history of the universe, but it is difficult to resist the impression that the inflationary theory, as it now stands, is *ad hoc*, speculation built on speculation (GUTs), and in conflict with the evidence.

Prior to 10^{-35} sec the physics becomes extremely speculative and even unknown. It is hypothesized that at the Planck time, 10^{-43} sec, gravitation and the grand unified force separate. One class of theories describing this post-Planckian period which paves the way for inflation is known as supersymmetry or SUSY GUTs and requires whole new families of as yet undiscovered exotic particles. Chaotic inflationary theories, on the other hand, hold that the universe at this time was a pastiche of regions possessing different physical properties, those of our region having been just right for inflation to occur. In order to describe the interval prior to the Planck time, a quantum theory of gravity is needed which will unify gravitation and the grand unified force into a single superforce. But no such theory, presumptuously called a Theory of Everything (TOE), exists. Quantum cosmologists who theorize about this domain have invested a great deal of energy in the study of superstring theory, which posits fundamental particles having length, but no width, and existing in 10 or 26 dimensions. Needless to say, such theories have yielded no testable predictions.

For good reviews of the mutual relevance of particle physics and cosmology, see David N. Schramm, 'Cosmology and GUTs: The Matter of the Universe', *Advances in Space Research*, 3 (1984), 419–30; A. Salam, 'Astroparticle Physics', in M. Caffo, R. Fanti, G. Giacomelli, and A. Renzini (eds.), *Astronomy, Cosmology and Fundamental Physics*, Third ESO/CERN Symposium, Astrophysics and Space Science Library, clv (Dordrecht: Kluwer, 1989), 1–22. On supersymmetry see Pierre Fayet, 'Unified Field Theories', in G. Setti and L. Van Hove (eds.), *Large Scale Structure of the Universe, Cosmology and Fundamental Physics*, Proceedings of the First ESO/CERN Symposium (Geneva: ESO and CERN, 1984), 35–71. For a list of offerings of particle physics to cosmology, ranked for plausibility and desirability, see John Ellis, 'A Brilliant Past in Front of Us', in Setti and Van Hove (eds.), *Large Scale Structure of the Universe*, 435–50. On inflation, see Alan Guth, 'Inflationary Universe: A Possible Solution to the Horizon and Flatness Problem', *Physical Review*, D23 (1981), 347–56; A. D. Linde, 'The

Present Status of the Inflationary Universe Scenario', *Comments on Astrophysics*, 10/6 (1985), 229–37; L. A. Khalfin, 'Limitations on Inflationary Models of the Universe', in M. A. Markov, V. A. Berezin, and V. P. Frolov (eds.), *Quantum Gravity* (Singapore: World Scientific, 1988), 891–909. On superstring theory, consult M. Green, J. Schwartz, and E. Witten, *Superstring Theory*, 2 vols. (Cambridge: Cambridge University Press, 1987). For readable accounts of quantum cosmology, see Paul Davies, *Superforce* (London: Heinemann, 1984) and John Barrow, *Theories of Everything* (Oxford: Clarendon Press, 1991). For a complaint about the speculative character of recent cosmology, see Tony Rothman and George Ellis, 'Has Cosmology Become Metaphysical?' *Astronomy*, 15 (1987), 6.

b. During the 1980s ever more precise measurements of the radiation background confirmed its remarkable isotropy. The launch of the Cosmic Background Explorer (COBE) satellite in Nov. 1989 afforded a breakthrough in precision measurements. Preliminary data gathered by COBE established isotropy to better than one part in 20 000 on all angular scales larger than 7° (G. F. Smoot *et al.*, 'Preliminary Results from the COBE Differential Microwave Radiometers: Large Angular Scale Isotropy of the Cosmic Microwave Background', *Astrophysical Journal Letters*, 371 (1991), L1–5). Indeed, the radiation background seemed *too* isotropic, since conventional theories of galaxy formation require perturbations in the background as the imprint of primordial density fluctuations which grew into galactic structures. Since gravitational accretion proceeds very slowly in a low-density universe dominated by baryonic matter, fluctuations so tiny as to satisfy these upper limits could not have produced the observed structure in the universe if galaxies were composed solely of baryonic matter (M. L. Wilson and Joseph Silk, 'On the Anisotropy of the Cosmological Background Matter and Radiation Distribution', i: 'The Radiation Anisotropy in a Spatially Flat Universe', *Astrophysical Journal*, 243 (1981), 14–25; M. L. Wilson, 'On the Anisotropy of the Cosmological Background Matter and Radiation Distribution', ii: 'The Radiation Anisotropy in Models with Negative Spatial Curvature', *Astrophysical Journal*, 273 (1983), 2–15; cf. Naoteru Gouda and Naoshi Sugiyama and Misao Sasaki, 'Constraints on Open Universe Models from Quadrupole Anisotropy of the Cosmic Microwave Background', *Astrophysical Journal Letters*, 372 (1991), L49–52), unless specific conditions are attached to a baryonic dark matter theory. Finally at a meeting of the American Physical Society the announcement was made that COBE had detected anisotropies in the radiation background (Joseph Silk, 'Cosmology back to the Beginning', *Nature*, 356 (1992), 741–2; see also 'Big Bang Brouhaha', *Nature*, 356 (1992), 731), a finding which has been hailed as of comparable importance to the discovery of the cosmic microwave background itself. COBE measured temperature fluctuations in the radiation background over angular scales of $10°-90°$ of a mere $5 \pm (1.5) \times 10^{-6}$, so that the microwave background is isotropic to better than one part in 100 000. The fluctuation amplitudes are consistent with those predicted by an inflationary, cold dark matter model with critical density, but only if one rejects the technique of biasing the distribution of luminous matter relative to dark matter (i.e. maintaining that baryonic matter tends to clump more than non-baryonic, thereby permitting vast quantities of dark matter to be more widely distributed than the galaxies). But, as we point out in our discussion of the density of the universe, the dark matter must be so distributed if the dynamical properties of galaxies are not to be upset. Hence, Silk admits 'such a model is a disaster on scales of a few megaparsecs, where it predicts excessive gravitational power. Correct the small scale problem, and cold dark matter fails on large scales.' (Silk, 'Cosmology back to the Beginning', 742.) In fact, no model succeeds completely in providing a description of initial conditions which yield both the small- and large-scale structure of the universe as we know it. (For an intriguing tableau weighting and assessing the probabilities of various models, see P. J. E. Peebles and Joseph Silk, 'A Cosmic Book of Phenomena', *Nature*, 346 (1990), 234.)

c. See now also Allan Sandage and G. A. Tammann, 'Steps toward the Hubble Constant', vii: 'Distances to the NGC 2403, M101, and the Virgo Cluster Using 21 Centimeter Line Widths Compared with Optical Methods: The Global Value of H_0', *Astrophysical Journal*, 210 (1976), 7–24; viii: 'The Global Value', *Astrophysical Journal*, 256 (1982), 339–45; ix: 'The Cosmic Value of H_0 Freed from All Local Velocity Anomalies', *Astrophysical Journal*, 365 (1990), 1–12.

d. There has been some controversy over the value of the Hubble constant. Part of the difficulty involves differing distance estimates to galactic structures. But in their most recent work, Sandage and Tammann were able to obtain on the basis of six independent methods a distance determination to the Virgo cluster of 'unprecedented accuracy' of $D = 21.9 \pm 0.9\,\mathrm{Mpc}$. Combining the velocity of the Virgo cluster relative to the fundamental cosmic frame v_{cosmic} (Virgo) = 1144 $\pm$ $18\,\mathrm{km\,s}^{-1}$ with this distance gives the Hubble constant to be $H_0 = 52 \pm 2\,\mathrm{km\,s}^{-1}\,\mathrm{Mpc}^{-1}$. The margin of error is so small because of the unprecedented accuracy of the distance and cosmic velocity estimates. (Sandage and Tammann, 'Steps toward the Hubble Constant', ix. 10.)

e. Estimates of the age of the universe are quite rough, and q_0 may be even more insignificant than previously thought. In a more recent review of the cosmic parameters, Sandage and Tammann, noting that if

H_0 is in the range $H_0 = 50 \pm 10 \, \text{km s}^{-1} \, \text{Mpc}^{-1}$, then $H_0^{-1} = 19.6 \times 10^9$ years, opt for an age of the universe $T_0 = 18.3 \pm 3$ billion years on the basis of the age of the globular cluster stars of 16 billion years (Allan Sandage and G. A. Tammann, 'The Dynamical Parameters of the Universe: H_0, q_0, Ω_0, Λ, and K', in Setti and Van Hove (eds.), *Large Scale Structure of the Universe*, 131). Sandage's colleague A. Dressler later estimated $T_0 \sim 15$ billion years; but he was using a value of $H_0 \approx 60 - 80 \, \text{km s}^{-1} \, \text{Mpc}^{-1}$, which is excluded by the most recent, more accurate determinations (A. Dressler, 'Cosmological Parameters of the Universe', in Caffo *et al.* (eds.), *Astronomy, Cosmology and Fundamental Physics*, 23). Part of the difficulty in obtaining more accurate estimates of T_0 is uncertainty over the age and especially the gestation time of the globular clusters.

f. During the 1980s, quantum models, taking as their point of departure Edward Tryon's suggestion in 1973 that the universe might be a vacuum fluctuation in a wider universe (Edward Tryon, 'Is the Universe a Vacuum Fluctuation?' *Nature*, 246 (1973), 396–7), became oft-suggested alternatives to an absolute beginning of the universe. These will be discussed in Essays IV and V. On wave-functional models of the universe's origin, see Essays XI and XII.

g. So it seemed; but Hoyle, ever the fighter, has stayed in the ring: see H. C. Arp, G. Burbidge, F. Hoyle, J. V. Narlikar, and N. C. Wickramasinghe, 'The Extragalactic Universe: An alternative View', *Nature*, 346 (1990), 807–12.

h. Actually, these are not synonymous, since models which have a positive spatial curvature and are thus closed can still expand for ever if one adopts a positive cosmological constant, as in the Lemaître model, or an unconventional spacetime topology. For a helpful chart of alternatives, see C. Misner, K. S. Thorne, and J. A. Wheeler, *Gravitation* (San Francisco: W. H. Freeman, 1973), 747.

i. Dressler emphasizes that attempts to measure q_0 directly are frustrated by evolutionary corrections to luminosity and, hence, there are no direct measures of q_0 to date that deserve to be taken very seriously (Dressler, 'Cosmological Parameters of the Universe', 32–4).

j. See also Sandage and Tammann's 'Steps toward the Hubble Constant', vii–ix, where q_0 takes on negligible values. 'Hence, we are forced to decide that $q_0 < \frac{1}{2}$ and therefore that it seems inevitable that the Universe will expand forever.' (Ibid. vii. 23.) In their helpful review article of the dynamical parameters of the universe, they arrive at a determination of Ω via three methods (the perturbational effect of the Virgo cluster on the Local Group, measurements of the luminosity density of galaxies, and the Big Bang nucleosynthesis constraints) and conclude, 'It is quite impressive that three independent routes toward

Ω_0 give so similar results. A value of $\Omega_0 = 0.12$, corresponding to a strongly open Friedman Universe with $q_0 = 0.06$, *satisfies all three determinations*.' (Sandage and Tammann, 'The Dynamical Parameters of the Universe', 137.) Similarly, Dressler finds that the present data favour $\Omega \approx 0.1 - 0.2$. (Dressler, 'Cosmological Parameters of the Universe', 23.)

k. See now also the review article by P. J. E. Peebles, 'The Mean Mass Density of the Universe', *Nature*, 321 (1986), 27–32, where he argues that either the standard picture of galaxy formation is wrong or else the density of the universe is about 0.1. Dressler observes that studies even out to scales of 5–10 Mpc still yield $\Omega = 0.1 - 0.2$; if closure density is to be reached, the remaining matter would have to be scattered over scales of 20 Mpc or more. (Dressler, 'Cosmological Parameters of the Universe', 35.) Dynamical analyses of galactic structures on ever larger scales consistently determine $\Omega \lesssim 0.2$. (Matthew Colless and Paul Hewett, 'The Dynamics of Rich Clusters', i: 'Velocity Data', *Monthly Notices of the Royal Astronomical Society*, 224 (1987), 453–72; Ann I. Zabludoff, John P. Huchra, and Margaret J. Geller, 'The Kinematics of Abell Clusters', *Astrophysical Journal Supplement*, 74 (1990), 1–36.) This meshes, as we have seen, with the density requirements for the nucleosynthesis of the light elements. Since inflationary scenarios predict $\Omega \approx 1.0$, Dressler asserts that 'the low value of Ω_0 appears incompatible with the inflation paradigm of the early Big Bang'. (Dressler, 'Cosmological Parameters of the Universe', 23.) None the less, theorists have attempted to preserve inflation by proposing that the universe is dominated by a 'missing mass' of non-baryonic matter (it must be non-baryonic to elude the nucleosynthesis constraints) which will serve to bring Ω to the critical value. Two broad types of theory are proposed: (1) the 'hot' dark matter theory (HDM), according to which neutrino-like particles have a mass $m \approx 10\,\text{eV}$ to $30\,\text{eV}$ and exist in such abundance to make up for the deficit in baryonic matter. Since the primordial neutrino gas does not interact with radiation, density fluctuations in it could have developed before the decoupling that spawned the microwave radiation background without leaving an imprint on it and so served as the large-scale cosmic skeleton which would be fleshed out by the later accretion of baryonic matter. (Y. B. Zeldovich, J. Einasto, and S. F. Shandarin, 'Giant Voids in the Universe', *Nature*, 300 (1982), 407–13.) The HDM theory has, however, fallen into disfavour because the neutrino-formed galactic structures would look different from those in the real universe. (Peebles and Silk, 'A Cosmic Book of Phenomena', 233–9; Carlos S. Frenk, Simon D. M. White, and Marc Davis, 'Massive Neutrinos and Galaxy Formation', in Setti and Van Hove (eds.), *Large Scale Structure of the Universe*,

257–65.) (2) The other alternative posits 'cold' dark matter (CDM) composed of exotic particles of weakly interacting matter (WIMPs) which are also conceived to have decoupled earlier than baryonic matter and left a negligible imprint on the radiation background. (Carlos S. Frenk, Simon D. M. White, Marc Davis, and George Efstathiou, 'The Formation of Dark Haloes in a Universe Dominated by Cold Dark Matter', *Astrophysical Journal*, 327 (1988), 507–25; *idem*, 'Galaxy Clusters and the Amplitude of Primordial Fluctuations', *Astrophysical Journal*, 351 (1990), 10–21.) It seemed at first that a CDM scenario could account for the observed structure of the universe, but as structure has been discovered on ever larger scales, it has become evident that CDM cannot plausibly explain it. Tóth and Ostriker have also recently argued that standard CDM models with $\Omega = 1$ can be ruled out because mass in galactic discs would have accreted too greatly in the last five billion years, whereas no such problem arises in an open universe. (G. Tóth and J. P. Ostriker, 'Galactic Disks, Infall and the Global Value of Ω', *Astrophysical Journal*, 389 (1992), 5–26.) Worst of all, perhaps, we have seen that while the perturbations in the microwave background detected by COBE can be deduced from CDM models with critical density (Nicola Vittorio and Joseph Silk, 'Fine Scale Anisotropy of the Cosmic Microwave Background in a Universe Dominated by Cold Dark Matter', *Astrophysical Journal Letters*, 285 (1984), L39–43; J. R. Bond and G. Efstathiou, 'Cosmic Background Radiation Anisotropies in Universes Dominated by Non-baryonic Dark Matter', *Astrophysical Journal Letters*, 285 (1984), L45–8), this is the case only if one abandons biasing, in which case the theory fails on large scales. CDM thus finds itself in a dilemma: it needs biasing to explain why the critical mass is not associated with galaxies, but is more widely distrubuted in lower-density interstices. But it must give up biasing, if it is to explain how the intensity of the primordial fluctuations is compatible with a universe of critical density. Thus, whether biasing is retained or abandoned, the model runs foul of observational astronomy. And, of course, the unanswered question in all this is *where* is this 'missing mass'? The dark matter associated with galactic clusters, even if 10 times the luminous matter, still brings Ω up to only 0.1. Therefore, if $\Omega = 1.0$ there must exist in the universe *one hundred times* as much dark matter as visible matter, which is, as Sandage and Tammann muse, 'a bizarre requirement'. (Sandage and Tammann, 'The Dynamical Parameters of the Universe', 137.) Martin Rees puts it bluntly: 'There is no astronomical evidence which supports a value $\Omega = 1$.' (Martin J. Rees, 'Concluding Comments: A Cosmologist's View', in Setti and Van Hove (eds.), *Large Scale Structure of the Universe*, 430.) Only partly tongue-in-cheek, Joseph Silk complains,

'If the hidden mass is completely unobservable, anything is permissible. Suggestions for the missing mass have ranged from snowflakes to rocks, planets, black holes, even to excess issues of the *Astrophysical Journal*.' '. . . in the absence of observable evidence, any suggestions must lack credibility'. (Silk, *The Big Bang*, 388.) It is difficult to resist the conclusion that the critical mass is not truly 'missing', since it is not to be expected, i.e. the inflationary theory which generated this expectation is wrong.

l. For a recent review of the evidence from nucleosynthesis of light elements, see Keith A. Olive, David N. Schramm, Gary Steigman, and Terry P. Walker, 'Big Bang Nucleo-Synthesis Revisited', *Physics Letters*, B236 (1990), 454–60; they confirm that baryons contribute only 0.02–0.11 to the critical density. Advocates of HDM and CDM attempt to subvert this consideration by positing mass in non-interacting form, so that nucleosynthesis constraints are not violated. But while some Soviet experiments a number of years ago seemed to suggest that the neutrino might have a mass, subsequent experiments have failed to confirm those findings, and the customary picture of the neutrino as a particle with zero rest-mass remains. As for cold dark matter, the sober fact is that there is no evidence that WIMPs even exist. CDM theorists are constrained to invent both particles and masses for them in the absence of any experimental data. Moreover, the observational evidence suggests that the dark matter associated with galactic clusters is baryonic in nature. (J. A. Tyson, F. Valdes, and R. A. Wenk, 'Detection of Systematic Gravitational Lens Galaxy Image Alignments: Mapping Dark Matter in Galaxy Clusters', *Astrophysical Journal Letters*, 349 (1990), L1–14.) Sandage and Tammann's conclusion stands: 'Plainly, there is no astronomical evidence of any nonbaryonic mass in the universe.' (Sandage and Tammann, 'The Dynamical Parameters of the Universe', 138.)

m. Sandage and Tammann point out that a universe with $\Omega = 0.1$ can actually be closed if $H_0 = 55 \, \text{km s}^{-1} \, \text{Mpc}^{-1}$ and the age of the universe is 22 billion years. But then closure does not imply that the expansion will eventually cease and contraction begin. For low-density universes that are spatially closed can exist if Λc^2 values are about $7 \times 10^{-35} \, \text{sec}^{-2}$, and such models, in spite of closure, will expand for ever. In fact, even high-density closed universes will expand for ever, so long as $\Lambda c^2/3H_0^2 > 0.5$. Indeed, they argue, for all values of $H_0 > 50$, the high-density case requires a positive Λ to satisfy the time-scale argument that $H_0 T_0 > 1$. Their conclusion is worth quoting: 'The one philosophical consequence of the high-density case is that all models in the observed parameter space have positive curvature (hence finite volume and therefore finite mass). They will all expand forever. Hence, the one certain conclusion is that in all models of either high or low density, $\dot{R}$ is always positive—the Universe will not stop its expansion. *This means it has happened only once*. The creation event was unique. (Sandage and Tammann, 'The Dynamical Parameters of the Universe', 144.)

II

Infinity and the Past
QUENTIN SMITH

Introductory Note

This essay aims to refute the argument that the past is necessarily finite. I consider the arguments offered by Craig in Essay I as well as some related arguments offered by other philosophers, such as G. J. Whitrow. I conclude that the past can be either infinite or finite and that empirical evidence needs to be introduced to decide the issue. Craig's argument for a finite past and a Creator that is based on Big Bang cosmology is addressed in my Essays IV and VI. In Essays IV and VI, I agree that Big Bang cosmology warrants the conclusion that the past is finite but I argue that the beginning of the universe is not caused by God or anything else.

In this essay I do not define 'set of past events' or 'series of past events' but leave the reader to understand this notion on an intuitive basis. A relevant definition would be as follows. Let an event be *a complex that includes whatever happens at one time*. (Since we are dealing with an a priori argument, we need not define 'a time' in terms of Einstein's relativity theory as 'a time relative to a given reference frame'.) Let each time be an extended interval of some given length, say an interval of 1 second. The set of past events form a series (which has sequentially ordered members), since past events are sequentially ordered by the relation *is later than*. We may say that the set of past events is infinite if there are an infinite number of events that are past, with each such event being of 1 second's duration and being later than some other event.

By defining an event in this way we avoid trivial arguments to an infinite past. For example, if an event were defined as a maximal complex of simultaneous instantaneous point-events, it would follow that the series of past events would be infinite even if the universe began to exist 5 minutes ago. (There are an infinite number of instants, each of zero duration, in any temporal extended interval of time.) Furthermore, if an event were

First pub. in *Philosophy of Science*, 54 (1987), 63–75. c. the Philosophy of Science Association, 1987.

I am grateful to William Vallicella and a reader for *Philosophy of Science* for their helpful criticisms of an earlier version of this paper.

defined as a maximal complex of simultaneous and temporally extended occurrences of any length whatsoever, the series of past events would be also infinite if the universe began to exist 5 minutes ago. Within this 5-minute interval, there are an infinite number of sequentially ordered events of decreasing temporal length, corresponding to the series $4, \frac{1}{2}, \frac{1}{4}, \frac{1}{8}, \ldots$ In addition, if an event was not defined as a maximal complex of simultaneous occurrences, i.e. as a complex of whatever occurs at one time, but instead as an occurrence that may be simultaneous with other occurrences, it would follow that the set of past events is infinite if there is a first instant of time, this instant is past, and there are an infinite number of distinct but simultaneous occurrences at this instant.

The definition of an event as *a maximal complex of whatever occurs during a temporal interval of 1 second* is suitable for the discussion in this essay but it is not one of the orthodox definitions of 'event' employed by philosophers or scientists and I shall not employ this unorthodox definition in my other essays in this book. For example, in Essay IV I employ the definition of 'an event' in the Theory of Relativity to refer to a point in a spacetime manifold.

Recently there has been a growing tendency to argue that the past is necessarily finite. Writers who argue this, G. J. Whitrow, William Lane Craig, Pamela Huby, David A. Conway, and others, acknowledge the formal correctness of contemporary set theory derived from Cantor's writings but deny that such a theory is applicable to the past. I believe their arguments are based on several errors, which I shall expose in the following.

I. FIRST ARGUMENT

In *Time and the Universe* Whitrow argues that the series of past events if infinite must be an actual rather than a potential infinity, and that an actual infinity of elapsed events is an impossibility.[1] He begins by defining the phrases 'actual infinity' and 'potential infinity' in terms of future events.[2] The future is potentially infinite in that

[1] G. J. Whitrow, 'Time and the Universe', in J. T. Fraser (ed.), *The Voices of Time* (New York: George Braziller, 1966).
[2] Ibid. 567.

(1) for any event in the future of E there will occur future events, and (2) any event in the future of E is separated from E by a finite number of intermediate events. Condition (1) asserts that the future is infinite, and condition (2) that this infinity is merely potential. An actual infinity would obtain if there were events separated from E by an infinite number of intermediate events.

Whitrow then argues that an infinite past is an actual infinity, and consequently is impossible:

If all the events in a temporal chain culminating in the present are infinite in number, then, because these events actually occurred, the infinity concerned must be an actual, not merely a potential, infinity.

Consequently, if the chain of events forming the past of E is infinite, there must have occurred events that are separated from E by an infinite number of intermediate events. For, if not, then any event in the past of E would be separated from E by only a finite number of intermediate events. This would mean that the set of past events would, like the set of future events, constitute only a potential infinity, whereas it must constitute an actual infinity. It thus follows that, if the past of E contains an infinite number of events in a temporal chain culminating in E, there must have occurred events O in the past of E that are separated from E by an infinity of intermediate events. But this conflicts with our condition that an infinite future with respect to any event, in this case O, is a potential infinity, for E is an event that occurs and O has already occurred. Even if, in this context, we are prepared to forgo the Law of Contradiction, we are still confronted with the same insoluble problem that arose earlier in our discussion: when, in the temporal chain from O to E, does the total number of events that have occurred since O become infinite? . . . We conclude that the idea of an elapsed infinity of events presents an insoluble problem to the mind.[3]

This argument is based on a fallacy of equivocation with respect to the phrases 'actual infinity' and 'potential infinity'. Whitrow's proof that the past if infinite is 'actually infinite' is based on a different sense of 'actually infinite' than that belonging to his proof that an 'actually infinite past' is impossible. The former proof utilizes 'actually infinite' to mean an infinity of events that *have really occurred*. Whitrow writes, 'because these events actually occurred, the infinity concerned must be an actual, not merely a potential, infinity'. In this sense of 'actuality', actuality is opposed to 'potentiality' in the sense of *able to occur but not yet having occurred*.

---

[3] Ibid. 567–8.

However, in his proof that an actually infinite past is impossible, Whitrow uses 'actually infinite' in the sense of a *series of events some of which are separated from E by an infinite number of intermediate events*. Once this equivocation is recognized, Whitrow's argument loses any sense of plausibility it might have had. For if the past is an 'actual infinity' in the sense of being an infinity of events that *have really occurred*, it does not follow that it is also an 'actual infinity' in the sense that *some past events are separated from the present event by an infinite number of intermediate events*. It is quite possible for there to be an infinite number of events that have really occurred such that each of these events is separated from the present event by a *finite* number of intermediate events.

To make this clear, it can be observed that if past events are infinitely numerous, then they form a set that is open on one end and closed on the other (the present event being the closure). They would correspond to the set of negative numbers:

Past Events	Present Event
... −4 −3 −2 −1	0

This set has the cardinal number $\aleph_0$ (aleph zero) and the order type of regression ω^*. Note that no matter which number one takes in this set, there will be a finite number of numbers intermediate between it and 0. Suppose we take the number -128, then there are -127 numbers between it and 0. Suppose we take a trillion times -128, there still is some finite number between it and 0, and so on for any other number in this set. And yet there is not a finite but an *infinite number* of negative numbers in this set; before each negative number, there is another negative number.

Accordingly, we can conceive of past events as forming an infinite set of events that have 'actually occurred' without being presented with the 'insoluble problem' to which Whitrow refers, namely, that if some past events are separated by an infinite number of events from the present event, when in the chain of past events did the events start being separated from the present event by an infinite rather than finite number of intermediate events?

2. SECOND ARGUMENT

Apart from Whitrow's equivocation upon 'actual infinity' and 'potential infinity', there is another and deeper fallacy underlying his argument, a fallacy also committed by Pamela Huby[4] and William Lane Craig.[5] The error in question appears in the commonly stated argument that

(1) $\aleph_0$ events have occurred before the present event

entails

(2) Events separated from the present event by $\aleph_0$ events have occurred,

which in turn entails

(3) From one of the events separated from the present event by $\aleph_0$ events the present event could not have been reached.[6]

The fallacy lies in the belief that (1) entails (2). It does not; for $\aleph_0$ events could have occurred before the present event such that no one of these events is separated from the present event by $\aleph_0$ events. It is this state of affairs that I discussed in the last section in connection with the set of negative numbers with the order type ω^*. The number of these negative numbers is $\aleph_0$, but no one of these negative numbers is separated from o by an $\aleph_0$ number of numbers.

It might be thought that the entailment of (2) by (1), accepted as self-evident by the above authors, could be proven by the following argument (suggested to me by William Vallicella). The set of negative numbers with order type ω^* by which the past is represented can be mapped on to the set with order type $\omega^* + \omega^*$:

$$\ldots -8 \; -6 \; -4 \; -2 \; + \ldots -7 \; -5 \; -3 \; -1 \quad \text{o}.$$

This set has the same members as the set of negative numbers in their natural order, and consequently, by the axiom of extensionality,

---

[4] Pamela Huby argues that past time must be finite for the same reason that the number of objects in space must be finite; namely, because every object in space (or time) must be 'a finite distance only from every other object. But between any object and any other object there can then be only a finite number of objects, and therefore, however vast the total number of objects may be, it will still be finite.' See Pamela Huby, 'Kant or Cantor: That the Universe, if Real, Must Be Finite in Both Space and Time', *Philosophy*, 46 (1971), 127. This implies the false assertion that if the total number of objects is infinite, there will be an infinite number of objects between some two of these objects.

[5] Craig reaffirms Whitrow's argument that 'if the chain of events prior to *E* is infinite, then there must be an event *O* that is separated from *E* by an infinite number of intermediate events'. See William Lane Craig, *The Kalām Cosmological Argument* (New York: Harper & Row, 1979), 200.

[6] This error appears in G. J. Whitrow, 'On the Impossibility of an Infinite Past', *British Journal for the Philosophy of Science*, 29 (1978), 43; and *The Natural Philosophy of Time*, 2nd edn. (Oxford: Clarendon Press, 1980), 31–2.

is identical with this set. Now the set with the order type $\omega^* + \omega^*$ contains members separated from other members by an infinite number of intermediate members; for example, -4 is separated from -3 by $\aleph_0$ members. It follows that the set with order type ω^* also contains some members infinitely distant from other members. For instance, the number in the set with order type ω^* that corresponds (in the one-to-one mapping of the two sets) to -4 in the set with order type $\omega^* + \omega^*$ is infinitely distant from the number in the set with order type ω^* that corresponds to -3 in the set with order type $\omega^* + \omega^*$.

I believe this argument can be contested in two areas. First, it assumes that the issue of whether some members in a set are infinitely distant from others is logically independent of the order type of the set. More specifically, it assumes that if two sets have the same cardinality, $\aleph_0$, and are identical with one another, then irrespective of differences in their order type if one of the sets has infinitely removed members then so must the other. However, it can be proven that whether some members of some set S_1 are infinitely removed from one another is determined by the ordered position of the members in S_1, and that the ordinal properties of these members are logically independent of the ordinal properties of any set S_2 to which S_1 corresponds or with which it is identical. Take, for example the set $\omega^* + \omega^*$; the number -4 is infinitely removed from -3. However, if I reorder this set, so that it possesses the order type ω^*, then -4 is no longer infinitely removed from -3 but is immediately adjacent to it. This shows that the property of being infinitely or finitely removed from another member of the set is an ordinal property of -4.

Next, note that if some infinite set S_1 is mapped on to set S_2, the ordinal property of a member x_1 of S_1 is logically independent of the ordinal property of the member y_1 of S_2 to which x_1 corresponds. This can be proven by a number of instances; thus, the set N of positive whole numbers can be mapped on to the set R of rational numbers; in this case, the number 1 in set N has the ordinal property of being the first member of N. However, the member in the set R to which 1 corresponds does not have the ordinal property of being the first member of R, for set R by definition has no first member.

Furthermore, each number in set N but the first member has the ordinal property of being the immediate successor of some other member of N; however, none of the numbers in R to which the numbers in N correspond have such an ordinal property; for a rational number by definition has no immediate successor.

The above considerations show that the ordinal properties of -4 in the set with the order type $\omega^* + \omega^*$ are logically independent of the ordinal properties of the members of the set with order type ω^* to which -4 corresponds. Accordingly, the fact that -4 has the ordinal property of being infinitely removed from some other member of the set with order type $\omega^* + \omega^*$ does not entail that the member of the set with order type ω^* to which -4 corresponds has the ordinal property of being infinitely removed from some members of the set with order type ω^*.

3. THIRD ARGUMENT

Another thesis common to several of the contemporary 'anti-infinitists' is the belief that an actual or completed infinity cannot be instantiated in the past since the set of past events is never completed but is always being added to. David Conway[7] expresses his difficulties about this: 'The notion of a completed infinite series [of past events], i.e. an infinite series all of whose members are already given, a series which is now infinite, infinite without the potential addition of further members, is indeed a strange notion.'[8] This argument is developed at greatest length by William Lane Craig, who claims that 'it would be impossible to add to a really existent actual infinite, but the series of past events is being increased daily'.[9] Craig sets forth his argument in terms of an example of an actually infinite collection of material objects, such as library books. Suppose that there is a library with an actually infinite collection of books on its shelves.

Suppose further that each book in the library has a number printed on its spine so as to create a one-to-one correspondence with the natural numbers. Because the collection is actually infinite, this means that *every possible* natural number is printed on some book. Therefore, it would be impossible to add another book to this library. For what would be the number of the

---

[7] David A. Conway, 'Possibility and Infinite Time: A Logical Paradox in St Thomas' Third Way', *International Philosophical Quarterly*, 14 (1974).

[8] Ibid. 206–7.

[9] Essay I, Sect. 2.1.1.

new book? Clearly there is no number available to assign to it. Every possible number already has a counterpart in reality, for corresponding to every natural number is an already existent book. Therefore, there would be no number for the new book.[10]

Craig does not spell out how this argument would apply to past events, but it is implicitly clear what he has in mind. If the collection of past events is actually infinite and has the cardinal number $\aleph_0$ and the order type ω^*, then corresponding to each negative number there is a past event. Since *every possible* negative number is assigned to some past event in this collection, it is impossible to add a new past event to it. For what could be the negative number of this event? It could not have a negative number, for *all* the negative numbers have been exhausted. But as a matter of fact the collection 'of past events is being increased daily', new past events *are* being added to the collection; so it follows that the collection of past events cannot have an infinite number. It must have some finite number.

Let us divide the above considerations into two separate arguments, one being that an actually infinite collection cannot be added to, the other being that if all possible negative numbers have been assigned to past events, no new event can be added to this collection. That the first argument is fallacious is apparent if we take any infinite collection of existing items, say, of books, and match them one-to-one with all positive whole numbers greater than or equal to 10. Such a collection is indeed a 'really existent actual infinite', for the books really exist and there is an actually infinite number of numbers in the series 10, 11, 12, ..., that corresponds to the collection of books. Now add nine books to the collection, matching them with the first nine integers; what has occurred is that a really existent actual infinite has been added to.

In regard to past events, match those that have occurred at some time t_1 with all the negative numbers greater than or equal to -10; this is an actually infinite collection of past events. Then match events that newly become past from t_2 to t_{11} with the negative numbers less than -10; the result is that an actually infinite collection of past events has been added to from t_2 to t_{11}.

The second argument is the one upon which Craig relies most heavily: if all possible negative numbers have been matched with past events, no new past events can be assigned to this collection. However, new assignments can be made if with the arrival of each new event in the past, each negative number is reassigned by being matched with the event immediately earlier than the event to which it had been assigned; such that -3 is reassigned to the event to which -2 formerly had been assigned, and -2 to the event to which -1 had been assigned, and so on for all the negative numbers greater than -3. This leaves -1 free to be matched with the event that has newly become past.

To the objection that this leaves some previously past event without a negative number assigned to it there is the following response: Let us call the time before some instance of the above-described reassignment t_1, and the time of the reassignment t_2. At t_2 there is a past event belonging to the collection of past events that had not belonged to this collection at t_1. However, at t_2 there is not a greater number of events belonging to this collection than at t_1, for the addition of the one event at t_2 to the infinite collection that had existed at t_1 results in a collection with the same number of members as the collection that existed at t_1, this number being $\aleph_0$. This is true because $\aleph_0$ plus 1 equals $\aleph_0$. Consequently, since there are $\aleph_0$ past events at both times, and since there are $\aleph_0$ negative numbers, there is no past event at either time that is unmatched with a negative number.

The collection of past events at t_1 is a proper subset of the collection of past events at t_2. Craig feels that the equivalence between an infinite set and a proper subset of that set as applied to real things and events is 'just not believable'.[11] It is only unbelievable, however, if one presupposes erroneously that the definition of an infinite set of real things or events is the same as the definition of a finite set of real things or events; namely, that a set necessarily has more things or events belonging to it than any proper subset of itself. If one does not make this false presupposition, then the equivalence in question is perfectly believable.

4. FOURTH ARGUMENT

Craig and Whitrow among others believe that the 'Tristram Shandy paradox' is sufficient to demonstrate the impossibility of an infinite

[11] Ibid.

past. This paradox, earlier discussed by Russell in reference to the future,[12] is based on Sterne's novel in which a character named Tristram Shandy is writing his autobiography so slowly that it takes him a year to record the events of a single day. Craig applies this story to the past and, relying in part on an argument developed by David Conway,[13] he purports to uncover a contradiction in the idea that the past is actually infinite:

suppose Tristram Shandy has been writing from eternity past at the rate of one day per year. Would he now be penning his final page? Here we discern the bankruptcy of the principle of correspondence in the world of the real. For according to that principle, Russell's conclusion would be correct: a one-to-one correspondence between days and years could be established so that given an actual infinite number of years, the book will be completed. But such a conclusion is clearly ridiculous, for Tristram Shandy could not yet have written *today's* events down. In reality he could never finish, for every day of writing generates another year of work. But if the principle of correspondence were descriptive of the real world, he should have finished—which is impossible. . . . But now a deeper absurdity bursts into view. For if the series of past events is an actual infinite, then we may ask, why did Tristram Shandy not finish his autobiography yesterday or the day before, since by then an infinite series of events had already elapsed? No matter how far along the series of past events one regresses, Tristram Shandy would have already completed his autobiography. Therefore, at no point in the infinite series of past events could he be finishing the book. We could never look over Tristram Shandy's shoulder to see if he were now writing the last page. For at any point an actual infinite sequence of events would have transpired and the book would have already been completed. Thus, at no time in eternity will we find Tristram Shandy writing, which is absurd, since we supposed him to be writing from eternity. And at no point will he finish the book, which is equally absurd, because for the book to be completed he must at some point have finished. What the Tristram Shandy story really tells us is that an actually infinite temporal regress is absurd.[14]

It is not clear at first glance why Craig believes the Tristram Shandy story to result in this 'absurdity', so it is best to reconstruct the logic of this story and try to pinpoint where the 'absurdity' is supposed to arise.

[12] Bertrand Russell, *The Principles of Mathematics* (New York: W. W. Norton, 1938), sect. 340, pp. 358–9.
[13] Conway, 'Possibility and Infinite Time', 201–8.
[14] Essay I, Sect. 2.1.1.

(1) Tristram Shandy has been writing his autobiography at every moment in the past, and it takes him one year to write about one day.

This entails that

(2) The temporal distance between any past day and the later time at which it is recorded increases with passage of time.

And this in turn entails that

(3) There is no later day finitely distant from any earlier day at which all prior days have been written about.

Now,

(4) The present day is finitely distant from any past day.

Therefore,

(5) At the present day all past days will not have been written about. Tristram Shandy's autobiography will *not* have been completed.

Nevertheless,

(6) The number of days written about is the same as the number of years elapsed prior to the present ($\aleph_0$), for in each year Tristram Shandy had written about one day.

At this point, we can see that Craig is tacitly appealing to this supposed contradiction: If in relation to any present day there are an infinite number of past days and an infinite number of past days written about, then in relation to any present there are no past days unwritten about—which contradicts (5).

However, it is false that the proposition 'The number of past days written about is the same as the number of past days' entails 'There are no past days unwritten about'. For, the number of past days written about is a proper subset of the infinite set of past days, and a proper subset of an infinite set can be numerically equivalent to the set even though there are members of the set that are not members of the proper subset. Just as the infinite set of natural numbers has the same number of members as its proper subset of even numbers, yet has members that are not members of this proper subset (these members being the odd numbers); so the infinite set of past days has the same number of members as its proper subset of days written about, yet has members that are not members of this proper subset (these members being the days unwritten about).

In conclusion, the fact that the number of past days written about corresponds to the number of past days does *not* entail that at each point in the past Tristram Shandy has completed his autobiography. Rather, at no point in the past, and at no present, will Tristram Shandy's autobiography be complete. The story of Tristram Shandy is internally consistent and so is the idea of an actually infinite past.

5. FIFTH ARGUMENT

In chapter 6 of *Our Knowledge of the External World*, Russell writes: 'classes which are infinite are given all at once by the defining property of their members, so that there is no question of "completion" or of "successive synthesis"'.[15] Some 'anti-infinitists' have inferred from this that the past cannot be infinite since events are not given all at once but successively. Craig writes that if the past is infinite then

> time and the events in it are like an actual infinite; the whole class of events and moments are given simultaneously, as Russell would say. . . . But, of course, such a picture is a crude caricature of time, for events in time, unlike events in space, exist serially. . . . The collection of all past events . . . is formed by successive addition or, to use Kant's phrase, successive synthesis.[16]

This argument is based on a confusion of givenness *in thought* with givenness *in reality*. The infinite class of events is given simultaneously *in thought*, but it is given successively *in reality*. In the thought of all events, all events are thought of 'all at once', rather than 'one at a time'. But that does not entail that all events *existed* all at once; rather they existed one at a time.

The idea behind the quoted passage from Russell's work is that classes that are infinite are given *in thought* all at once by the defining property of their members, so that there is no question of completion or of successive synthesis *in thought*. This definition of

---

[15] Bertrand Russell, *Our Knowledge of the External World* (New York: Mentor Books, 1960).
[16] William Lane Craig, *The* Kalām *Cosmological Argument* (New York: Harper & Row, 1979), 203 n. 25.

the manner of givenness *in thought* of infinite classes implies nothing about how these classes are given *in reality*. Whereas *every* infinite class is given simultaneously in our thought, *some* of the classes are also given simultaneously in reality (for example, the class of material objects if it is infinite), and *others* are given successively in reality (for example, the class of events in time if it is infinite).

6. SIXTH ARGUMENT

By acknowledging that events are given successively *in reality*, are we not admitting that *in reality* they can never add up to an infinite collection? Is it not true that for an infinite class to be given at all, whether in thought or reality, it must be given all at once?

The reply is that the collection of events cannot add up to an infinite collection in a finite amount of time, but they do so add up in an infinite amount of time. And since it is coherent to suppose that in relation to any present an infinite amount of time has elapsed, it is also coherent to suppose that in relation to any present an infinite collection of past events has already been formed by successive addition.

This suffices to disprove Kant's thesis that an infinite series 'can never be completed through successive synthesis';[17] for, although such a series can never be completely synthesized in a finite time, it can be completely synthesized in an infinite time.

I have explained that this infinite synthesis is to be understood in terms of the collection of negative numbers. But doubts may arise in connection with the idea of a *successive synthesis* corresponding to the collection of negative numbers. It seems, first of all, that a successive synthesis corresponding to the collection of negative numbers must be a potential infinite, not an actual infinite; this is because the collection of negative numbers itself cannot be completely synthesized. David Conway concludes from this that the representation of past events as corresponding to the negative numbers does not show that past events form an actual infinite:

we cannot understand the completeness of the series [of past events] on the analogy to the negative number series, since no one imagines that an infinite number of negative numbers is 'given', that, e.g., each number has actually been already written down prior to 'arriving at' -1. Rather, the

---

[17] I. Kant, *Kritik der reinen Vernunft* (Berlin, 1960), A426/B454.

latter series is infinite in the sense that there is an unending number of potential additions to it.[18]

Craig asserts, in a similar vein, that 'we cannot conceive of anyone writing down all the negative numbers from eternity past so that he ends at -1'.[19]

Moreover, there is the further problem that the negative numbers in being counted are counted *in reverse* to the order in which the past events existed. This problem is brought out by Huby and Whitrow, the latter writing:

A potentially infinite sequence of future events can be enumerated as 1, 2, 3, . . . , and so on indefinitely. Similarly, it has been argued that an infinite sequence of past events can be associated with the sequence of negative integers ending with -1 and that this demolishes Kant's objection to the possibility of an infinite sequence of past events. However, we can only enumerate the events in such a sequence by counting backwards, that is by beginning with -1 instead of ending with it. That is the reverse of the way in which the events would actually occur and yields only a potentially infinite sequence.[20]

These objections can be dealt with in turn. Certainly Conway is right in believing that no one imagines that an infinite series[21] of negative numbers *has been* written down. But he is wrong if he believes, with Craig, that no one can conceive *the possibility* of this series being written down. It may be the case that in a finite period of time this series cannot be written down,[22] but it certainly is the case that it could be written down in an infinite period of time. It is coherent to suppose that in relation to any present event, an infinite series of past events has already elapsed, and that each number in the negative number series has been written down at a time corresponding to each one of these past events.

P. F. Strawson believes that this is impossible 'because we think of the process of counting as having to *start* at some time'.[23] But

---

[18] Conway, 'Possibility and Infinite Time', 207.

[19] Craig, *The Kalām Cosmological Argument*, 203 n. 25.

[20] G. J. Whitrow, *The Natural Philosophy of Time*, 2nd edn. (Oxford: Clarendon Press, 1980), 31.

[21] Negative numbers form a series as well as a collection (set), a series being a collection of *sequentially ordered* members. Collections that are not series are exemplified by the collection of grains of sand.

[22] Can any infinite series of numbers be written down in a finite time? For a discussion of an issue directly related to this, namely, whether an infinite task can be completed in a finite time, see the articles in Wesley C. Salmon (ed.), *Zeno's Paradoxes* (Indianapolis: Bobbs-Merrill, 1970).

[23] P. F. Strawson, *The Bounds of Sense* (London: Methuen, 1966), 176.

this belief in the sense that it is true is irrelevant, and in the sense that it is relevant is false. The *empirically observable* processes of counting with which we are familiar all start at some time, but that has little bearing upon the issue of whether all *logically possible* processes of counting must start at some time. If we define a logically possible process of counting as a *synthetic series of acts of counting*, then it makes sense to conceive each past event to correspond to one act of counting, such that each earlier past event is correlated with an act of counting a greater negative number and each later past event is correlated with an act of counting a smaller negative number. In relation to any present event, the immediately preceding event can be conceived as correlating to an act of counting the number -1, such that at the time of this immediately preceding event the series of counting acts *is being* brought to completion, and at the time of the present event the series of counting acts *has already been* brought to completion.

These reflections enable the second objection, Whitrow's, to be disposed of. It may be true in the empirical sense that 'we' can only enumerate the series of past events by counting backwards from -1, and that such an enumeration yields only a potential infinite. But what we can or cannot do given our empirical limitations is not essentially relevant to the issue of whether it is *logically possible* to enumerate the series of past events in accordance with the negative number series. It may be the case that *we* must start at -1 and can only count some ways backwards, but a *logically possible counter* could have been counting at every moment in the past *in the order* in which the past events occurred. And this logically possible counter in relation to any present would have completely counted the negative numbers.

I conclude that these arguments have given us no reason to believe that an actually infinite past is logically impossible.[24]

---

[24] Some theses assumed in or related to this paper are defended elsewhere. The assumption that the past, present, and future are mind-independent elements of time has been defended in Quentin Smith, *The Felt Meanings of the World* (West Lafayette, Ind.: Purdue University Press, 1986), sect. 38, and in Quentin Smith, 'The Mind-Independence of Temporal Becoming', *Philosophical Studies*, 47 (1985), 109–20. The thesis that the past may be infinite, but is not necessarily so, implies that time may have begun, an idea elaborated upon in Quentin Smith, 'On the Beginning of Time', *Noûs*, 19 (1985), 578–84. If the universe began, say, with the Big Bang, it is not necessary that time began also; an infinitude of empty time may have elapsed before the universe. See Quentin Smith, 'Kant and the Beginning of the World', *New Scholasticism*, 59 (1985), 339–46.

III

Time and Infinity

WILLIAM LANE CRAIG

It is gratifying to see that the ancient *kalām* cosmological argument, so long neglected by post-Enlightenment philosophers, is once again receiving a measure of the philosophical scrutiny it deserves.[1] In this discussion, I wish to examine recent critiques offered of two *kalām*-style arguments for the beginning of the universe.

I. ARGUMENT FROM THE IMPOSSIBILITY OF THE EXISTENCE OF AN ACTUAL INFINITE

One of the *kalām* arguments which I have defended is the argument based on *the impossibility of the existence of an actual infinite*. While potential infinites can exist, no actual infinite can exist, for such a

A longer version of this essay was first pub. in *International Philosophical Quarterly*, 31/4, Issue 124 (Dec. 1991), 387–401.

[1] For an exposition of the argument in its historical context, see William Lane Craig, *The Cosmological Argument from Plato to Leibniz* (London: Macmillan, 1980); Harry Austryn Wolfson, 'Patristic Arguments against the Eternity of the World', *Harvard Theological Review*, 59 (1966), 354–67, and *The Philosophy of the* Kalām (Cambridge, Mass.: Harvard University Press, 1976); and, most recently, H. A. Davidson, *Proofs for Eternity, Creation and the Existence of God in Medieval Islamic and Jewish Philosophy* (New York: Oxford University Press, 1987). Recent discussions of the philosophical arguments for a beginning of the universe include William Wainwright, 'Critical Notice of *The* Kalām *Cosmological Argument*', *Noûs*, 16 (1982), 328–34; J. L. Mackie, *The Miracle of Theism* (Oxford: Clarendon, 1982); Richard Sorabji, *Time, Creation and the Continuum* (Ithaca, NY: Cornell University Press, 1983); David A. Conway, ' "It would have Happened Already"; On One Argument for a First Cause', *Analysis*, 44 (1984), 159–66; William Lane Craig, 'Professor Mackie and the Kalām *Cosmological Argument*', *Religious Studies*, 20 (1985), 367–75; Robin Small, 'Tristram Shandy's Last Page', *British Journal for the Philosophy of Science*, 37 (1986), 213–16; Quentin Smith, Essay II; Ellery Eells, 'Quentin Smith on Infinity and the Past', *Philosophy of Science*, 55 (1988), 453–5; Stewart C. Goetz, 'Craig's Kalām *Cosmological Argument*', *Faith and Philosophy*, 6 (1989), 99–102; William Lane Craig, 'The Kalām *Cosmological Argument* and the Hypothesis of a Quiescent Universe', *Faith and Philosophy*, 8 (1991), 104–8; Graham Oppy, 'Craig, Mackie, and the Kalām *Cosmological Argument*', *Religious Studies*, 27 (1991), 189–97.

real existence would entail absurdities. Because an actually infinite number of things cannot exist, the series of past events must be finite in number and, hence, the universe began to exist.

Several critics have followed Cantor's lead in charging that the Aristotelian position that only potential, but no actual, infinites exist in reality is incoherent because a potential infinite presupposes an actual infinite. Rucker claims that there must be a 'definite class of possibilities' which is actually infinite in order for the mathematical intuitionist to regard the natural number series as potentially infinite through repetition of certain mathematical operations.[2] Similarly, Sorabji asserts that Aristotle's view of the potentially infinite divisibility of a line entails that there is an actually infinite number of positions at which the line could be divided.[3] More recently, Prevost argues that what are to us potential infinites must be for an omniscient God actual infinites, since God knows the places where a division potentially may be made.[4]

If this line of argument is successful, it is, indeed, a *tour de force*, since it would show mathematical thought from Aristotle to Gauss to be, not merely mistaken or incomplete, but incoherent in this respect. But the objection does not seem to be successful. For the claim that a substance is, say, potentially infinitely divisible does not entail that the substance is potentially divisible *here* and *here* and *here* and . . . Potential infinite divisibility (the property of being susceptible of division without end) does not entail actual infinite divisibility (the property of being composed of an infinite number of points where divisions can be made). The argument that it does seems to be guilty of a modal operator shift, invalidly inferring from

(1) Possibly, there is some point at which *x* is divided

to

(2) There is some point at which *x* is possibly divided.

It is therefore coherent to maintain that a substance is potentially infinitely divisible without holding that there are an infinite number of positions where it could be divided.

[2] Rudolf v. B. Rucker, 'The Actual Infinite', *Speculations in Science and Technology*, 3 (1980), 66.
[3] Sorabji, *Time, Creation and the Continuum*, 210–13, 322–4.
[4] Robert Prevost, 'Classical Theism and the *Kalām* Principle', in William Lane Craig and M. McLeod (eds.), *Rational Theism: The Logic of a Worldview* (Lewiston, NY: Edwin Mellen, 1990), 113–25.

To illustrate: William Alston has recently defended the view that God's knowledge is non-propositional in nature.[5] His is a simple, intuitive knowledge that embraces all truth. Finite creatures break up the whole of what God knows into propositions which they know. But the fact that God's simple intuition can be broken down into a potentially infinite number of propositions does not entail that what God knows is an actually infinite number of propositions. His simple, intuitive knowledge can be endlessly expressed in propositional forms without there being an actually infinite number of propositions which He knows. In the same way one can admit potential infinities of extendability or divisibility without entailing actual infinites of positions of extension or division.

Rucker also argues that there are probably in fact physical infinities.[6] If one says, for example, that time is potentially infinite, then Rucker will reply that the modern, scientific world-view sees past, present, and future as merely different regions coexisting in spacetime. If one says that any physical infinity exists only as a temporal (potentially infinite) process, Rucker will rejoin that it is artificial to make physical existence a by-product of human activity. If there are, for example, an infinite number of bits of matter, this is a well-defined state of affairs which obtains right now regardless of our apprehension of it. He concludes that it seems quite likely that there is some form of physical infinity.

His conclusion, however, clearly does not follow from his arguments, since all they prove at best is the possibility of physical infinity. Time and space may well be finite. But could they be potentially infinite? Concerning time, Rucker is simply incorrect in saying that 'the modern, scientific world-view' precludes a theory of time according to which temporal becoming is a real and objective feature of the universe. Following McTaggart, contemporary philosophers of space and time distinguish between an A-theory of time, according to which events are temporally ordered by tensed determinations of past, present, and future, and temporal becoming is an objective feature of physical reality, and a B-theory of time, according to which events are ordered by the tenseless relations of earlier than, simultaneous with, and later than, and temporal becoming is subjective and mind-dependent. Although some thinkers

---

[5] William Alston, 'Does God Have Beliefs?' *Religious Studies*, 22 (1986), 287–306.
[6] Rucker, 'The Actual Infinite', 69.

have carelessly asserted that relativity theory has vindicated the B-theory over against its rival, such claims are untenable. One could harmonize the A-theory and relativity theory in at least three different ways:[7] (1) Distinguish metaphysical time from physical or clock time and maintain that while the former is A-theoretic in nature, the latter is a bare abstraction therefrom, useful for scientific purposes and quite possibly B-theoretic in character, the element of becoming having been abstracted out. (2) Relativize becoming to reference frames, just as is done with simultaneity. (3) Select a privileged reference frame to define the time in which objective becoming occurs, most plausibly the cosmic time which serves as the time parameter for hypersurfaces of homogeneity in spacetime in the General Theory of Relativity. And concerning space, to say space is potentially infinite is not to say it depends on human activity (nor again, that there are actual places to which it can extend), but simply that space expands limitlessly as the distances between galaxies increase with time. As for the number of bits of matter, there is no incoherence in saying there is a finite number of bits or that matter is capable of only a finite number of physical subdivisions, though mathematically one could proceed potentially *ad infinitum*.

Thus, the attempts to prove that an actual infinite must or does exist appear to be unsuccessful. But what about arguments that an actual infinite cannot exist?

Smith focusses on one particular argument I gave in support of the premiss that an actual infinite cannot exist. I argued that if an actually infinite number of concrete objects could exist, then it would be impossible to add to them, which is absurd. Such a contention must seem patently false to anyone familiar with trans-finite arithmetic, for it is mathematical commonplace to add to the infinite. But it must be kept in mind that the argument concerns concrete objects and ontological possibility, not mathematical objects

[7] Solution (1) is preferred by Mary Cleugh, *Time and its Importance in Modern Thought* (London: Methuen, 1937), 29–67, and perhaps by Peter Kroes, *Time: Its Structure and Role in Physical Theories* (Dordrecht: D. Reidel, 1985), p. xiii; solution (2) by H. Stein, 'On Einstein–Minkowski Space-Time', *Journal of Philosophy*, 65 (1968), 5–23, and by Milic Capek, 'The Inclusion of Becoming in the Physical World', in M. Capek (ed.), *The Concepts of Space and Time* (Dordrecht: D. Reidel, 1976), 501–24; and solution (3) by Richard Swinburne, *Space and Time*, 2nd edn. (London: Macmillan, 1981), 185–202, and by G. J. Whitrow, *The Natural Philosophy of Time*, 2nd edn. (Oxford: Clarendon, 1980), 283–307, 371.

and logical possibility. The realm of abstract objects is not for me a realm of Platonic existents, but a realm of thought and imagination. I thus find myself entirely sympathetic to the view of mathematicians such as Abraham Robinson, who on the one hand accepts as logically consistent axiomatized infinite set theory and yet supports on the other hand disbelief in the reality or objectivity not only of set theory, but also of all other infinite mathematical structures.[8]

Consider, then, a library with an actually infinite number of books, each of which has a number printed on its spine so as to create a one-to-one correspondence between the books and the natural numbers. In such a case, it would seem impossible to add another book to the collection, for all the natural numbers have been *used up*, and there remains no number not already concretely represented which could be assigned to the new book.

Smith offers a twofold response: (1) An actual infinite can be added to by, for example, matching each object in the collection with the natural numbers greater than 10, then adding 10 new objects to the collection. (2) If the objects are already matched in a one-to-one correspondence with all the natural numbers, then when new members are added to the collection, simply renumber the objects beginning with the most recent addition.

But (1) simply violates the initial conditions laid down in the argument. We are to imagine a series of consecutively numbered books beginning at 0 and increasing infinitely, not a series of books numbered from some finite number. Moreover, since any collection with $\aleph_0$ members can be numbered from 0, the argument can be made to apply to any actually infinite collection of objects by stipulating that they be so numbered. (2) Once the objects are numbered as stipulated, reassigning the numbers to begin with the proposed addition seems impossible. As I argued in my earlier work,

It might be suggested that we number the new book '1' and add one to the number of every book thereafter. This is perfectly successful in the mathematical realm, since we accommodate the new number by increasing all the others out to infinity. But in the real world this could not be done. For an actual infinity of objects already exists that completely exhausts the natural number system—every possible number has been instantiated in

---

[8] Abraham Robinson, 'Metamathematical Problems', *Journal of Symbolic Logic*, 38 (1973), 500–16.

reality on the spine of a book. Therefore, book 1 could not be called book 2, and book 2 be called book 3, and so on, to infinity. Only in a potential infinite, where new numbers are created as the collection grows, could such a re-count be possible. But in an actual infinite, all the members exist in a determinate, complete whole, and such a re-count would necessitate the creation of a new number. But this is absurd, since every possible natural number has been used up.[9]

I thus attempted to anticipate the objection pressed by Smith; so far as I can see, however, he takes no cognizance of this and thus leaves this reasoning unrefuted.

But let us push the analysis a step further. The argument might be thought to confuse numbers with numerals, such that one might hold that the proposed acquisition to the library can be made to correspond to the number 1 even though it cannot be assigned the numeral '1', since that numeral has already been used. But I do not think the argument makes this equivocation. It is not just the numeral '1' that is correlated with a book already in the collection, but also the number 1 and every other natural number as well which the numerals serve to represent. This can be clearly seen by making the Platonic assumption that numbers are actually existing entities, every bit as real as concrete objects. Is it possible to add a new integer to the series of natural numbers? Of course not, for the natural number series is determinate and complete. When we perform calculations with natural numbers, for example, $1 + 1 = 2$, we are not increasing the store of natural numbers or creating new abstract objects—there is, after all, only one number 1. But just as we cannot add more numbers to the natural numbers, so we cannot add more books to an infinite collection of books each of which stands in a one-to-one correspondence with the natural numbers. Now I do not in fact believe that numbers are actually existing objects, but clearly books are. Therefore, just as the collection of natural numbers could not be added to, neither could an actually infinite collection of books—which is absurd.

Oddly enough, the above argument for the impossibility of the real existence of an actual infinite is, in my opinion, one of the most tentative I presented, and yet Smith in his refutation leaves entirely out of account what I consider to be the stronger arguments which show the counter-intuitive absurdities—and, in the end, logical

---

[9] Essay I, Sect. 2.1.1.

contradictions—which would result if an actually infinite number of things could exist. Smith, like Mackie,[10] offers only the passing comment that once we understand that an infinite set has a proper subset which has the same number of members as the set itself, the purportedly absurd situations become 'perfectly believable'.[11]

But to my mind, it is precisely this feature of infinite set theory which, when translated into the realm of the real, yields results which are perfectly unbelievable. Confronted with such results, my sympathies are with Wittgenstein, when he remarked: 'I would say, "I wouldn't dream of trying to drive anyone from this paradise." I would do something quite different: I would try to show you that it is not a paradise—so that you'll leave of your own accord. I would say, "You're welcome to this; just look about you." '[12]

The proponent of the *kalām* argument has two options open to him at this point. On the one hand, the proponent of *kalām* can argue that if an actual infinite were to exist, then the Principle of Correspondence would be valid with respect to it and that if an actual infinite were to exist and the Principle of Correspondence were to be valid with respect to it, then the various counter-intuitive situations would result. Therefore, if an actual infinite were to exist, the various counter-intuitive situations would result. But because these are absurd and so really impossible, it follows that the existence of an actual infinite is impossible. Smith and Mackie's response does nothing to prove that the envisioned situations are not absurd, but only reiterates, in effect, that if an actual infinite were to exist and the Principle of Correspondence were valid with respect to it, then the relevant situations would result, which is not in dispute. At any rate, I have also pointed out that the real existence of an actual infinite entails not just counter-intuitive absurdities, but, in the cases of inverse mathematical operations, logical contradictions. Here the case against the real existence of the actual infinite becomes decisive.

On the other hand, the proponent of *kalām* might call into question the premiss that if an actual infinite were to exist, then the Principle of Correspondence would be valid with respect to it. As I point out in Essay I,[13] there is no reason to think that the principle

---

[10] Mackie, *The Miracle of Theism*, 93.

[11] Essay II, end of Sect. 3.

[12] Ludwig Wittgenstein, *Lectures on the Foundations of Mathematics*, ed. Cora Diamond (Sussex: Harvester Press, 1976), 103.

[13] Sect. 2.1.1.

is universally valid. It is merely a convention adopted in infinite set theory. Now, necessarily, if an actual infinite were to exist, then either the Principle of Correspondence or Euclid's maxim that 'The whole is greater than its part' would apply to it. But since the application of either of these two principles to an actual infinite results in counter-intuitive absurdities, it is plausible that if the existence of an actual infinite were possible, then if an actual infinite were to exist, neither of these two principles would be valid with respect to it. It therefore follows that the existence of an actual infinite is impossible. When Smith and Mackie assert that given the Principle of Correspondence, the resulting situations are perfectly believable, the Euclidean can with equal justification retort that when his maxim is granted, the relevant resulting situations are perfectly believable. But both sides cannot be correct. All each side has shown is that if an actual infinite were to exist and his respective principle were valid with respect to it, then certain situations would in fact follow. But neither has shown that those situations are in themselves believable or possible. Each regards the other's resulting situations as unbelievable and the principle generating them as false. The proponent of *kalām* may agree with both of their critiques, concluding that since the counterfactual that 'If an actual infinite were to exist, then neither principle would be valid with respect to it' is necessarily false, it follows that the existence of an actual infinite is impossible.

2. ARGUMENT FROM THE IMPOSSIBILITY OF THE SUCCESSIVE FORMATION OF AN ACTUAL INFINITE

2.1. *The Tristram Shandy Paradox*

A second *kalām* argument which I defended is the argument based on *the impossibility of forming an actual infinite by successive addition*. If an actual infinite cannot be formed by successive addition, then the series of past events must be finite, since that series is formed by successive addition of one event after another in time.[14]

---

[14] Cantor himself held that the argument against the infinity of the past is sound. In a letter of 1887 he wrote: 'When it is said that a mathematical proof for the beginning of the world in time cannot be given, the emphasis is on the word "mathematical", and to this extent my view agrees with that of St Thomas. On the other hand, a mixed mathematical–metaphysical proof of the proposition might well

The problem of forming an infinite series of past events by successive addition is vividly illustrated by the Tristram Shandy paradox, according to which a man writing his autobiography from eternity at the rate of one day per year of writing would finish by the present moment.

Critics of the argument agree that Tristram Shandy cannot complete his autobiography by the present day.[15] But they contend that there is no reason to think that if he has been writing from eternity then he should be finished by the present moment. But the reason he should be finished with his autobiography by the present moment seems to be that given by Russell: in order to write about an actually infinite number of days at the rate of one day per year all one needs is an actually infinite number of years. Since according to the Principle of Correspondence there is a year corresponding to every day in the past, there is plenty of time available to write about all the days. There is no day of the past to which a past year does not correspond; therefore, assuming that his goal is to record every day of his life, Tristram Shandy should finish the job in an infinite number of years, which, by now, he has had.

But the problem with this argument seems to be that while an infinite number of years is a *necessary* condition of recording an infinite number of days at the rate of one day per year, it is not a *sufficient* condition. What is also needed is that the days and years be arranged in a certain way such that every day is succeeded by a year in which to record it. But then it will be seen that Tristram Shandy's task is inherently paradoxical; the absurdity lies not in the infinity of the past but in the task itself. The argument's critics would thus far seem to be vindicated.

But does not then another paradox arise? I disputed Russell's conclusion that no part of the autobiography will remain unwritten if Tristram Shandy writes for ever, since as time goes on 'Tristram Shandy would only get farther and farther behind so that instead of finishing his autobiography, he will progressively approach a state in which he would be *infinitely* far behind.'[16] Should we not,

be produced on just the basis of the true theory of the transfinite, and to this extent I depart from St Thomas, who defends the view: *Mundum non semper fuisse, sola fide tenetur, et demonstrative probari non potest.*' (Cantor, in H. Meschkowski, *Probleme des Unendlichen: Werk und Leben Georg Cantors* (Braunschweig: Freidrich Vieweg, 1967), 125–6.) I am indebted to Robin Small for this reference.

[15] So Sorabji, Conway, Small, and Smith.
[16] Essay I, Sect. 2.1.1.

therefore, have argued, not that Tristram Shandy would have completed his autobiography by now, but on the contrary that he would now be infinitely far behind? For if Tristram Shandy has lived for an infinite number of years, then he has recorded an equally infinite number of past days. Given the thoroughness of his autobiography, these days are all consecutive days. At any point in the past or present, therefore, Tristram Shandy has recorded a beginningless, infinite series of consecutive days. But now the question inevitably arises: *Which* days are these? Where in the temporal series of events are the days recorded by Tristram Shandy at any given point? The answer can only be that *they are days infinitely distant from the present.* For there is no day on which Tristram Shandy is writing which is finitely distant from the last recorded day. This may be seen through an incisive analysis of the Tristram Shandy paradox given by Robin Small. He asks,

in the case involving past time, what day corresponds to the latest complete year of writing? Clearly the most recent day whose events can have been written down during this year is a day which took place at least a year ago. But the most recent day whose events can have been written down in the year preceding this one is a day which took place at least two years ago. Therefore, the most recent day whose events can have been written down in the latest complete year is the day after one which took place at least two years ago. And this argument can be repeated again and again. We find then that we cannot locate the day which corresponds to the latest year of writing, or for that matter to any other year of writing. It is therefore *not* the case that 'a one-to-one correspondence between days and years could be established'.[17]

The succinctness of this paragraph belies its insightfulness. Small points out that if Tristram Shandy has been writing for one year's time, then the most recent day he could have recorded is one year ago. But if he has been writing two years, then that same day could not have been recorded by him. For since his intention is to record *consecutive* days of his life, the most recent day he could have recorded is the day immediately after a day at least two years ago. This is because it takes a year to record a day, so that to record two days he must have two years. Similarly, if he has been writing three years, then the most recent day recorded could be no more recent than three years and two days ago. In other words, the longer he has written the further behind he has fallen. In fact, the recession

[17] Small, 'Tristram Shandy's Last Page', 214–15.

into the past of the most recent recordable day can be plotted according to the formula (present date − *n* years of writing) + *n* − 1 days. But what happens if Tristram Shandy has, *ex hypothesi*, been writing for an infinite number of years? The first day of his autobiography recedes to infinity, that is to say, to a day infinitely distant from the present. Nowhere in the past at a finite distance from the present can we find a recorded day, for by now Tristram Shandy is infinitely far behind. The beginningless, infinite series of days which he has recorded are days which lie at an infinite temporal distance from the present. In other words, the Tristram Shandy paradox serves to vindicate G. J. Whitrow's contention that an infinite past entails infinitely distant events!

But it might be rejoined that according to our formula, Tristram Shandy's most recently recorded day was an infinite number of days after that first infinitely distant day. Does this not imply that he would therefore have reached the present day or at least some day in the finite past? No, not at all; for adding an infinite number of days to a day infinitely distant does not guarantee arriving at the present day or a day in the finite past. Indeed, he *cannot* have reached any such day, for in order to record that day he would have to have been writing less than a certain number of years; for example, if today is 31 December 1987, then in order to record 5 January 1983, he would have to have been writing less than 5 years. Therefore, as Small perceives, no matter how finitely far we regress into the past we can never find Tristram Shandy's most recently recorded day. They all exist in some mysterious way beyond the frontier of past infinity.

Small's conclusion that the recorded days and the years of writing do not correspond does not, however, follow. The number of recorded days is $\aleph_0$ and the number of years of writing is similarly $\aleph_0$, so that a one-to-one correspondence exists. The fact that we cannot locate the days is merely an anthropocentric limitation and does not affect the cardinality of the set of such days any more than my inability to locate all the beetles that have ever crawled over the face of the earth affects the cardinality of the set of such beetles.

Small's analysis serves to expose the flaw in the reasoning of Eells,[18] who argues that because there will always be a finite number of days between any given day and the day on which Tristram

---

[18] Eells, 'Quentin Smith on Infinity and the Past', 453–5.

Shandy finishes writing about that day, we discover as we go back in time that the temporal distance between a day and the day he finishes writing about it becomes shorter and shorter. As we continue to look back in time, we must eventually find Tristram Shandy writing about the day on which he is then currently writing, and earlier than this time we must find him writing about his future. Indeed, only after infinitely many years of writing about his future could he have begun to write about his past! Therefore, there is no logical contradiction in the story, nothing that contradicts the logical possibility of an infinite past, for no day of Tristram Shandy's life will remain unwritten, including the present. But Eells errs in assuming that there is always a finite number of days between any given day and the day Tristram Shandy finishes writing about it; for if he has been writing for an infinite number of days, that distance will be infinite. As we look back in time we will never come to the day on which he is writing about that day and before which he was writing about his future. The days during which he wrote about his future are in an infinite series temporally prior to the infinite past. That this is the case is evident from the fact that Tristram Shandy could die, say, the day after tomorrow. In that case he could not prior to some point in the finite past have been writing from eternity about his future, since the days of his future were only finite in number. Only if there were infinitely many days after that turn-around point could he have been for infinite time writing about his future.

Not only does the problem now arise of how those infinitely distant days could have once been present, but even more seriously, Eells has achieved logical consistency only at the cost of metaphysical absurdity: for how can Tristram Shandy record *future* days of his life, of which he knows nothing? The task of slowly writing one's autobiography is evidently a coherent one; but if it becomes paradoxical when carried out for infinite time, then the solution is not to posit the additional absurdity of making records of the future, but to deny the metaphysical possibility of infinite past time.

What seems to follow from the Tristram Shandy story is that an infinite series of past events is absurd. For there is no way to traverse the distance from an infinitely distant event to the present, or, more technically, for an event which was once present to recede to an infinite temporal distance.

2.2. *The Paradox Deepened*

In Essay I I discussed what seemed to be a 'deeper absurdity' revealed by the Tristram Shandy story, namely, that wholly apart from the conditions stipulating the rate of writing, if Tristram Shandy were going to complete his book by the present moment, then he would always have at any moment in the finite past completed it.

Smith distinguishes in this connection between givenness in thought and givenness in reality. 'The infinite class of events is given simultaneously *in thought*, but it is given successively *in reality*.'[19] This is, of course, a distinction which I want to insist on and is perhaps best captured by distinguishing with Popper between the *set* of temporal events and the *series* of temporal events.[20] Although Popper mistakenly utilizes this distinction to claim that the set of past events is actually infinite while the series of past events is potentially infinite,[21] it seems to me that the proponent of this argument could maintain that something like the opposite is true: the set of future events may be actually infinite, but the series of future events can be only potentially infinite (in the sense that the number of events later than any point in the past is finite but continually increasing; if we use the present as our point of departure, then there simply exist no future events, much less a series of them). So the question is not whether an actual infinite can be given in thought in the sense of an abstract object like a set, but whether an actual infinite can be sequentially and completely instantiated in reality as a temporal series. I do not think that it can.

But Smith retorts, 'the collection of events cannot add up to an infinite collection in a finite amount of time, but they do so add up in an infinite amount of time'.[22] With regard to the series of past events, 'It is coherent to suppose that in relation to any present event, an infinite series of past numbers has already elapsed, and

---

[19] Essay II, Sect. 5.
[20] Karl Popper, 'On the Possibility of an Infinite Past: A Reply to Whitrow', *British Journal for the Philosophy of Science*, 29 (1978), 47–8.
[21] See my critique in 'Whitrow and Popper on the Impossibility of an Infinite Past', *British Journal for the Philosophy of Science*, 30 (1979), 165–70.
[22] Essay II, Sect. 6. Since Smith refers in this connection to the paradoxes of motion formulated by Zeno, I refer the reader to my discussion of these in app. 2 of my Kalām *Cosmological Argument*.

that each number in the negative number series has been written down at a time corresponding to each one of these past events.'[23]

This familiar rejoinder to the *kalām* argument seems, however, to be question-begging. For the argument can be restated in terms of time itself. If we divide time into temporal segments of equal duration, say, hours, then, if the past is actually infinite, before the present hour could arrive an infinite number of previous hours would have to have successively elapsed, which, according to the argument, is absurd. Now clearly it would be nonsensical to reply that it is only impossible for them to elapse *in a finite time*, for the argument concerns time itself. It is thus question-begging to explain how one purportedly infinite collection (the series of past events) can be formed by successive addition merely by correlating it with another purportedly infinite collection (the series of past hours) also formed by successive addition. For the question is then merely moved back a notch: how can the *latter* collection be given successively? I do not therefore see that Smith has done anything to render more intelligible the formation in reality of an infinite past by successive addition.

Conway and Sorabji have, however, addressed the argument more directly, maintaining that there is no reason to think that Tristram Shandy would at any point have already finished and, hence, that the successive formation of a temporal series of order type ω^* is impossible.[24] Since I have elsewhere commented briefly on Sorabji's remarks,[25] I shall restrict my discussion to Conway's objection.

Conway's perplexity with this version of the Tristram Shandy paradox stems largely, I think, from his incorrect rendering of it. In his analysis, the nub of the argument lies in the conditional

(3) If an infinite number of pages had been written by yesterday, then Tristram Shandy will have finished by yesterday.

But Conway's conditional is quite ambiguous, and the arguments which he suggests in support of it have no apparent relevance to the reasoning behind the Tristram Shandy paradox. Rather, the con-

---

[23] Essay II, Sect. 6.

[24] Conway, 'It would have Happened Already', 159–66; Sorabji, *Time, Creation and the Continuum*, 222.

[25] See William Lane Craig, 'Critical Notice of *Time, Creation, and the Continuum*', *International Philosophical Quarterly*, 25 (1985), 319–26.

ditional at the heart of this version of the paradox is something more like this:

(4) If Tristram Shandy would have finished his book by today, then he would have finished it by yesterday,

and the truth of this conditional would seem to be guaranteed by the Principle of Correspondence. It is on the basis of this principle that the defender of the infinite past seeks to justify the intuitively impossible feat of someone's counting down all the negative numbers and ending at -1. Since the negative numbers can be put into a one-to-one correspondence with the series of, say, past hours, someone counting from eternity would have completed his count-down. But by the same token, Tristram Shandy at any point in the past should have already completed his book, since by then a one-to-one correspondence exists between, say, each page of writing and a past hour. In this version of the story, having infinite time does seem to be a sufficient condition of finishing the job. Having had infinite time, Tristram Shandy should have already completed his task.

Such reasoning in support of the finitude of the past and the beginning of the universe is not mere armchair cosmology. P. C. W. Davies, for example, utilizes this reasoning in explaining two profound implications of the thermodynamic properties of the universe:

The first is that the universe will eventually die, wallowing, as it were, in its own entropy. This is known among physicists as the 'heat death' of the universe. The second is that the universe cannot have existed for ever, otherwise it would have reached its equilibrium end state an infinite time ago. Conclusion: the universe did not always exist.[26]

The second of these implications is a clear application of the reasoning that underlies the 'deeper absurdity' revealed by the Tristram Shandy paradox: even if the universe had infinite energy, it would in infinite time come to equilibrium; since at any point in the past infinite time has elapsed, a beginningless universe would have already reached equilibrium, or as Davies puts it, it would have reached equilibrium an infinite time ago. Therefore, the universe began to exist, *quod erat demonstrandum.*[27]

[26] Paul Davies, *God and the New Physics* (New York: Simon & Schuster, 1983), 11.
[27] See the similar reasoning of John Barrow and Frank J. Tipler, *The Anthropic*

3. CONCLUSION

It seems to me, therefore, that both the argument based on the impossibility of an actual infinite and the argument based on the impossibility of forming an actual infinite by successive addition have yet to be shown unsound. Therefore these arguments—amazing as it may seem—do seem to furnish philosophical grounds for holding to the finitude of the past and the beginning of the universe.

Cosmological Principle (Oxford: Clarendon Press, 1986), 601–8, against steady state cosmologies on the ground that any event which would have happened by now would have already happened before now if the past were infinite.

IV

The Uncaused Beginning of the Universe
QUENTIN SMITH

Introductory Note

At this point, the debate between Craig and myself shifts from the argument about the Cantorian infinite to a discussion of Big Bang cosmology. In Essay IV, I take up the second half of Craig's theistic argument in Essay I, the half consisting of his argument for God's existence based on the empirical evidence of Big Bang cosmology. (Readers who are interested in pursuing the debate between Craig and myself about the Cantorian infinite and the past may consult Craig's article 'Time and Infinity' (*International Philosophical Quarterly*, 31 (1991), 387–410), which is an extended version of his Essay III, my 'Reply to Craig: The Possible Infinitude of the Past' (*International Philosophical Quarterly*, 33 (Mar. 1993), 109–115), and Craig's rejoinder, 'Smith on the Finitude of the Past', *International Philosophical Quarterly*, 33 (Mar. 1993), 225–31.

This essay relates to the discussion of Big Bang cosmology in Craig's Essay I in two ways. I further defend the thesis that Big Bang cosmology warrants the belief that the past history of the universe is finite. However, Essay IV also challenges Craig's argument that the Big Bang is caused by God; I argue the Big Bang has no cause.

My presentation of Big Bang cosmology begins on a fairly technical level, so readers unfamiliar with the Einstein equation, the Friedman solutions to the Einstein equation, and the singularity theorems should consult Appendix IV.2 to this essay before they begin reading the essay itself. Appendix IV.2 gives a non-technical account of these mathematical ideas.

My purpose in this paper is to argue that there is sufficient evidence at present to warrant the conclusion that the universe probably began

Portions of this essay were first pub. in 'The Uncaused Beginning of the Universe', *Philosophy of Science*, 55 (1988), 39–57. c. Philosophy of Science Association, 1988.

to exist over 10 billion years ago, and that it began to exist without being caused to do so. I believe accordingly that the positions held by many if not most contemporary philosophers concerning this issue are unjustified, for their beliefs typically fall into one of three mutually exclusive categories, (1) the universe is probably infinitely old, (2) the universe began to exist and its beginning was caused by God, and (3) insufficient evidence is available to enable us to decide about whether the universe began to exist or is infinitely old.

I. THE PREDICTION OF A SPACETIME SINGULARITY IN OUR PAST

Most philosophers today are aware that the Big Bang cosmological theory has superseded the steady state theory, but a great number of these philosophers erroneously believe either that there is probably an infinite number of cycles of expansion and contraction of the universe, or that there is insufficient evidence to decide between the infinitely oscillating model and the theory that there was an earliest or single expansion, or that there is a first expansion that needs to be explained by introducing divine causality. That these beliefs are unfounded becomes apparent once the evidence for the prediction of a singularity in our past by the Big Bang cosmological model is adequately clarified.

The most important but by no means the only observational evidence for the Big Bang theory is the red-shift of the light from distant galactic clusters, first discovered by Slipher and Hubble, which indicates the universe to be expanding uniformly in all directions.[1] This suggests that there is some time in the past when all the galactic clusters, or all the materials in these clusters, were

---

[1] See Stromberg's summary of V. M. Slipher's measurements in G. Stromberg, 'Analysis of Radial Velocities of Globular Clusters and Non-galactic Nebulae', *Astrophysical Journal*, 61 (1925), 353–62; also see E. Hubble, 'A Relation between Distance and Radial Velocity among Extra-galactic Nebulae', *Proceedings of the National Academy of Sciences*, 15 (1929), 168–73. Other observational evidence that supports Big Bang cosmology includes the background microwave radiation of 2.7 K, which is a remnant of the intense heat generated at an early stage of the expanding universe. This radiation was first discovered (and initially measured to be 3.5 K) by A. Penzias and R. Wilson ('A Measure of Excess Antenna Temperature at 4080 Mc/s', *Astrophysical Journal*, 142 (1965), 419–21). A third major set of data supporting the Big Bang cosmology is the abundance of helium 4, deuterium, helium 3, and lithium 7, the formation of which is predicted to occur in the first minutes of the Big Bang.

arbitrarily close together, and that this time represents the beginning of the universe.

This is more exactly understood in terms of the models of the universe provided by the so-called Friedman solutions of the field equations of Einstein's General Theory of Relativity (GTR). The field equations show that the metric of spacetime is dependent upon the matter present in that spacetime.[2] The field equations can be solved for the universe as a whole if figures reflecting the observed values of the universe are introduced. Since the universe is isotropic (the same in all directions) and homogeneous (the matter is evenly distributed), it is described by the Robertson–Walker metric,[3] which applied to the field equations enable them to be reduced to (with the cosmological constant λ omitted):

$$-3d^2a/dt^2 = 4\pi G(p + 3P/c^2)a$$
$$3(da/dt)^2 = 8\pi Gpa^2 - 3kc^2.$$

a is the scale factor representing the radius of the universe at a given time. da/dt is the rate of change of a with time; it is the rate at which the universe expands or contracts. d^2a/dt^2 is the rate of change of da/dt, that is, the acceleration of the expansion or the deceleration of the contraction. G is the gravitational constant and c the speed of light. P is the pressure of matter and p its density. k is a constant which takes one of three values: 0 for a flat Euclidean

---

[2] According to John Wheeler ('From Relativity to Mutability', in J. Mehra (ed.), *The Physicist's Conception of Nature* (Boston: D. Reidel, 1973), 220), the simplest expression of the Einstein equations is

(curvature of space time) =
8π(density of the mass-energy present in that spacetime).

More completely, it can be said that the field equations relate the metric tensor $g_{\mu\nu}$ and its derivatives, which describe the geometry of spacetime, to the energy–momentum tensor $T_{\mu\nu}$, which is determined by the distribution of the mass and energy in that spacetime. These equations enable paths in spacetime (specifically, geodesic paths) to be calculated. The formula summarizing the 10 field equations is

$$R_{\mu\nu} - (1/2)Rg_{\mu\nu} + \lambda g_{\mu\nu} = -(8\pi G/c^2)T_{\mu\nu}.$$

The terms on the left-hand side are composed of $g_{\mu\nu}$ and its derivatives, and also of the constant λ. G is the constant of gravitation and c the velocity of light.

[3] The Robertson–Walker metric is determined by a, the radius of the universe at a certain time, and by the curvature of spacetime. The metric of a homogeneous and isotropic universe is

$$ds^2 = dt^2 - (1/c^2)a^2d\sigma^2,$$

where ds is the spacetime interval between two events, $d\sigma$ is the line element of a space of constant curvature, and c the velocity of light.

space (in which case the universe is open, that is, expands for ever), −1 for a hyperbolic space (in which case the universe is also open), or +1 for a spherical space (in which case the universe is closed, that is, will contract).

What is important to note about these Friedman equations is that if p, the density of matter in the universe, is positive, then the right side of the first equation is positive, and this entails that d^2a/dt^2, the acceleration of the expansion or the deceleration of the contraction, cannot be zero. d^2a/dt^2 must be negative, which means that the acceleration of the expansion is decreasing or that the acceleration of the contraction is increasing. In a word, if there is matter present in the universe, then the universe must be either expanding or contracting with a varying acceleration.

This explanation should be modified in one respect, crucial to understanding the Theory of Inflation. Instead of saying that 'If p, the density of matter in the universe, is positive, then the right side of the first equation is positive', a more complete statement would be that 'If $(p + 3P/c^2)$ is positive, then the right side of the equation is positive'. This is crucial to the inflationary era (from 10^{-35} seconds after the Big Bang singularity to 10^{-32} seconds after the singularity), when the pressure P of matter becomes negative. At this time, the acceleration of the expansion increases.

It is the case of expansion that interests us, since the universe is now expanding. If the acceleration of the expansion is always decreasing (except during the brief fraction of a second comprising the inflationary era), then this suggests that the further we go into the past the greater the increase of the acceleration and the smaller the scale factor a of the radius of the universe, until a time t_0 is reached when $a = 0$. As d^2a/dt^2 increases and a decreases, the density of matter p increases, until at t_0 the value of p is infinite. At this time the entire universe is squeezed into at least one point of infinite density, infinite temperature, and infinite curvature. We have reached a spacetime singularity.

I shall argue in the next sections that these considerations support the idea that there is an uncaused beginning of the universe. In the remainder of this section I shall discuss the issue of whether or not the singularity is real.

At first it was thought that the singularity predicted by the Friedman equations was fictitious, since its prediction depended upon the assumption that the universe is exactly homogeneous and

isotropic, whereas in reality it is only approximately so. Consider
an inexactly symmetrical contracting universe: as the radius of the
universe approaches zero the convergence of particles, due to small
perturbations, would not focus upon a single point; rather the
particles would rebound off one another and result in a 'bounce' of
the universe and a new phase of expansion. In the words of Lifschitz
and Khalatnikov, who developed one of the more recent arguments
for this scenario, the fluctuations 'exclude the possibility of the
existence of a singularity in the future of the contracting universe
and imply that the contraction of the universe (if this must in
general occur) must finally be turned into an expansion'.[4] This is
the basis of the idea of an oscillating universe according to which
the universe runs through successive cycles of expansion and contraction. The present phase of the expansion, accordingly, can be
understood as a result of a prior phase of contraction.

Before I explain how it can be proven that the above argument is
mistaken and that a singularity must occur even if the universe is
inexactly symmetrical, I shall show first that the assumption that
the singularity is fictitious and that the universe oscillates does not
render probable the idea that the universe is *infinitely* old.

Models of an oscillating universe usually predict that with each
new cycle there is an increase in the size of the radius of the
universe, amount of radiation present, and entropy.[5] Radiation
from previous cycles accumulates in each new cycle, and the
accompanying increase in pressure causes the new cycle to be
longer than the last one; the universe expands to a greater radius
and takes a longer time to complete the cycle. This disallows an
infinite regress into the past, for a regress will eventually arrive at a
cycle that is infinitely short and a radius that is infinitely small; this
cycle, or the beginning of some cycle with values approaching the
values of this cycle, will count as the beginning of the oscillating
universe.

The inference to a finite past can also be made from a measure of
the amount of radiation present in the universe; if there were an

---

[4] E. M. Lifschitz and I. M. Khalatnikov, 'Investigations in Relativist Cosmology',
Advances in Physics, 12 (1963), 207.
[5] The most widely discussed models have been developed in R. C. Tolman,
Relativity, Thermodynamics and Cosmology (Oxford: Clarendon Press, 1934), 440 ff.,
and in P. T. Landsberg and D. Park, 'Entropy in an Oscillating Universe', *Proceedings of the Royal Society of London*, A346 (1975), 485–95.

infinite number of previous cycles, an infinite amount of radiation would be present in the current cycle, but the amount measured is finite. Joseph Silk calculates that the amount of radiation observed in the present expansion allows there to be 'about 100 previous expansion and collapse cycles of the universe'.[6]

The conclusion that the past is finite also follows from facts about entropy; if an infinite number of previous cycles have elapsed, each with increasing entropy, then the present cycle would be in a state of maximum entropy—but in fact it is in a state of relatively low entropy.

John Wheeler sweeps away these objections to an infinitely oscillating universe by supposing that at the end of each contracting phase all the constants and laws of that cycle disappear and the universe is 'reprocessed probabilistically'[7] so as to acquire new constants and laws in the next cycle. No information about a previous cycle is passed on to the next cycle. Accordingly, no inference to a finite past can be made on the basis of present observations and the laws and constants that hold in the current cycle.

Now there is no reason to think that such a universe is logically impossible, but that is not germane to our present concern, which is to establish probabilist grounds for a belief in the finitude or infinitude of the universe's past. It is logically possible that at the point of onset of each new cycle all laws and constants are transformed, but since these occurrences cannot be predicted according to any known physical law, there is no reason to think that these transformations occur.

Indeed, there is a theoretical reason to prefer the finite oscillatory models to Wheeler's model (supposing that we must choose among oscillating models). The finite models, through being constructed in accordance with the known physical laws and constants, obey a principle related to the principle of induction; the related principle is that physical laws and constants originally inductively established for one domain of physical events should be applied to other domains of physical events if there is no observational evidence that events in these other domains differ in the relevant respects from those in the original domain. In the present context, the domains

[6] Joseph Silk, *The Big Bang* (San Francisco: W. H. Freeman, 1980), 311.

[7] C. W. Misner, K. S. Thorne, and J. A. Wheeler, *Gravitation* (San Francisco: W. H. Freeman, 1973), 1214.

are cycles; since there is no observational evidence that events in past cycles differ relevantly from those in our cycle, we are not justified in supposing that the laws and constants inductively established in our cycle do not apply to the events in previous cycles.

The issue of whether oscillating universes are finite or infinite in respect of the past lost much of its urgency in the middle and late 1960s, with the development of the Hawking–Penrose singularity theorems,[8] which entailed that an inexactly homogeneous and isotropic universe must have a singularity. A spacetime contains a singularity if (1) the spacetime satisfies the equations of GTR, (2) time travel into one's own past is impossible and the principle of causality is not violated (there are no closed timelike curves), (3) the mass density and pressure of matter never becomes negative,[9] (4) the universe is closed and/or there is enough matter present to create a trapped surface, and (5) the spacetime manifold is not too highly symmetric.[10]

It is reasonable to assume that all of these conditions, except perhaps (4), apply to our universe. Condition (4) might seem to be open to question if the universe is not closed and the condition of a trapped surface must obtain. A trapped surface is one from which light and matter cannot escape due to the intensity of the gravitational forces, such that under the influences of these forces the spacetime paths of radiation and matter within the trapped surface converge towards a singularity. If the singularity is in the past, the geodesics of the rays and particles stem from the singularity; if it is in the future, the geodesics aim towards it. In the case of our universe, there is a singularity in its past (be the universe open or closed) if there is enough matter present to create a trapped surface. And there *is* enough matter: 'Recent observations of the microwave background indicate that the universe contains enough matter to

---

[8] R. Penrose, 'Gravitational Collapse and Space-Time Singularities', *Physical Review Letters*, 14 (1965), 57–9; S. W. Hawking, 'Singularities in the Universe', *Physical Review Letters*, 17 (1966), 444–5.

[9] That is, the stress–energy tensor satisfies

$$(T_{\alpha\beta} - \tfrac{1}{2} g_{\alpha\beta}T)u^{\alpha}u^{\beta} \geq 0.$$

[10] That is, the spacetime is such that

$${}^{t}(a^{R}b)cd(e^{t}f)^{t^{c}t^{d}} \neq 0$$

holds at some point along each timelike or null geodesic. t^{c} is the tangent vector.

cause a time-reversed closed trapped surface. This implies the existence of a singularity in the past.'[11]

2. THE BEGINNING OF THE UNIVERSE DEFINED

In order to show how the foregoing considerations render probably true the idea that the universe spontaneously began, a precise definition of the beginning of the universe must be developed. That is the task of the present section.

There are at least three possible definitions that might seem to be consistent with the ideas of the last section. Either the universe began (1) at the singularity, (2) after the singularity, or (3) neither at nor after the singularity.

(1) If the universe were closed and perfectly homogeneous and isotropic, then the definition of the beginning of the universe 'at the singularity' would be relatively simple. At the first time, t_0, when $a = 0$, there exists a single point in which the entire universe is compressed, and the existence of this point counts as the beginning of the universe. This point exists for one instant before exploding in the Big Bang. However, since the universe is imperfectly homogeneous and isotropic, a more complicated definition is necessary. The imperfect symmetry implies that the universe began non-simultaneously at a series of points.[12] Moreover, it is necessary that these points be infinite in number if the universe is open and space is infinite, for in one point there can be compressed only a finite volume of space. Given these factors, the appropriate definition of the beginning of the universe is that it began at the earliest singularity, this singularity consisting of the existence at t_0 of the first point(s)

---

[11] S. W. Hawking and G. F. R. Ellis, *The Large Scale Structure of Space-Time* (Cambridge. Cambridge University Press, 1973), 3. The proof that the trapped surface created by this matter implies a singularity in our past (rather than in our future) is given on pp. 356-9.

[12] The less dense parts of the universe exploded from points first, followed by the more dense parts. See J. Barrow and J. Silk, *The Left Hand of Creation* (New York: Basic Books, 1983), 42. It should also be noted that if the universe is not sufficiently isotropic and homogeneous, some past-directed timelike geodesics will not end in singularities. Observational evidence, however, suggests that the universe is sufficiently symmetric so that all do end in singularities. See Hawking and Ellis, *The Large Scale Structure of Space-Time*, 358-9.

to explode in a Big Bang. The universe began at this time in the sense that this is the earliest time at which some part of the universe exists.

There is a serious problem with this definition. The universe is standardly defined as the set of events, each event being a point in a four-dimensional spacetime continuum, such that each event is characterized by four coordinates (x_1, x_2, x_3, t), the first three being spatial and the fourth temporal. But the singularity at t_0 is not in a three-dimensional space; it is in a space either of zero dimensions (if it is just one point), one dimension (if it is a series of points constituting a line or line segment) or two dimensions (if it is a series of points comprising a surface-like space). Accordingly, the singularity at t_0 is not a part of the universe and *a fortiori* not the earliest part of the universe. Rather it is a *source* of the universe. The universe itself began at some time after t_0, which leads us to the second definition of the beginning of the universe.[13]

(2) On this definition the beginning of the universe is the explosion of four-dimensional spacetime out of the earliest singularity, the singularity at t_0. In other words, the beginning of the universe is the Big Bang. The Big Bang is the first state of the universe.

The Big Bang occurs at $t > t_0$. However, there is not some instant at which the Big Bang occurs, for (assuming that time is dense or continuous) there is no earliest instant after the first instant t_0; for every instant $t_a > t_0$ there is another instant $t_b < t_a$. Accordingly, if the phrase 'the Big Bang' is to be used unequivocally, it must be used to designate a state occupying an interval that is the first interval of some length to elapse after t_0. Although on a priori grounds there is no non-arbitrary basis for selecting this length, there are empirical reasons for identifying the first post-t_0 interval of length 10^{-43} seconds as the time of the Big Bang. The earliest state of the universe that cosmologists have determined to be unpreceded by a state of a different kind is the state constitutive of the Planck era, which occupies the first post-t_0 interval of length 10^{-43} seconds. A cosmological state of some kind K is the only

---

[13] Although the space of the singularity is standardly defined as less than 3D, Roger Penrose has proposed a definition of the cosmological singularity as a 3D spacelike surface, in which case it could count as a part of the universe and thus as its beginning. See R. Penrose, 'Singularities in Cosmology', in M. S. Longair (ed.), *Confrontation of Cosmological Theories with Observational Data* (Boston: D. Reidel, 1974).

state at which all and only the types of particles and forces present in that state exist. It is speculated by many cosmologists that during the Planck era and only during the Planck era there existed only one type of force, the superforce, and one type of particle, the superparticle; the superforce is the force from which the gravitational, strong, weak, and electromagnetic forces subsequently separated due to symmetry-breaking, and the superparticle likewise became differentiated into the various types of bosons and fermions. Following the Planck era there is the GUT era from 10^{-43} to 10^{-35} seconds (the inflationary expansion occurs at 10^{-35} seconds at the end of the GUT era), the electroweak era from 10^{-35} seconds to 10^{-10} seconds, the free quark era from 10^{-10} seconds to 10^{-4} seconds, and so on up to the present.

In order to eliminate possible confusion about the identity of this Big Bang, it must be noted that it is the explosion of four-dimensional spacetime out of the point(s) that exist(s) at t_0; there are other explosions from points that exist at instants later than t_0. The Big Bang that explodes from the singularity at t_0 is the first Big Bang, and can be designated as 'the Big Bang$_1$'. It is the Big Bang$_1$ that is the beginning of the universe.

The Big Bang$_1$ is 'the first state of the universe' in two senses. In the first sense, the Big Bang$_1$ begins and ends before every other state of a different kind begins; in this sense, the first state of the universe is a non-overlapped state. In the second sense, the Big Bang$_1$ begins before every other Big Bang begins; in this sense the first state of the universe is a partially overlapped state, for it is likely that other Big Bangs begin before the Big Bang$_1$ ends.

(3) A third possible definition is that the universe began with the Big Bang$_1$ at the earliest interval of 10^{-43} seconds, but that it did not begin at or after the earliest singularity. This does not entail that the universe began before this singularity, for it is possible for the universe to begin neither before, at, nor after this singularity; this possibility is actual if there is no time at which the singularity exists. The concept of a singularity, on this view, is a limiting concept that refers to nothing existent. The prediction by the Friedman equations of a time t_0 when $a = 0$ is interpreted as a prediction of a limit to time and to the radius of the universe. 't_0' does not refer to a time but expresses a concept of an ideal limit that past times can approach with arbitrary closeness but can never reach; every actual time t is such that $t > t_0$. The same holds for the

concept of $a = 0$, and the concepts of infinite density, temperature, and curvature.

An alternative explication of the definition of the universe as beginning 'neither before, at, nor after this singularity' is implied by some remarks of Richard Swinburne. According to this explication, there is no time at which the singularity exists, but there is time, empty time, prior to the Big Bang$_1$. If t_0 is the time at which the first singularity would have existed had it existed, then the Friedman equations can be taken as predicting that 'the Universe must have come into being after t_0'.[14] Paul Fitzgerald interprets Swinburne to mean something that Fitzgerald takes to be absurd, that 'the universe popped into being, preceded by a finite empty lapse of time!'[15] But this is a misinterpretation of Swinburne, for Swinburne argues that it is logically necessary for past time to be infinite,[16] and therefore that if the universe began at $t > t_0$ there must be an infinite amount of empty time prior to the Big Bang$_1$ or to t_0.

I have elsewhere shown that Swinburne's and others' putative proofs of the necessary infinitude of time are fallacious, so Swinburne's characterization of the beginning of the universe need not be accepted.[17] But that is not to say that it is logically impossible for the Big Bang$_1$ to have occurred after an infinite (or finite) period of empty time, as G. J. Whitrow[18] and others have argued it to be; this *is* logically possible.[19] What is pertinent to my present investigation is that this is *improbable*, given that the empirically established cosmological theories that predict a beginning of the universe do so by predicting a beginning of time (this will be proven in the next section).

If we reject the 'empty time' explication of the third definition of the beginning of the universe, that leaves us with two seemingly viable definitions, the second and the third as originally explicated.

[14] R. Swinburne, *Space and Time*, 2nd edn. (New York: St Martin's Press, 1981), 254.

[15] P. Fitzgerald, 'Discussion Review: Swinburne's *Space and Time*', *Philosophy of Science*, 43 (1976), 635.

[16] Swinburne, *Space and Time*, 172–3.

[17] Quentin Smith, 'On the Beginning of Time', *Noûs*, 19 (1985), 579–84.

[18] G. J. Whitrow, *The Natural Philosophy of Time*, 2nd edn. (Oxford: Clarendon Press, 1980), 32.

[19] I have shown that it is logically possible for there to be empty time before the Big Bang$_1$ in Q. Smith, 'Kant and the Beginning of the World', *New Scholasticism*, 59 (1985), 339–46.

I shall assume the second definition to be the correct one, as it treats the singularity as real and thus complies with the Hawking–Penrose singularity theorems. The few objections that have been made to the reality of the singularity since the development of the singularity theorems have been based for the most part on philosophical grounds, and do not seem very convincing.[20] In any case, I will show that the conclusion that the universe spontaneously began follows no less if the third definition is used.

3. ARGUMENTS THAT THE FIRST SINGULARITY AND THE BIG BANG₁ ARE UNCAUSED

The idea that the Friedman equations and the Hawking–Penrose singularity theorems predict an uncaused beginning of the universe is resisted by many philosophers. W. H. Newton-Smith writes:

supposing that the Big Bang emerged from a singularity of infinite density, it is hard to see what would constitute a reason for denying that singularity itself emerge from some prior cosmological goings-on. And as we have reasons for supposing that macroscopic events have causal origins, we have reason to suppose that some prior state of the universe led to the production of this particular singularity.[21]

This argument fails on several accounts. Note first that

(1) We have reason for supposing that macroscopic events have causal origins

entails

(2) We have reason for supposing that the cosmological singularity has a causal origin

only given the additional premiss

(3) The cosmological singularity is a macroscopic event,

which is false, for the singularity, far from being a macroscopic event, is infinitely smaller than the smallest *microscopic* event that physicists have yet detected. Moreover, the singularity is not even

---

[20] A frequent objection is that singularities involve infinite values and that infinities cannot be real. See e.g. Craig, Essay I, Sect. 2.2. I have rebutted Craig's and others' arguments against infinite realities in Essay II.

[21] W. H. Newton-Smith, *The Structure of Time* (London: Routledge & Kegan Paul, 1980), 111.

an *event*, that is, a point in four-dimensional spacetime; it is not a part of but a boundary or edge of the four-dimensional spacetime continuum.

Furthermore, it belongs analytically to the concept of the cosmological singularity that it is not the effect of prior physical events. The definition of a singularity that is employed in the singularity theorems entails that it is *impossible* to extend the spacetime manifold beyond the singularity. The definition in question is based on the concept of inextendible curves, a concept that has been most completely and precisely explicated by B. G. Schmidt.[22] In a spacetime manifold there are timelike geodesics (paths of freely falling particles), spacelike geodesics (paths of tachyons), null geodesics (paths of photons), and timelike curves with bounded acceleration (paths along which it is possible for observers to move). If one of these curves terminates after a finite proper length (or finite affine parameter in the case of null geodesics), and it is impossible to extend the spacetime manifold beyond that point (for example, because of infinite curvature), then that point, along with all adjacent terminating points, is a singularity. Accordingly, if there is some point p beyond which it is possible to extend the spacetime manifold, beyond which geodesics or timelike curves can be extended, then p by definition is not a singularity.

This effectively rules out the idea that the singularity is an effect of some prior natural process. A more difficult question is whether or not the singularity or the Big Bang probably is an effect of a supernatural cause, God. I will consider first the question of whether the Big Bang$_1$ is probably supernaturally caused. This fits in with W. L. Craig's argument for a divine causality of the beginning of the universe, for Craig rejects the singularity as unreal and treats the Big Bang as the first physical state (Craig does not distinguish among the several Big Bangs).[23] Craig's argument includes the steps:

(4) We have reason to believe that all events have a cause.

(5) The Big Bang is an event (or a set of events).

(6) Therefore, we have reason to believe that the Big Bang has a cause.

---

[22] B. G. Schmidt, 'A New Definition of Singular Points in General Relativity', *General Relativity and Gravity*, 1 (1971), 269–80.

[23] William Lane Craig, *The* Kalām *Cosmological Argument* (New York: Harper & Row, 1979).

Additional steps are introduced to show that the cause of the Big Bang probably is a personal Creator.

An argument of this sort avoids the problems in Newton-Smith's argument, for it does not argue from macroscopic events to something that is neither macroscopic nor an event, but argues from events in general to another event or set of events, the Big Bang. Furthermore, it does not violate the singularity theorems in supposing that the spacetime manifold is extended beyond a singularity.

Nevertheless, the argument fails because its first premiss, (4), is false. Craig writes of (4): 'Constantly verified and never falsified, the causal proposition may be taken as an empirical generalization enjoying the strongest support experience affords.'[24] However, quantum-mechanical considerations show that the causal proposition is limited in its application, if applicable at all, and consequently that a probabilistic argument for a cause of the Big Bang cannot go through. It is not relevant to the demonstration of this fact whether the causal relation be analysed in terms of physical necessity or in terms of the regular but non-necessary conjunction of events of a certain kind. Either analysis may be assumed. It is sufficient to understand causality in terms of a law enabling single predictions to be deduced, precise predictions of individual events or states. That there are uncaused events in this sense follows from Heisenberg's uncertainty principle, which states that for conjugate magnitudes such as the position q and momentum p of a particle, it is impossible in principle to measure both simultaneously with precision. If p lies within a certain interval of length Δp, and q lies within a certain interval of length Δq, then if Δp is made very small (measured exactly), Δq cannot at the same time be made very small (measured exactly). Exactly put, the product of Δp and Δq cannot be made smaller than Planck's constant h divided by 4π, so that

(7) $\Delta p \cdot \Delta q \geqslant h/4\pi$

Now if the initial conditions such as p and q of a particle x cannot all be known precisely at time t_1, then the subsequent conditions of x at time t_2 cannot be precisely predicted. The prediction of the conditions of the conjugate magnitudes of x at t_2 must be statistical and indeterministic. For example, the position of x at t_2 is represented in terms of various possible positions each with a different probability

---

[24] Essay I, Sect. 3.

value, such that none of these values is able to arbitrarily approach 1. These predictions are effected by means of the Schrödinger wave function ψ; the square of the amplitude of ψ at each point of ψ determines the probability distribution of condition q of x at t_2. If by $d(q,t_2)$ we mean the probability distribution of q at t_2, this can be calculated as

(8) $d(q,t_2) = |\psi(q,t_2)|^2$.

Equations (7) and (8) at most tend to show that acausal laws govern the *change of condition* of particles, such as the change of particle x's position from q_1 to q_2. They state nothing about the causality or acausality of absolute beginnings, of beginnings of the existence of particles. Consequently, supposing that with suitable additional premises we can draw inferences from (7) and (8) to the whole universe, the only relevant argument that we could show to be unsuccessful is

(9) We have reason to believe that all changes of condition are caused.

(10) Therefore, we have reason to believe that all changes of condition of the universe as a whole are caused.

And the failure of this argument does not entail the failure of

(11) We have reason to believe that all beginnings of existence are caused.

(12) Therefore, we have reason to believe that the beginning of the universe's existence is caused.

Thus if the latter argument is to be refuted, it is necessary to find premises more relevant to absolute beginnings than (7) and (8).

Such premises can be obtained on the basis of Heisenberg's uncertainty relation

(13) $\Delta E \cdot \Delta t \geq h/4\pi$,

where E = energy, t = time and h = Planck's constant. This relation implies that if the energy of a particle is measured precisely, so that ΔE is made very small, the time at which the particle possesses this energy can be known only imprecisely, so that Δt is very large. Now if Δt is small enough, ΔE becomes so large that it becomes impossible in principle to determine if the law of energy conservation is violated. During the interval of time

(14) $\Delta t \simeq (h/4\pi)\Delta E$

this law is inapplicable and consequently an amount of energy ΔE can spontaneously come into existence and then (before the interval has elapsed) cease to exist. There is observational evidence, albeit indirect, that this uncaused emergence of energy or particles (notably virtual particles) frequently occurs. It appears, then, that the argument (11)–(12) is unsuccessful and that the crucial step in the argument to a supernatural cause of the Big Bang, or more exactly, of the Big Bang$_1$, is faulty.

It might be objected that quantum acausality applies only on the microscopic level and not to macroscopic cosmological states or beginnings. This could be granted without detriment to my argument, for the physical processes constitutive of the Big Bang$_1$ (which I have defined to occur during the Planck era) are one and all microscopic and occur at dimensions where quantum-mechanical principles unquestionably apply.

It might then be objected that the Big Bang$_1$ is the beginning of the existence of four-dimensional spacetime itself, and that the uncaused events I have specified involve merely the beginnings of existence *within* four-dimensional spacetime. Surely 'There are some uncaused beginnings of existence within spacetime' is irrelevant to and thus cannot increase the probability value of 'The beginning of the existence of spacetime itself is uncaused'.

My response is that if this is so (and I will provide reasons for doubting that this is the case in Section 4 with my discussion of the vacuum fluctuations models of the universe), then the same holds for the parallel argument for a supernatural cause of four-dimensional spacetime; for 'There are some caused beginnings of existence within spacetime' or even 'All beginnings of existence within spacetime are caused' would by the same token be irrelevant to and thus fail to increase the probability of 'The beginning of the existence of four-dimensional spacetime is caused'.

I conclude that quantum-mechanical considerations show that the argument to a divine cause of the Big Bang$_1$ based on the causal principle (4) is unsuccessful.

But this does not end the matter, for it is still open to a defender of the theistic argument to claim that I have no right to introduce quantum-mechanical acausality into the discussion, since the Big Bang cosmological model is based on GTR and GTR presupposes a causal determinism to operate in the domain of its application.

I shall not decide whether this objection is valid, but will instead show that even if it is valid it still can be proven that we have no reason to think the Big Bang$_1$ is caused, supernaturally or otherwise. This can be demonstrated solely on the basis of the GTR Big Bang cosmological model itself.

Let us begin by assuming what I have already maintained to be the case, that the singularity at t_0 is real. Given that, we can note that the classical notions of space and time and all known laws of physics (since they are formulated on a classical spacetime background) break down at the singularity, and consequently it is impossible to predict what will emerge from the singularity. This impossibility is not due to our ignorance of the correct theory but is a limitation upon possible knowledge that is similar but additional to the limitation entailed by the quantum-mechanical uncertainty principle. The former limitation is due to the causal structure of spacetime that is postulated by GTR; the interaction region postulated by GTR can be bounded not only by an initial surface on which data are given and a final surface on which measurements are made *but also by a hidden surface*. A hidden surface is one about which any possible observer can have only limited information, such as (in the case of black holes) the mass, angular momentum, and charge. This surface 'emits with equal probability all configurations of particles compatible with the observers limited knowledge'.[25] A surface close to the Big Bang$_1$ singularity, a surface at the Planck time 10^{-43} seconds, is a hidden surface. The singularity hidden by this surface 'would thus emit all configurations of particles with equal probability'.[26] If we assume with Craig that the singularity is unreal, and that the first physical state is the Big Bang$_1$, then the hidden surface is not taken to be subsequent to the singularity; instead of the particles being regarded as randomly and spontaneously being emitted from the singularity, they are regarded as randomly and spontaneously being emitted from nothing at all. This means, precisely put, that if the Big Bang$_1$ is the first physical state, then every configuration of particles that does constitute or might have constituted this first state is as likely on a priori grounds to constitute it as every other configuration of particles. In either case, the constitution of the Big Bang$_1$ is impossible in principle to

[25] S. W. Hawking, 'Breakdown of Predictability in Gravitational Collapse', *Physical Review*, D14 (1976), 2460.
[26] Ibid. 2463.

predict and thus is uncaused (for 'uncaused' minimally means 'in principle unpredictable').

The singularity itself is also regarded in the GTR-based Big Bang theory as uncaused, although for a different reason. It is defined as a point beyond which spacetime curves cannot be extended, and thus which cannot have causal antecedents.

In sum, then, we may say that although the GTR-based Big Bang theory does suppose causality to operate in its domain of application, it also supposes that there is a limit to its operation; it represents causality as breaking down at the initial physical states, the singularity and the Big Bang. Consequently, this theory cannot be used to support the thesis that the initial physical states are probably caused and that this cause is God.

4. QUANTUM GRAVITY AND THE UNCAUSED BEGINNING OF THE UNIVERSE

There is a serious lacuna in my foregoing account of the beginning of the universe; I have been presuming that the GTR-based Big Bang theory has an unlimited application and therefore applies to the extreme conditions of the universe during the Big Bang$_1$. In fact, GTR fails to apply when quantum-mechanical interactions predominate, and these predominate when the temperature is at or above 10^{32} K, when the density is at or above 10^{94} gm cm^{-3}, and when the radius of curvature becomes of the order of 10^{-33} cm. Since these conditions obtain during the Planck era at the first 10^{-43} seconds after the singularity, the GTR-based Big Bang theory cannot be used as a reliable guide in reconstructing the physical processes that occurred during this time and *a fortiori* cannot be used as a reliable basis for predicting that the density, temperature, and curvature reached infinite values prior to this time. Accordingly it seems that the foregoing probabilistic argument to an uncaused beginning of the universe is in jeopardy.

I believe, however, that there are three reasons for a continued support of the idea that the universe spontaneously began to exist. To comprehend these reasons, we must first observe that the reason why GTR is inapplicable during the Planck era is that the theory of gravity in GTR is unable to account for the quantum-mechanical behaviour of gravity during this era. A new *quantum theory of gravity* is needed. Although such a theory has not yet been developed, there are some general indications of what it may predict. It is in terms of these indications that our three reasons are to be understood.

First, it is thought that a quantum theory of gravity may show gravity to be repulsive rather than attractive under conditions that obtain during the Planck era. During this time regions of negative energy density may be created by the forces and particles present, and these regions lead to a gravitational repulsion. This suggests that any given finite set of past-directed timelike or null geodesics will not converge in a single point but will be pushed apart, as it were, by the repulsive gravitational force. This possibility is consistent with an oscillating universe, for as each contracting phase ends gravity becomes repulsive and prevents converging geodesics from terminating in a point; gravity repels them so that they enter a new expanding phase.

But this way of avoiding the singularity predicted by the Hawking–Penrose theorems does not give us a universe that is infinitely old. For—and this is the first of the three reasons I want to mention—this oscillating quantum-gravitational universe would still be subject to the same problems that were discussed in Section 1, namely, increase in radius, length of cycle, radiation, and entropy with each new cycle. Consequently, this theory does no more than push the cosmological singularity further into the past, at a time just before (or at) the beginning of the first cycle when the radius of the universe is zero (or near zero).

The second reason is that there is a way in which the Hawking–Penrose theorems' prediction of a singularity at the beginning of the present expansion can be made consistent with a quantum theory of repulsive gravity. These theorems do not *define* a singularity as that wherein curvature, density, and temperature are infinite and the radius is zero. A singularity is defined as a point or series of points beyond which the spacetime manifold cannot be extended. Consequently, if the effects of quantum gravity prevent a build-up of temperature, density, and curvature to infinite values, and a decrease of radius to zero, this need not mean there is no singularity at the beginning of the present expansion. The singularity could occur with *finite* and *non-zero* values.

The third reason is that the most theoretically developed attempts to account for the past of the universe on the basis of specifically quantum-mechanical principles have represented the universe as spontaneously beginning at the onset of the present expansion. These theories are collectively known as the 'vacuum fluctuation models of the universe'. The models developed by Tryon, Brout *et al.*, Grishchak and Zeldovich, Atkatz and Pagels, and Gott[27] picture the universe as emerging spontaneously from an empty background space, and the model of Vilenkin[28] depicts it as emerging without cause from nothing at all.

The first vacuum fluctuation model was developed by Edward Tryon in 1973. A vacuum fluctuation is an uncaused emergence of energy out of empty space that is governed by the uncertainty relation $\Delta E \cdot \Delta t \geq h/4\pi$, and which thus has zero net value for conserved quantities. Tryon argues that the universe is able to be a fluctuation from a vacuum in a larger space in which the universe is embedded since it does have a zero net value for its conserved quantities. Observational evidence (Tryon claims) supports or is consistent with the fact that the positive mass-energy of the universe is cancelled by its negative gravitational potential energy, and that the amount of matter created is equal to the amount of antimatter. (But this last point is inconsistent with current Grand Unified Theories.)

A disadvantage of Tryon's theory, and of other theories that postulate a background space from which the universe fluctuates, is that they explain the existence of the universe but only at the price of introducing another unexplained given, namely, the background space. This problem is absent from Vilenkin's theory, which represents the universe as emerging without a cause 'from literally *nothing*'.[29] The universe appears in a quantum tunnelling from nothing at all to de Sitter space. Quantum tunnelling is normally understood in terms of processes *within* spacetime; an electron, for example, tunnels through some barrier if the electron lacks sufficient energy to cross it but nevertheless still does cross it. This is possible because the above-mentioned uncertainty relation allows the electron

---

[27] E. P. Tryon, 'Is the Universe a Vacuum Fluctuation?' *Nature*, 246 (1973), 396–7; R. Brout, F. Englert, and E. Gunzig, 'The Creation of the Universe as a Quantum Phenomenon', *Annals of Physics*, 115 (1978), 78–106; L. P. Grischak and Y. B. Zeldovich, 'Complete Cosmological Theories', in M. J. Duff and C. J. Isham (eds.), *Quantum Structure of Space and Time* (Cambridge: Cambridge University Press, 1982), 409–22; D. Atkatz and H. Pagels, 'Origin of the Universe as a Quantum Tunneling Event', *Physical Review*, D25 (1982), 2065–73; J. R. Gott, 'Creation of Open Universes from de Sitter Space', *Nature*, 295 (1982), 304–7.
[28] A. Vilenkin, 'Creation of Universes from Nothing', *Physical Letters*, 117B (1982), 25–8.
[29] Ibid. 26.

to acquire spontaneously the additional energy for the short period of time required for it to tunnel through the barrier. Vilenkin applies this concept to spacetime itself; in this case, there is not a state of the system before the tunnelling, for the state of tunnelling is the first state that exists. The state of tunnelling thus is the analogue of the Big Bang$_1$ in the third definition of the beginning of the universe offered in Section 2, for it is the first state of the universe and there is no time before this state. The equation describing this state is a quantum-tunnelling equation, specifically the bounce solution of the Euclidean version of the evolutionary equation of a universe with a closed Robertson–Walker metric.[30] The universe emerged from the tunnelling with a finite size ($a = H^{-1}$) and with a zero rate of expansion or contraction ($da/dt = 0$). It emerged in a symmetric vacuum state, which then decays and the inflationary era begins; after this era ends, the universe evolves according to the standard Big Bang model.

These quantum-mechanical models of the beginning of the universe are explanatorily superior in one respect to the standard GTR-based Big Bang models; they do not postulate initial states at which the laws of physics break down but explain the beginning of the universe in accordance with the laws of physics. The GTR-based theory predicts a beginning of the universe by predicting initial states at which the laws of the theory that are used to predict these states break down. The singularity and the explosion of four-dimensional spacetime from the singularity obey none of the laws of GTR that are obeyed by states within the universe or subsequent states of the universe. In contrast, the quantum-mechanical theories represent the universe as coming into existence via the same laws that processes within the universe obey. Instead of an exploding singularity, there is a quantum fluctuation or tunnelling that is analogous to the fluctuations or tunnellings within the universe and that obeys the same acausal laws as the latter fluctuations or tunnellings.[31]

[30] The bounce solution is $a(t) = H^{-1} \cos(Ht)$. See ibid.

[31] For further discussion of the vacuum fluctuation theories of the beginning of the Universe, see Quentin Smith, 'World Ensemble Explanations', *Pacific Philosophical Quarterly*, 67 (1986), 73–86, esp. 81–4. Other pertinent cosmological discussions can be found in Quentin Smith, 'The Anthropic Principle and Many-Worlds Cosmologies', *Australasian Journal of Philosophy*, 63 (1985), 336–48, and Quentin Smith, *The Felt Meanings of the World: A Metaphysics of Feeling* (West Lafayette, Ind.: Purdue University Press, 1986), ch. 6.

This review of the role of quantum mechanics in accounts of the beginning of the universe strongly suggests that the probabilistic argument to an uncaused beginning of the universe, although more complicated than we had been supposing in Sections 1–3, still goes through. Its conclusion is summarized in this disjunctive statement: it is probably true that *either* the universe began without cause at the beginning of this expansion (1) subsequent to a singularity of infinite density, temperature and curvature, and zero radius, or (2) at a singularity with finite and non-zero values, or (3) in a vacuum fluctuation from a larger space or a tunnelling from nothing, *or* the universe spontaneously began to exist at the beginning of some prior expansion phase under conditions described in (1), (2), or (3).

APPENDIX IV.1. *Some New but Unsound Arguments for an Infinitely Old Oscillating Universe*

A main argument of Essay IV is that there is no reason to think that our universe is an infinitely old oscillating universe and that it is probably true the universe has an uncaused beginning. This claim has subsequently been challenged by some, e.g. Smith and Weingard,[1] Gale,[2] and Sullivan,[3] and I shall here further defend this thesis. I shall show that these counter-arguments, if successful, establish only that

(1) It is epistemically possible that there is an infinitely old oscillating universe

and that no bridge is supplied from (1) to

(2) It is probably true that there is an infinitely old oscillating universe.

Since (2) cannot be inferred from (1), and these counter-arguments establish only (1), any argument for (2) that is based on these counter-arguments is unsound because invalid.

[1] Gerrit Smith and Robert Weingard, 'Quantum Cosmology and the Nature of the Universe', *Philosophy of Science*, 57 (1990), 663–7.
[2] George Gale, 'Cosmological Fecundity: Theories of Multiple Universes', in John Leslie (ed.), *Physical Cosmology and Philosophy* (New York: Macmillan, 1990), 189–206.
[3] T. D. Sullivan, 'Coming to Be without a Cause', *Philosophy*, 65 (1990), 261–70.

One main argument I used against an infinitely old oscillating universe is that entropy increase prevents the past number of cycles from being infinite. This is the main point addressed by the defenders of an infinite oscillatory past history.

I shall consider the first the argument of Geritt Smith and Robert Weingard, who have criticized this essay for failing to take into account certain possibilities. It seems to me, however, that all of the points of Smith and Weingard can be acknowledged consistently with claiming that there is no reason to think it probably true that there is an infinitely old oscillating universe. They offer a quantum model of a universe in a pure state, in which the entropy is zero. This universe is an infinitely old oscillating universe. But is this model physically realistic? Smith and Weingard admit that it is not:

Although this model contains an infinite number of cycles, it is too simple or unrealistic to be a counterexample to Quentin Smith's argument concerning entropy increase in successive cycles. In fact, because the state of the universe is a pure state, its total entropy is zero. Rather, we think the model is relevant to Smith's argument in a heuristic manner.[4]

Thus, the model does not describe our universe and therefore we do not have a model corresponding to the actual universe that implies an infinite past. Smith and Weingard add that if we have 'a smearing of the singularity as illustrated here, we think it is not clear what sort of information can pass from cycle to cycle'[5] and thus that the problem of entropy increase may be avoided. But this at best shows it is epistemically possible that the problem of entropy increase is avoided. Smith and Weingard allude to our lack of knowledge of a realistic model incorporating quantum gravity, but this shows merely that it is consistent with our limited present knowledge about quantum models that there be a realistic quantum model that implies our universe has an infinite oscillatory past. Of course, it is also consistent with our limited knowledge of a theory of quantum gravity that there be a realistic quantum model that has a finite past. If we remain in the area of epistemic possibility we remain in the area of sheer speculation and virtually anything goes. However, if we stick to our best confirmed currently developed

---

[4] Smith and Weingard, 'Quantum Cosmology and the Nature of the Universe', 666.
[5] Ibid.

theories, then we should conclude (as I have argued in this essay) that the past history of our universe is finite.

Smith and Weingard allude to a possibility I have not considered, that 'space is infinite, space would contract from $t = -\infty$, but before reaching infinite density, space would start expanding again to $t = +\infty$'.[6] Smith and Weingard do not elaborate on which model they have in mind, but this sentence brings to mind the de Sitter model of a universe. However, de Sitter's universe is empty of matter and therefore does not describe our universe.

George Gale criticizes my article 'World Ensemble Explanations'[7] for maintaining the similar claim that there is no reason to think it probable that there are an infinite number of past oscillations. Gale writes: 'given the robust models presented by Linde and Markov, there is plenty of life left in the oscillating universe model. This strongly disputes Smith's claim to the contrary that "at present the evidence favors the conclusion that there is no WE [World Ensemble] composed of oscillating worlds".'[8]

Let us consider first Markov's model. Markov supposes that at the time of maximal contraction in the oscillatory cycle the energy–momentum tensor of matter becomes $T_{\mu\nu} \sim g_{\mu\nu}p_P$ and the universe enters a de Sitter phase. Following this, it exponentially expands and then changes over to a normal Friedman expansion. The problem is that this violates the second law of thermodynamics, that entropy always increases. Markov's model requires that entropy decrease at the end of each contraction, which makes his model physically impossible. Markov suggests a way around this problem, that we know nothing about the physical laws governing the bounce that occurs at the end of the collapse:

At this stage we have no quantum description of the bounce phenomenon and of the consequence which will arise in this situation (surely different from the classical one) in the universe at its now expansion state that follows the bounce.

And what if the time description of the bounce is in principle impossible? If the concept of time is meaningless in this case is there any hope that in such scenario (where matter is transformed into a vacuum state of the metric) the universe must appear only in the form and amount which it

---

[6] Ibid. 664.
[7] Quentin Smith, 'World Ensemble Explanations', *Pacific Philosophical Quarterly*, 67 (1986), 73–86.
[8] Gale, 'Cosmological Fecundity', 201.

had before, only with the same quantitative parameters? [If so 'the entropy problem is solved'.][9]

Clearly, Markov's argument is hardly anything more than 'we don't know what happens at the bounce; therefore virtually anything is possible; therefore, it is even possible that there is no entropy increase'. This is far from showing that this is physically possible, i.e. consistent with some known law of physics. At best, it is merely epistemically possible.

Gale also refers to Linde's model. Linde supposes that in the final stages of collapse there would be a gravitational confinement phase. He states that 'in the gravitational confinement phase (if such a phase can exist) *any* particle would have infinite energy'.[10] This would imply that 'no real particle excitations can exist and the entropy of the universe vanishes'.[11] This gives us a vacuum state in curved space in which no free particles exist and in which there are no inhomogeneities of the curvature tensor.[12] If matter and radiation disappear, the disorder of matter and radiation disappear. By this means, it is possible for entropy to be generated anew in each new cycle, which avoids the problem of entropy increase in each cycle.

However, this theory rests on a conjecture about a gravitational confinement phase that Linde offers no evidence is a physical possibility. There is an extant and confirmed theory, for example, of the strong force's confinement phase, in which it would take infinite energy to free up a single quark. This would imply that no quarks exist as free particles. But Linde does not show how a gravitational confinement phase is physically possible. Indeed he admits that we will 'just *assume* that at sufficiently large curvature $R \gtrsim G^{-1}$ in quantum gravity there occurs a phase transition to the gravitational confinement phase. We are unable to prove or disprove this conjecture at present, since it would require a much deeper

[9] M. A. Markov, 'Entropy in an Oscillating Universe in the Assumption of Universes "Splitting" into Numerous Smaller "Daughter" Universes', in M. A. Markov *et al.* (eds.), *Proceedings of the Third Seminar on Quantum Gravity* (Singapore: World Scientific, 1985), 3–15, esp. 10. Also see M. A. Markov, *JETP Letters*, 36 (1982), 265–7.

[10] A. D. Linde, in G. W. Gibbons *et al.* (eds.), *The Very Early Universe* (Cambridge: Cambridge University Press, 1982), 205.

[11] Ibid.

[12] A. D. Linde, 'The Inflationary Universe', *Reports on Progress in Physics*, 47 (1984), 978.

understanding of quantum gravity than we have at present.'[13] In short, Linde's argument is that this is epistemically possible, not physically probable, which is insufficient to establish the probabilistic claim (2) about an infinite past of oscillating universes.

T. D. Sullivan presents a different sort of argument against Essay IV, although he too is concerned to deny the thesis of an uncaused beginning of the universe. Sullivan argues that Hawking now rejects the hypothesis of a Big Bang singularity and *therefore* that Hawking rejects the idea of an uncaused beginning of the universe:

Smith notes that some philosophers resist the implication of an uncaused beginning of the universe, but it is not only philosophers who resist the conclusion. Hawking himself now resists the implication. 'It is perhaps ironic that, having changed my mind, I am now trying to convince other physicists that there was in fact no singularity at the beginning of the universe—as we shall see later, it can disappear once quantum effects are taken into account.'[14] With Hawking himself against the putative implication of the Hawking–Penrose singularity theorems, why give up on causality?[15]

I think Sullivan misunderstands Hawking's quantum theory. Hawking's denial of the singularity is not a denial that the universe began to exist without a cause. It merely denies that the first state of the universe has the property of being lawless. Hawking's new theory is that there is an uncaused first state of the universe, but that it is governed by a law, the so-called 'wave function of the universe' and therefore is not a singularity. His new theory is that 'the ordinary laws of science . . . hold everywhere, including at the beginning of time'.[16] Hawking emphasizes that there is not only no singularity but also no naturally unexplained boundary conditions that provide room for a supernatural causal explanation of these conditions. If his new theory were true, 'there would be no singularity at which the laws of science broke down and no edge of space-time at which one would have to appeal to God . . . to set the

---

[13] Ibid.
[14] Quoted from S. W. Hawking's *A Brief History of Time* (New York: Bantam Books, 1988), 50.
[15] Sullivan, 'Coming to Be without a Cause'. For further criticism of Sullivan, see Quentin Smith, 'Can Everything Come to Be without a Cause?' *Dialogue: Canadian Philosophical Review*, forthcoming.
[16] Hawking, *A Brief History of Time*, 133.

boundary conditions for space-time'.[17] Thus, Hawking's new theory does not count as evidence against an uncaused beginning of the universe, but as evidence for it. (For a more detailed discussion of Hawking's quantum cosmology, see Part 3 of this book, 'Theism, Atheism, and Hawking's Quantum Cosmology'.)

More pertinent to the idea of oscillating universes is that Hawking and Luttrell[18] have formulated an equation that some have interpreted as corresponding to an oscillating universe. But this interpretation cannot be sustained. Hawking and Luttrell note:

Although the classical solution will collapse to a singularity of zero radius, the particular class of solutions that correspond to the wave function will have a bounce at or near the right-hand branch of the curve $V = 0$. At $V = 0$ however the WKB approximation will break down and the concept of time will become ill-defined. It does not really have any meaning therefore to say whether or not the universe collapses to zero radius but the wave function remains regular at $a = 0$.[19]

Since an oscillating universe is physically possible only if its description is physically meaningful, it follows that it is not physically possible.

There are other theories of oscillating universes not mentioned by Smith and Weingard or Gale, but they also remain on the shelf of epistemic possibility. For example, in a 1991 article in *Nature* Sikkema and Israel seem to present a theory of how a bounce is possible. But do they? They trace the contracting phase through the merging of the galaxies and the supermassive black holes that lie at the centre of the galaxies. And then comes the 'proof' of the bounce: 'For the subsequent quantum phase no reliable theory currently exists, and we enter the realm of speculation. We proceed by assuming the existence of a consistent quantum theory of gravity that permits non-singular passage of a sufficiently homogeneous contracting structure through an epoch of maximal compression.'[20] This in effect is an admission that they have no theory of how it is physically possible for there to be a bounce. In addition, they announce an implication of their theory, based on the above-mentioned assumptions, is that it allows for a 'finite number of

<hr>

[17] Ibid. 136.
[18] J. C. Luttrell and S. W. Hawking, 'Higher Derivatives in Quantum Cosmology', i: 'The Isotropic Case', *Nuclear Physics*, B247 (1984), 250–60.
[19] Ibid. 260.
[20] A. E. Sikkema and W. Israel, 'Black-Hole Mergers and Mass Inflation in a Bouncing Universe', *Nature* (3 Jan. 1991), 46.

cycles of descending mass to a first "baby" universe large enough to spawn mini black holes'.[21] Thus, these authors have given us no reason to believe in an infinitely old oscillating universe.

In short, the putative 'evidence' for an infinitely old oscillating universe is no evidence at all but merely assertion that this is epistemically possible since we do not yet have a complete theory of quantum gravity and therefore that 'anything goes'. These assertions do not make it reasonable to believe that there probably is an infinite oscillatory past. Rather, the evidence and the extant and confirmed theories point in the other direction and have been doing so since at least the late 1960s.

In conclusion, I would suggest two things. I would endorse the idea that it is metaphysics, not physics, that partly motivates (against all evidence) the attraction to an infinitely old oscillating universe. I partly agree with MacRobert: 'The idea of an oscillating universe, in which the Big Bang resulted from the recollapse of a previous phase of the universe, gained currency merely because it avoided the issue of [divine] creation—not because there was the slightest evidence in favor of it '[22] I only partly endorse this because I think another reason is an unjustified allegiance to the principle that everything that begins to exist has a cause of its beginning to exist. Many people share C. D. Broad's feeling: 'I must confess that I have a very great difficulty in supposing that there was a first phase in the world's history . . . I can not really *believe* in anything beginning to exist without being *caused* . . . by something else which existed before and up to the moment when the entity in question began to exist.'[23] However, I believe the principle that everything that begins to exist has a cause is not true a priori and is not supported by the empirical evidence, as I shall further argue in Essay VI of this book. The fact of the matter is that the most reasonable belief is that we came from nothing, by nothing and for nothing. With some exceptions, such as Stephen Hawking, thinkers are too often adversely affected by Heidegger's dread of 'the nothing'. We should instead acknowledge our foundation in nothingness and feel awe at the marvellous fact that we have a chance to participate briefly in this incredible sunburst that interrupts without reason the reign of non-being.

[21] Ibid.

[22] MacRobert, *Sky and Telescope* (Mar. 1983), 213.

[23] C. D. Broad, 'Kant's Mathematical Antinomies', *Proceedings of the Aristotelian Society*, 40 (1955), 1–22.

APPENDIX IV.2.   *A Non-technical Explanation of the*
Mathematical Basis of Big Bang Cosmology

In this appendix I shall explain the mathematical foundations of
Big Bang cosmology in four 'easy to follow' steps, so that readers
unfamiliar with the mathematical technicalities of Big Bang cosmology can follow the argument of this essay.

1. The first step is the introduction of the so-called Einstein
equation, which is the heart of Einstein's General Theory of Relativity. Einstein's equation says, in simplified terms, that the geometry
(curvature) of spacetime is determined by the distribution of mass
and energy in spacetime. The equation may be simplified as

(curvature of spacetime) = 8π(density of matter).

This equation suggests that if the matter in the universe is sufficiently dense, then the curvature of spacetime will become so
great that it eventually curves to a point, as at the tip of a cone.
The history of a particle or light ray is a path in spacetime, and if
spacetime eventually curves to a point then these spacetime paths
will converge and intersect at the point. If this intersection occurs
at some time in the future, the point of intersection would seem to
constitute the end of spacetime. If the intersection occurs in the
past, such that the spacetime paths emerge from the point of
intersection and gradually curve away from each other, the point of
intersection would appear to constitute the beginning of spacetime.
This possibility leads to a discussion of the next relevant aspect of
Big Bang cosmology.

2. Einstein's equation admits of many solutions and it is an
empirical question which solution describes our universe. The
Friedman solutions (first obtained by Friedman in 1922 and
1924) are the ones thought to apply to our universe. His solutions
describe a universe that is perfectly isotropic (it looks the same in
every direction) and perfectly homogeneous (matter is evenly distributed). If we apply to Einstein's equation a metric that describes
a perfectly isotropic and homogeneous universe, the Friedman
solutions are obtained, which in a simplified form read

−3(acceleration of expansion or deceleration of contraction of
the universe) = 4π(density of matter).

The Friedman solutions tell us that if there is matter evenly
distributed throughout the universe, then the universe must be

expanding at a decreasing rate or contracting at an increasing rate (except at the instant, if any, at which the expansion stops and changes to a contraction). To see this, note that the right side of the above (simplified) equation represents the density of matter multiplied by 4π. If there is matter present in the universe, then the matter density of the universe is positive. This implies that the right side of the equation, 4π(density of matter), will be positive. This in turn implies that the value for the acceleration of the expansion or the deceleration of the contraction will be negative. This is because the acceleration of the expansion or the deceleration of the contraction is multiplied by -3 and the result must be equal to the positive number represented by the right side of the equation. If the value of the expansion's acceleration is negative, this means that the universe is expanding at an ever decreasing rate. If the value of the contraction's deceleration is negative, this means that the universe is contracting at an ever increasing rate. This result is of momentous significance, for it implies that if the universe contains evenly distributed matter then its existence is temporally limited. If the universe is contracting at an ever increasing rate, then it cannot contract *for ever* but must eventually reach an endpoint, when it curves to a point and its radius becomes zero. If the universe is expanding at an ever decreasing rate, then it cannot have been expanding *for ever* but must have begun expanding at some time in the past, when its radius began extending from zero.

Let us further consider the case of expansion, since the universe is now expanding. The further we trace the universe into the past, the faster we find its rate of expansion. As the rate of expansion increases, the curvature of the universe and the density of matter increase and the radius of the universe decreases, until a time is reached when the curvature of the universe is infinite, the density of matter infinite and the radius of the universe is zero. Due to the infinite curvature, the past-directed spacetime paths of particles converge, such that each spacetime path ends at some point at which other spacetime paths also end. If the Friedman equations describe a spherical universe, the universe is finite in extent and consequently the past-directed spacetime paths all intersect in one point. All of matter is squeezed into this one point, which has zero spatial dimensions. This point exists instantaneously before exploding in the Big Bang. The instantaneously existing point is a singularity, which means that it is an end-point of spacetime; there is no earlier

time than the instant of the singularity for it itself is the first instant
of time. On the other hand, if the universe is flat (uncurved) or
hyperbolic (curved like a saddle) it is infinite in extent, which
implies that the past-directed spacetime paths end in a spatially
one-dimensional singularity. Only a finite volume of space can be
compressed into a point; consequently, if there are an infinite
number of spatial volumes of any given finite size (which there
would be if the universe were flat or hyperbolic), then there must
be an infinite number of points constitutive of the singularity.
These points exist instantaneously (at the first instant of time) and
then explode in an infinitely extended Big Bang.

However, Friedman's solutions to Einstein's equation do not by
themselves show that our universe began in a Big Bang singularity.
There exists a certain discrepancy between his solutions and the
global features of our universe, a discrepancy that might seem to
render inapplicable their prediction of a Big Bang singularity. The
statement and resolution of this problem leads to the third aspect of
Big Bang cosmology that is pertinent to my argument.

3. Friedman's solutions are based on the assumption that the
universe is perfectly isotropic and homogeneous. But this assumption
is inconsistent with observational evidence, which reveals the universe to consist of clusters or superclusters of galaxies separated by
vast stretches of empty or near empty space. The universe is
isotropic and homogeneous only when averaged over distances of
billions of light years. (For example, we may assume that different
cubic regions of space differ in mass by less than 1 per cent only if
these regions are taken to be 3 billion light years or more in
diameter.) This might suggest that the prediction of a Big Bang
singularity is inapplicable to the universe since this prediction is
based on the assumptions of perfect homogeneity and isotropy. The
assumption of perfect isotropy entails that the relative motion of
any pair of particles is purely radial and the assumption of perfect
homogeneity entails there are no pressure gradients. The fact that
our universe is imperfectly isotropic and homogeneous entails that
past-directed spacetime paths of particles exhibit transverse velocities
and clusterings that make up clumpings of matter. This suggests
that the paths will miss each other instead of converging at a point.
This is turn suggests that the present expansion phase of the
universe results from a 'bounce' that terminated a prior contracting
phase of the universe. But this suggestion of an oscillating universe

was contradicted in the late 1960s by the Hawking–Penrose singularity theorems, which demonstrate that under certain conditions imperfectly isotropic and homogeneous universes also orginate in a Big Bang singularity. Precisely put, the theorems state that a singularity is inevitable given the following five conditions:

(1) Einstein's General Theory of Relativity holds true of the universe.
(2) There are no closed timelike curves (i.e. time travel into one's past is impossible and the principle of causality is not violated).
(3) Gravity is always attractive.
(4) The spacetime manifold is not too highly symmetric; i.e. every spacetime path of a particle or light ray encounters some matter or randomly oriented curvature.
(5) There is some point p such that all the past-directed (or future-directed) spacetime paths from p start converging again. This condition implies that there is enough matter present in the universe to focus every past-directed (or future-directed) spacetime path from some point p.

The solutions for the Hawking–Penrose theorems show, as Hawking notes, that 'in the general case there will be a curvature singularity that will intersect every world line. Thus general relativity predicts a beginning of time.'[1]

4. The last aspect of Big Bang cosmology that I address is Hawking's *principle of ignorance*, which states that singularities are inherently chaotic and unpredictable. In Hawking's words,

A singularity is a place where the classical concepts of space and time break down as do all the known laws of physics because they are all formulated on a classical space-time background. In this paper it is claimed that this breakdown is not merely a result of our ignorance of the correct theory but that it represents a fundamental limitation to our ability to predict the future, a limitation that is analogous but additional to the limitation imposed by the normal quantum-mechanical uncertainty principle.[2]

[1] S. W. Hawking, 'Theoretical Advances in General Relativity', in H. Woolf (ed.), *Some Strangeness in the Proportion* (Reading, Mass.: Addison-Wesley, 1980), 149.
[2] Hawking, 'Breakdown of Predictability in Gravitational Collapse', *Physical Review*, D14 (1976), 2460.

One of the quantum-mechanical uncertainty relations concerns the position q and momentum p of a particle. This relation states that $\Delta p \cdot \Delta q \geq h/4\pi$, which implies that if the position of a particle is definitely predictable then its momentum is not, and vice versa. The principle of ignorance is stronger in that it implies that one can definitely predict neither the position nor the momentum of any particle emitted from a singularity. In fact, this principle implies that none of the physical values of the emitted particles are predictable. According to this principle, the Big Bang singularity would emit all configurations of particles with equal probability. Thus, there will be a chaotic outpouring from the singularity and the first post-singularity state of the universe will be a state of maximal chaos (complete disorder or entropy).

V

The Caused Beginning of the Universe
WILLIAM LANE CRAIG

I. INTRODUCTION

Quentin Smith has recently argued that there is sufficient evidence at present to warrant the conclusions that (*a*) the universe probably began to exist and that (*b*) it began to exist without being caused to do so.[1] While I am inclined to agree with (*a*),[2] it seems to me that Smith has overstated the case for (*b*).

As part of his argument for (*b*), Smith takes on the task of disproving what we may call the theistic hypothesis, that the beginning of the universe was caused by God. It is apparently Smith's contention that the theist who believes in divine *creatio ex nihilo* must fly in the face of the evidence in order to do so. But is this in fact the case? As I read him, Smith's refutation of the theistic hypothesis basically falls into two halves: (i) there is no reason to regard the theistic hypothesis as true, and (ii) it is unreasonable to regard the theistic hypothesis as true. Let us, therefore, examine each half of his refutation in turn.

2. NO REASON TO REGARD THE THEISTIC HYPOTHESIS AS TRUE

In order to show that there is no reason to think that God caused the beginning of the universe, Smith attacks the universality of the causal principle, variously construed. After arguing that 'it belongs analytically to the concept of the cosmological singularity that it is

Forthcoming in *British Journal for the Philosophy of Science*.

[1] Quentin Smith, Essay IV.
[2] See William Lane Craig, *The* Kalām *Cosmological Argument*, Library of Philosophy and Religion (London: Macmillan, 1979), 65–140.

not the effect of prior physical events' and that 'This effectively rules out the idea that the singularity is an effect of some prior natural process,'[3] Smith turns to the 'more difficult question' of whether the singularity or the Big Bang is the effect of a supernatural cause. He presents the following argument (which he incorrectly attributes to me) as a basis for inferring a supernatural cause of the universe's origin:

(1) We have reason to believe that all events have a cause.

(2) The Big Bang is an event.

(3) Therefore, we have reason to believe that the Big Bang has a cause.

While admitting that this argument does not violate singularity theorems, since the cause is not conceived to be a spatiotemporal object, Smith maintains that the argument fails because (1) is false. Quoting me to the effect that 'the causal proposition may be taken as an empirical generalization enjoying the strongest support experience affords', Smith rejoins that quantum-mechanical considerations show that the causal principle is limited in its application, so that a probabilistic argument for a cause of the Big Bang cannot succeed. For according to Heisenberg's uncertainty principle, it is impossible to predict precisely the conditions of the values of momentum or position of some particle x at some time t_2 on the basis of our knowledge of the conditions of x at t_1. Since it is sufficient to understand causality in terms of a law enabling precise predictions of individual events to be deduced, it follows from Heisenberg's principle that there are uncaused events in this sense.[4]

---

[3] Essay IV, Sect. 3.

[4] Notice that Smith's claim that such events are uncaused is predicated on the very dubious equivalence between 'unpredictability in principle' and 'uncausedness', an equivalence which I shall criticize in the text. If all that quantum indeterminism amounts to is 'uncausedness' in the sense of 'unpredictability in principle', then the demonstration that quantum events are uncaused in this sense fails to confute the causal proposition at issue in the first premiss of the *kalām* argument, unpredictability being an epistemic affair which may or may not result from an ontological indeterminism. For clearly, it would be entirely consistent to maintain determinism on the quantum level even if *we* could not, even in principle, predict precisely such events. In this paper, however, I shall not assume some controversial 'hidden variables' view, but shall for the sake of argument go beyond Smith and assume that indeterminism does hold on the quantum level. For discussion see A. Shimony, 'Metaphysical Problems in the Foundations of Quantum Mechanics', *International Philosophical Quarterly*, 18 (1978), 3–17; A. Aspect and P. Grangier, 'Experiments on Einstein–Podolsky–Rosen Type Correlations with Pairs of Visible Photons', in R. Penrose and C. J. Isham (eds.), *Quantum Concepts in Space and Time* (Oxford:

Therefore, the causal proposition is not universally applicable and may not apply to the Big Bang.

But what exactly is the causal proposition which is at issue here? The proposition which I enunciated was not (1), as Smith alleges, but rather

(1') Whatever begins to exist has a cause.

The motions of elementary particles described by statistical quantum-mechanical laws, even if uncaused, do not constitute an exception to this principle. As Smith himself admits, these considerations 'at most tend to show that acausal laws govern the *change of condition* of particles, such as the change of particle x's position from q_1 to q_2. They state nothing about the causality or acausality of absolute beginnings, of beginnings of the existence of particles.'[5]

Smith seeks to rectify this defect in his argument, however, by pointing out that the uncertainty relation also permits energy or particles (notably virtual particles) to 'spontaneously come into existence' for a very brief time before vanishing again. It is therefore false that 'all beginnings of existence are caused' and, hence, 'the crucial step in the argument to a supernatural cause of the Big Bang . . . is faulty'.[6]

But as a counter-example to (1'), Smith's use of such vacuum fluctuations is highly misleading. For virtual particles do not literally come into existence spontaneously out of nothing. Rather the energy locked up in a vacuum fluctuates spontaneously in such a way as to convert into evanescent particles that return almost immediately to the vacuum. As John Barrow and Frank Tipler comment, 'the modern picture of the quantum vacuum differs radically from the classical and everyday meaning of a vacuum—nothing. . . . The quantum vacuum (or vacua, as there can exist many) states . . . are defined simply as local, or global, energy minima ($V'(0) = 0$, $V'''(0) > 0$).'[7] The microstructure of the quantum vacuum is a sea of continually forming and dissolving particles which borrow energy from the vacuum for their brief existence. A quantum

Clarendon Press, 1986), 1–15, and S. Bhave, 'Separable Hidden Variables Theory to Explain Einstein–Podolsky–Rosen Paradox', *British Journal for the Philosophy of Science*, 37 (1986), 467–75.

[5] Essay IV, Sect. 3.

[6] Ibid.

[7] J. Barrow and F. J. Tipler, *The Anthropic Cosmological Principle* (Oxford: Clarendon Press, 1986), 440.

vacuum is thus far from nothing, and vacuum fluctuations do not constitute an exception to the principle that whatever begins to exist has a cause. It seems to me, therefore, that Smith has failed to refute premiss (1′).

Let us pursue Smith's argument further, however. He proceeds to argue that there is no reason to think that the causal principle applies to the Big Bang, whether one adopts a model based exclusively on the General Theory of Relativity or whether one uses a model adjusted for quantum effects during the Planck era. Consider on the one hand a model in which quantum physics plays no role prior to 10^{-43} second after the singularity. Since the classical notions of space and time and all known laws of physics break down at the singularity, it is in principle impossible to predict what will emerge from a singularity. If we regard the Big Bang as the first physical state,[8] then the particles that constitute that state must be regarded as being randomly and spontaneously emitted from nothing at all. Smith states,

This means, precisely put, that if the Big Bang is the first physical state, then every configuration of particles that does constitute or might have constituted this first state is as likely on a priori grounds to constitute it as every other configuration of particles. In [this] case, the constitution of the Big Bang$_1$ is impossible in principle to predict and thus is uncaused (for 'uncaused' minimally means 'in principle unpredictable').[9]

Moreover, since the singularity is a point beyond which spacetime curves cannot be extended, it cannot have causal antecedents.

On the other hand, consider a model in which quantum processes do predominate near to the Big Bang. If the defender of the causal principle maintains that the proposition

(4) There are some uncaused beginnings of existence *within* spacetime

is irrelevant to and thus cannot increase the probability of

(5) The beginning of the existence of spacetime itself is uncaused,

---

[8] Smith's own view that the universe began to exist at $t > t_0$ and that the state of affairs existing at t_0 in 0, 1, or 2 dimensions is the source of the universe contradicts his claim that the universe began to exist uncaused, for on his view the universe did not come from nothing but is causally connected to the singularity, whose existence remains unexplained. Smith rejects Newton-Smith's demand for a cause of the singularity. Therefore, his argument for (*b*) fails.

[9] Essay IV, end of Sect. 3.

then Smith will respond that the same holds for the parallel argument for a supernatural cause of four-dimensional spacetime. For the proposition

(6) All beginnings of existence *within* spacetime are caused

would by the same token be irrelevant to and thus not increase the probability of

(7) The beginning of the existence of four-dimensional spacetime is caused.

So whether one adopts an unmodified relativistic model or a quantum model, there is no reason to postulate a cause, natural or supernatural, of the Big Bang.

Is this a sound argument? It seems to me not. To pick up on a point noted earlier, Smith's argument throughout his paper appears to be infected with positivism, so that it is predicated upon a notion of causality that is drastically inadequate. Smith assumes uncritically the positivistic equation between predictability in principle and causation.[10] But this verificationist analysis is clearly untenable, as should be obvious from the coherence of the position that quantum indeterminacy is purely epistemic, there existing hidden variables which are in principle unobservable, or even the more radical position of die-hard realists who are prepared to abandon locality (and perhaps even Special Relativity) in order to preserve the hidden variables. Clearly, then, to be 'uncaused' does not mean, even minimally, to be 'in principle unpredictable'.

This single point alone seems to me to vitiate Smith's entire argument for his conclusion (*b*) and against the theistic hypothesis in particular. For now we see that Smith's argument, even if successful, in no way proves that the universe began to exist without a cause, but only that its beginning to exist was unpredictable. What is ironic about this conclusion is that it is one with which the theist is in whole-hearted agreement. For since according to classical theism creation is a freely willed act of God, it follows necessarily that the beginning and structure of the universe were in principle unpredictable even though they were caused by God. The theist will therefore not only agree with Smith 'That there are uncaused

---

[10] For a good example of this equation, see H. Bondi, 'Why Mourn the Passing of Determinism?' in Alwyn van der Merwe (ed.), *Old and New Questions in Physics, Cosmology, Philosophy, and Theoretical Biology* (New York: Plenum Press, 1983), 77–82.

events *in this sense* follows from Heisenberg's uncertainty principle',[11] but even more will insist that such uncaused events are entailed by classical theism's doctrine of creation. He will simply deny that this is the relevant sense when we are enquiring whether the universe could have come into being uncaused out of nothing.

When we ask that question, we are asking whether the whole of being could come out of non-being; and here a negative answer seems obvious. Concerning this question, even genuine quantum indeterminacy affords no evidence for an affirmative response. For if an event requires certain physically necessary conditions in order to occur, but these conditions are not jointly sufficient for its occurrence, and the event occurs, then the event is in principle unpredictable, but it could hardly be called uncaused in the relevant sense. In the case of quantum events, there are any number of physically necessary conditions that must obtain for such an event to occur, and yet these conditions are not jointly sufficient for the occurrence of the event. (They are jointly sufficient in the sense that they are all the conditions one needs for the event's occurrence, but they are not sufficient in the sense that they guarantee the occurrence of the event.) The appearance of a particle in a quantum vacuum may thus be said to be spontaneous, but cannot properly be said to be absolutely uncaused, since it has many physically necessary conditions. To be uncaused in the relevant sense of an absolute beginning, an existent must lack any non-logical necessary or sufficient conditions whatsoever. Now at this juncture, someone might protest that such a requirement is too stringent: 'For how could *anything* come into existence without *any* non-logical necessary or sufficient conditions?' But this is my point exactly; if nothing existed—no matter, no energy, no space, no time, no deity—if there were absolutely nothing, then it seems unintelligible to say that something should spring into existence.

As for Smith's two cases, then, in the case of the classical relativistic theory, the fact that the universe originates in a naked singularity only proves that we cannot *predict* what sort of universe will emerge therefrom (and Smith does not claim otherwise), but it in no way implies that anything and everything can actually come into existence uncaused. Indeed, when we reflect on the fact that a physical state in which all spatial and temporal dimensions are zero

---

[11] Essay IV, Sect. 3.

is a mathematical idealization whose ontological counterpart is nothing, then it becomes clear why the universe is unpredictable and why its unpredictability in no way implies the possibility of its coming into being without a cause.[12] As for Smith's consideration that a singularity is a point beyond which spacetime curves cannot be extended, this only proves that the creation event cannot have been brought about by any natural cause; but it does not prove that a being which transcended space and time could not have caused it.

As for the quantum case, the problem with the inference from (4) to (5) is not that it moves from existents within the universe to the universe as a whole, but rather that Smith's faulty concept of causation makes the notion of 'uncaused' equivocal. For some beginnings of existence within spacetime are uncaused in the sense of being spontaneous or unpredictable, but one cannot conclude that therefore spacetime itself could come into being uncaused in the stronger sense of arising from nothing in the utter absence of physically necessary and sufficient conditions. But the inference from the necessity of causal conditions for the origin of existents in spacetime to the necessity of causal conditions for the origin of spacetime itself is not similarly equivocal. Indeed, our conviction of the truth of the causal principle is not based upon an inductive survey of existents in spacetime, but rather upon the metaphysical intuition that something cannot come out of nothing.[13] The proper inference, therefore, is actually from 'Whatever begins to exist has a cause' and 'The universe began to exist' to 'The universe has a cause', which is a logically impeccable inference based on universal instantiation. It seems to me, therefore, not only that Smith has

---

[12] At the risk of seeming repetitious, I cannot refrain from referring again to G. E. M. Anscombe, '"Whatever Has a Beginning of Existence Must Have a Cause". Hume's Argument Exposed', *Analysis*, 34 (1973–4), 150. As she points out, we can form various pictures in our minds and give them appropriate titles, e.g. 'Superforce Emerging from the Singularity', 'Gravitons Emerging from the Singularity', or 'Rabbits Emerging from the Singularity', but our ability to do that says absolutely nothing about whether it is ontologically possible for something to come into being uncaused out of nothing.

[13] My defence of the causal proposition as an 'empirical generalization enjoying the strongest support experience affords' cited by Smith was in its original context a last-ditch defence of the principle designed to appeal to the hard-headed empiricist who resists the metaphysical intuition that properly grounds our conviction of the principle. (Craig, *The Kalām Cosmological Argument*, 141–8.) It does seem to me that only an aversion to the theism implied by the principle in the present context would lead the empiricist to think that the denial of the principle is more plausible than the principle itself.

failed to show that the Big Bang does not require a supernatural cause, but that, on the contrary, we see from these considerations that if the universe did originate from nothing, then that fact does point to a supernatural cause of its origin.[14]

Hence, I conclude that Smith has failed to show that there is no reason to regard the theistic hypothesis as true.

3. UNREASONABLE TO REGARD THE THEISTIC HYPOTHESIS AS TRUE

If Smith is to prove his point (*b*), in any case, he has to do much more than show that there is no reason to adopt the theistic hypothesis. He has to show that in light of the evidence, the theistic hypothesis has now become unreasonable. Smith believes the evidence for vacuum fluctuation models of the origin of the universe is such as to render unreasonable the theistic hypothesis. For it is physically necessary for quantum effects to predominate near to the Big Bang, and quantum-mechanical models of the origin of the universe as or on the analogy of a vacuum fluctuation provide the most probable account of the beginning of the universe out of nothing.

Now Smith's line of reasoning raises some intriguing epistemological issues, to which, unfortunately, Smith gives no attention.[15] Under what circumstances would it be irrational to believe in supernatural *creatio ex nihilo*? Under what circumstances would it be rational? When is a supernatural explanation preferable to a naturalistic one and vice versa? Rather than seek to adjudicate these questions, let us assume for the sake of argument that it would be unreasonable, all things being equal, to posit a supernatural cause for the origin of the universe when a plausible empirical explanation is available or even likely to become available. Notice that in such a case the theistic hypothesis would not be *falsified*; it

---

[14] Such a cause would have to be uncaused, eternal, changeless, timeless, immaterial, and spaceless; it would, as I have argued elsewhere, also have to be personal and therefore merits the appellation 'God'. (Essay I, Sect. 4; William Lane Craig, 'The *Kalām* Cosmological Argument and the Hypothesis of a Quiescent Universe', *Faith and Philosophy*, 8 (1991), 492–503).

[15] See, however, T. V. Morris, 'Creation *ex Nihilo*: Some Considerations', in *Anselmian Explanations* (Notre Dame, Ind.: University of Notre Dame Press, 1987), 151–60, for some initial and interesting analysis of these issues.

would simply be *unreasonable*, all things being equal, to believe it. The question is, then, whether quantum-mechanical models of the origin of the universe as or on the analogy of a vacuum fluctuation are or are likely to become plausible empirical explanations.

The answer to the question whether such models now provide plausible empirical explanations for the universe's origin is, of course, no, since the theories are so undeveloped and there is no empirical evidence in their favour. It is remarkable that Smith has so high a degree of confidence in quantum fluctuation models that he thinks it unreasonable to believe in the theistic hypothesis, for this is tantamount to saying that in light of these theories it is no longer reasonable to hold to a Big Bang model involving a singularity. But these theories are so inchoate, incomplete, and poorly understood that they would hardly commend themselves to most scientists as more plausible than traditional Big Bang models. Of course, quantum effects will become important prior to the Planck time, but it is pure speculation that these will serve to avert the initial singularity.[16]

Smith's bold assertions on behalf of these models greatly overshoot the modest, in some cases almost apologetic, claims made by the proponents of the models themselves. Brout, Englert, and Gunzig, for example, advised: 'We present our work as a hypothesis . . . For the present all that can be said in favor of our hypothesis is that these questions can be examined and on the basis of the answers be rejected or found acceptable.'[17] Atkatz and Pagels offered the following justification: 'While highly speculative, we believe this idea is worth pursuing.'[18] All that Vilenkin claimed on behalf of his model was, 'The advantages of the scenario presented here are of

---

[16] See also the interesting discussion by J. Barrow and F. Tipler, 'Action Principles in Nature', *Nature*, 331 (1988), 31–4, in which they explain that since we have no tested theory of quantum gravitation to supersede General Relativity nor any observational evidence for the existence of matter fields which violate the strong or weak energy conditions, the initial cosmological singularity has not been eliminated. In fact, they point out that the finiteness of the action in Friedman models is due to the cosmological singularities. 'Thus in general there is a trade-off between space time singularities: a singularity in the action is avoided only at the price of a singularity in curvature invariants, and vice versa. In cosmology some sort of singularity seems inevitable.' (Ibid. 32–3.)

[17] R. Brout, F. Englert, and E. Gunzig, 'The Creation of the Universe as a Quantum Phenomenon', *Annals of Physics*, 115 (1978), 78, 98.

[18] D. Atkatz and H. Pagels, 'Origin of the Universe as a Quantum Tunneling Event', *Physical Review*, D25 (1982), 2072.

aesthetic nature.'[19] Other proponents of such models claim no more than that their model is *consistent* with observational data—and sometimes they do not even claim that much. In fact, it is ironic that—apparently unbeknownst to Smith—several of the original proponents of these models have, as we shall see, already abandoned the vacuum fluctuation approach to cosmogony as implausible and are seeking elsewhere for explanations of the universe's origin.

Are these models then likely to become plausible empirical explanations of the universe? Again the difficulty here is that it is just too early to be able to give an affirmative answer to this question. Such models are provocative and worth pursuing, but we simply do not yet know if they are likely to become plausible empirical explanations of the universe's origin. Indeed, there is some reason to doubt that such models can ever become plausible *empirical* explanations, since such models, by their very nature, tend to posit events which are in principle inaccessible to us, causally discontinuous with our universe, or lying beyond event horizons. According to Vilenkin, the only verifiable prediction made by his model is that the universe must be closed—a prediction which observational cosmology tends to falsify. Smith likes Gott's model because it makes empirical predictions.[20] But so far as I can see, his only prediction is that the universe is open, which is so general as to be useless in serving as evidence for the model. None of the proponents of such models has to my knowledge laid down conditions which would verify his theory. At present, then, such models are perhaps best viewed as naturalistic metaphysical alternatives to the theistic hypothesis.

But even so construed, their superiority to theism is far from obvious:

1. Such models make the metaphysical presupposition that the observed expansion of the universe is not, in fact, the expansion of the Universe-as-a-whole, but merely the expansion of a restricted region of it. Our expanding space is contiguous with some sort of wider space (whether a Minkowski space as in Brout, Englert, and Gunzig's model or a curved de Sitter space as in Gott's model) in which the quantum fluctuations occur in the spacetime geometry

[19] A. Vilenkin, 'Creation of Universes from Nothing', *Physical Letters*, B117 (1982), 27.

[20] Quentin Smith, 'World Ensemble Explanations', *Pacific Philosophical Quarterly*, 67 (1986), 73–86.

which 'pull' particles into existence out of the energy locked up in empty space. Thus, throughout this broader Universe-as-a-whole, which is considered to be a quantum-mechanical vacuum, fluctuations occur which blow up, rather like the ears on a Mickey Mouse balloon, into distinct material universes. But immediately the question arises, Why, since all the evidence we possess suggests that space is expanding, should we suppose that it is merely our *region* of space (and regions like it) that is expanding rather than *all* of space? This thesis would appear to be in violation of the Copernican Principle, which holds that we occupy no special place in the universe. This methodological principle, which underlies all of modern astronomy and astrophysics, would be violated because what we observe would not be typical of the universe at large. A violation of the principle in this case would appear to be entirely gratuitous. Moreover, it is not just that the postulate of a different wider space is required, but that a good deal of fine-tuning is necessary in order to get the space to spawn appropriate universes. But there is no independent reason to think that such a different wider space exists or, indeed, that a different wider space of any sort at all exists. In this sense, the postulation of such a wider space is an exercise in speculative metaphysics akin to the postulate of theism—except that theism enjoys in this case the advantage that there are at least putative independent reasons for accepting the existence of God.

Moreover, it is questionable whether the models at issue are anything more than mathematical constructs lacking any physical counterpart. For, as David Lindley points out, such models depend on the use of certain mathematical 'tricks' for their validity.[21] For example, quantities derived from the conformal factor most naturally belong to the geometrical side of Einstein's equation, but by being put on the other side of the equation they can be imagined to be part of the stress–energy tensor instead. This 'rather arbitrary procedure' allows one to think of the conformal factor as a physical field. But this seems to be a clear case of unjustified ontologizing of a mathematical notion into a physical entity. To make matters worse, proponents of such models then propose the trick of coupling these conformal fields dynamically to other more conventional physical components of the stress–energy tensor, such as the fields

---

[21] D. Lindley, 'Cosmology from Nothing', *Nature*, 330 (1987), 603–4.

associated with particles similar to gauge bosons in high-energy physics. In this way the conformal field can be made to generate regions of distorted geometry and a local density of particles. But what reason or evidence is there to regard such a procedure as anything more than mathematical legerdemain? As Barrow and Tipler point out, 'It remains to be seen whether any real physical meaning can be associated with these results.'[22]

Nor does the comparison of the universe's origin to the spontaneous production of a virtual particle serve to render these models plausibly realistic. For if this comparison is meant to be reasoning by analogy—and it apparently is[23]—then it seems extraordinarily weak, since the disanalogies between the universe and a virtual particle are patent. If we are to believe with Tryon that the universe literally *is* a virtual particle,[24] then this seems even more preposterous, since the universe has neither the properties nor the behaviour of a virtual particle. One might ask, too, why quantum fluctuations are not now spawning universes in our midst? Why do vacuum fluctuations endure so fleetingly rather than grow into mini-universes inside ours?

2. While there may be no difficulty in conceiving of a wider space, the postulation of such a space must also involve the postulation of a wider time, which raises profound metaphysical questions. The wider space cannot be conceived of as a static, timeless entity, for it is filled with successive events, and, moreover, time must, it seems, exist in order to speak meaningfully about the probability of a fluctuation's occurring. In another place, Smith recognizes that the wider space is a spacetime, and he construes this to mean that the time of the background space is like the trunk of the tree of time and the times of the emerging universes are like its branches.[25] But what is this metaphor supposed to mean? Perhaps it refers to the sort of 'route-dependence' (to borrow Bondi's phrase) of empirical or measured time such as is envisioned in the clock paradox in relativity theory, according to which moments in one

---

[22] Barrow and Tipler, *The Anthropic Cosmological Principle*, 441.

[23] E.g. Atkatz and Pagels write, 'The big bang is analogous to a single radioactive decay, on a huge scale' and 'In analogy to this decay process we assume that the Universe began as a classically stable, static configuration of spacetime.' (Atkatz and Pagels, 'Origin of the Universe as a Quantum Tunneling Event', 2065–6.)

[24] E. Tryon, 'Is the Universe a Vacuum Fluctuation?' *Nature*, 246 (1973), 396–7.

[25] Quentin Smith, 'A New Typology of Temporal and Atemporal Permanence', *Noûs*, 23 (1989), 310–12.

temporal series literally have no simultaneous counterparts in another temporal series as calculated by an observer in the latter series. This analogy seems particularly apt when diverging branches later reconverge, as in Gott's model. Events in the two universes cannot be temporally related to each other, despite their common origin and coalescence, presumably because an exchange of light signals is impossible between the branches. But then the relation of the trunk-time to the branching times seems to become unintelligible. For prior to the Planck time, photons did not even exist, and at some point the density of the universe and the curvature of spacetime become so extreme that one cannot speak of an exchange of signals prior to that point, so that empirical time is no longer meaningful. In order to have a tree of time in which branches sprout out of the trunk, one is forced, it seems, to posit a sort of metaphysical time not dependent upon the exchange of light signals. Smith, in fact, holds to the existence of metaphysical time and champions an A-theory of time, as his other publications make clear. But while it makes sense, perhaps, to speak of a B-theory of branching metaphysical time, I can see no such sense of an A-theory of branching metaphysical time. For on an A-theory of metaphysical time, moments of time elapse successively one after another completely independently of what may be happening physically. Since such time is not wedded to space, even the formation of new universes would not cause time to 'branch': the same metaphysical time in which the wider space endures measures all the separate universes as well. This seems to me a coherent picture, and the theist who regards God as temporal would no doubt conceive of Him as existing in this metaphysical time, which would measure the succession in God's contents of consciousness. But the problem is that it is not at all clear that this metaphysical time is of any use to quantum fluctuation models, which are predicated on the use of empirical time and therefore seem to face real difficulties in giving a coherent account of the relation of wider time to the various subtimes.

3. Our relatively young universe seems incompatible with vacuum fluctuation models. In the quantum-mechanical vacuum of the wider spacetime, sooner or later a fluctuation would appear which would produce an open universe. (If this is debarred, then the model stands in conflict with current observational cosmology.) Indeed, given an infinite past wider time, such a universe would have originated an infinite time ago. But then by now, the open universe

would have grown so large as to coalesce with all other fluctuation-formed universes and would have filled all of the wider space.[26] Gott attempts to avoid this difficulty by simply laying down conditions where the fluctuations are allowed to occur in the wider space.[27] For any universe-spawning event E, there must not exist another similar event E' in the past light cone of E. The volume of this region which must be free of events like E is infinite. In order to prevent any E' from occurring in this region, Gott stipulates that the probability of randomly producing events like E per unit 4-volume be infinitesimal. Since de Sitter space is infinite, one can thus construct a model of an infinite number of disjoint universes formed by fluctuations. But not only is this scenario extraordinarily *ad hoc*, but it does not seem even to avoid the difficulty.[28] For

---

[26] See the similar objection urged against the quantum tunnelling model of Atkatz and Pagels by M. Munitz, *Cosmic Understanding* (Princeton, NJ: Princeton University Press, 1986), 136, who observes, 'For if the actual closed universe arose by a process of quantum tunneling from a prior stable initial state, then the universe in its pre-creation state could not remain indefinitely long in that state, if indeed it is unstable with respect to quantum tunneling.' Atkatz and Pagels have admittedly no answer to the question of how the universe got into its pre-creation state. See also C. Isham, 'Creation of the Universe as a Quantum Process', in R. Russell, W. Stoeger, and G. Coyne (eds.), *Physics, Philosophy, and Theology: A Common Quest for Understanding* (Vatican City State: Vatican Observatory, 1988), 375–408. Cf. the analogous reasoning of Paul Davies, 'Cosmic Heresy?', *Nature*, 273 (1978), 336, that, given the evolution of intelligent life somewhere in a static universe, it becomes inexplicable why intelligent life should evolve at some time t_n rather than before. If intelligent life will evolve at all, it should by now have done so, filling all appropriate ecological niches in the universe, thereby preventing the evolution of the life-forms we now observe.

[27] J. R. Gott III, 'Creation of Open Universes from de Sitter Space', *Nature*, 295 (1982), 304–7.

[28] See a similar objection urged by Barrow and Tipler, *The Anthropic Cosmological Principle*, 602–7, who point out that although Gott's model posits a causal structure of the background space consisting of infinitely many non-intersecting, open bubble universes, there must be such a bubble universe in the past light cone of any event p, given a constant probability of bubble formation in the de Sitter space. Notice that because Gott's bubbles are potentially infinite in their expansion, so long as the bubbles are formed after p they will not intersect; the walls of each bubble reach spacelike infinity at an infinite time in the future. But since 'the volume of an open bubble becomes infinite in an infinite time', then given an infinite past prior to p, it follows that the past light cone of any p will already contain an open bubble universe that has already expanded to infinity. So also Isham, 'Creation of the Universe as a Quantum Process', 387, who writes, 'The existence within a single spacetime of infinitely many "conefuls" of matter might be theoretically acceptable if they did not interfere with each other. But this is far from being the case. . . . the matter emitted from a seed-point Y will eventually interact with that emerging from the point X. This is a rather peculiar picture, and not one that seems particularly consistent with

given infinite past wider time, each of the infinite regions of the de Sitter space will have spawned an open universe which will have filled the volume of that region completely, so that all the bubble universes will by now have coalesced and thus filled all of the de Sitter space. Hence, even Gott seems constrained to posit some origin of the wider de Sitter spacetime—but then we are right back to where we started.

4. It is obvious from what has been said above that vacuum fluctuation models have, in fact, nothing to do with the origination of the universe *ex nihilo*. They posit metaphysical realities of precise specifications in order to generate our universe. Some of them are really more closely related to inflationary scenarios than to cosmogony. Interpreted cosmogonically, vacuum fluctuation models constitute in the final analysis denials that the universe began to exist, for it is only our observable segment of the universe that had a beginning, not the Universe-as-a-whole. As Barrow and Tipler comment,

It is, of course, somewhat inappropriate to call the origin of a bubble Universe in a fluctuation of the vacuum 'creation *ex nihilo*,' for the quantum mechanical vacuum state has a rich structure which resides in a previously existing substratum of space-time, either Minkowski or de Sitter space-time. Clearly, a true 'creation *ex nihilo*' would be the spontaneous generation of everything—space-time, the quantum mechanical vacuum, matter—at some time in the past.[29]

Smith admits that 'A disadvantage of . . . theories that postulate a background space from which the universe fluctuates, is that they explain the existence of the universe but only at the price of introducing another unexplained given, viz., the background space.'[30] In that case, Smith has not only failed to carry his point (*b*) but (*a*) as well.

But Smith asserts that there are even more radical models of a quantum origin of the universe that do not postulate the existence of a wider space, but hold that the universe is the result of some sort of quantum transition out of nothingness into being. For example, in the Vilenkin model, the origin of the universe is

large-scale astronomical observations.' He notes that to avoid this difficulty, the background de Sitter spacetime must itself be in a state of expansion—which forces the origination question right back on us again.

[29] Barrow and Tipler, *The Anthropic Cosmological Principle*, 441.
[30] Essay IV, Sect. 4.

understood on the analogy of quantum tunnelling, a process in which an elementary particle passes through a barrier, though it lacks the energy to do so, because the uncertainty relation allows it to acquire spontaneously the energy for the period of time necessary for it to pass through the barrier. Vilenkin proposes that spacetime itself tunnels into existence out of nothing, except that in this case there is no prior state of the universe, but rather the tunnelling itself is the first state that exists.

Unfortunately, Smith seems to have misinterpreted in a literal way Vilenkin's philosophically naïve use of the term 'nothing' for the quantum-mechanical vacuum.[31] Be that as it may, if the quantum tunnelling is supposed to be literally from nothing, then such models seem to be conceptually flawed.[32] For as Thomas Aquinas saw,[33] creation is not properly any kind of a change or transition at all, since transition implies the existence of an enduring subject, which is lacking in creation. In a beginning to be out of nothing, there can be no talk whatsoever of transition, quantum or otherwise. It is therefore incoherent to characterize creation as a quantum transition out of nothingness.

Even more fundamentally, however, what we are being asked to believe is surely metaphysical nonsense. Though dressed up in the guise of a scientific theory, the thesis at issue here is a philosophical one, namely, can something come out of nothing? Concerning his own model, even Vilenkin admits, 'The concept of the universe being created from nothing is a crazy one.'[34] He tries to mitigate this craziness by comparing it to particle pair creation and annihilation— an analogy which we have seen to be altogether inadequate and in any case irrelevant to the Vilenkin model as Smith interprets it, since he supposedly lacks the embedding quantum-mechanical spacetime. The principle *ex nihilo nihil fit* seems to me to be a sort of metaphysical first principle, one of the most obvious truths we intuit when we reflect seriously. If the denial of this principle is the

[31] See J. Leslie, *Universes* (London: Routledge, 1989), 80, who writes, 'Often— for example in A. Vilenkin's influential writings—"nothing" appears to mean the spacetime foam mentioned just a moment ago.' So also W. Carroll, 'Big Bang Cosmology, Quantum Tunneling from Nothing, and Creation', *Laval théologique et philosophique*, 44 (1988), 59–75.

[32] This is emphasized by Carroll, 'Big Bang Cosmology, Quantum Tunneling from Nothing, and Creation', 68–75.

[33] Aquinas, *Summa contra Gentiles*, 2.17.

[34] Vilenkin, 'Creation of Universes from Nothing', 26.

alternative to a theistic metaphysic, then let those who decry the irrationality of theism be henceforth forever silent!

If this fourth criticism is on target, then vacuum fluctuation models say nothing against divine *creatio ex nihilo*, for even if some such model turns out to be correct, the theist will maintain that God created the wider spacetime from which our material universe emerged. It might be rejoined that there would then be no grounds for positing God as the creator of the embedding spacetime, since there is no scientific evidence that it began to exist. But as I have defended it in Essay I, divine *creatio ex nihilo* is grounded in revelation and philosophical argument, and the scientific evidence merely serves as empirical *confirmation* of that doctrine. The theist, after all, has no vested interest in denominating the Big Bang as the moment of creation. He is convinced that God created all of spacetime reality *ex nihilo*, and the Big Bang model provides a powerful suggestion as to when that moment was; on the other hand, if it can be demonstrated that our observable universe originated in a broader spacetime, so be it—in that case it was this wider reality that was the immediate object of God's creation. But unless the conceptual difficulties in such models can be overcome and some empirical evidence for them is forthcoming, the theist will probably apply Ockham's Razor and be content to regard the Big Bang as the creation event.

5. I earlier alluded to the fact that vacuum fluctuation models have been abandoned as plausible accounts of the origin of the universe by some of their principal expositors and to that extent are already somewhat dated. Brout, Englert, and Spindel of the Free University of Brussels, where much of the theoretical work on these models was done, have, for example, moved beyond such models and have criticized the attempt of some of their colleagues to refurbish the old, untenable models.[35] Noting that 'Theories of this type have not found wide acceptance,' Isham explains that their interest lies mainly in some of the rather general problems that they raise.[36] Rather the model that seems to have fired the imagination of many current theorists is the model of Hartle–Hawking based on the assigning of a wave function to the universe.[37]

---

[35] R. Brout and P. Spindel, 'Black Holes Dispute', *Nature*, 337 (1989), 215–16.
[36] Isham, 'Creation of the Universe as a Quantum Process', 387.
[37] J. Hartle and S. Hawking, 'Wave Function of the Universe', *Physical Review*, D28 (1983), 2960–75.

Unfortunately, models of this sort confront acute philosophical difficulties concerning the metaphysics of time.[38] For such models presuppose a geometrodynamical interpretation of spacetime that suppresses objective temporal becoming in favour of a Parmenidean static construal of the dynamics of spacetime in terms of positions on a leaf of history in superspace.[39] In the same way that spacetime serves as the dynamical arena of particle dynamics, so superspace is the dynamical arena of geometrodynamics. Given a spacetime manifold, classical geometrodynamics constitutes a rule for calculating and constructing a leaf of history that slices through superspace by mapping the points of spacetime on to the corresponding points in superspace; conversely, given the leaf of history in superspace one can construct a unique spacetime by a reverse mapping. There are innumerably more 3-geometries in superspace than lie on this particular leaf, but only those that do so lie comprise the building-blocks of the 4-geometry that is classical spacetime. Already this reduction of spacetime to a leaf of history in superspace has completed the Parmenidean reinterpretation of the dynamics of time along static lines, for the arena of geometrodynamics is not a superspacetime, but a super*space* alone. But the introduction of quantum theory into geometrodynamics—a move essential to wave-functional models of the origin of our 4-geometry—not only makes spacetime ontologically derivative from superspace, but, far more, actually expunges spacetime altogether, for quantum indeterminacy makes it impossible to distinguish sharply between 3-geometries on a leaf of history in superspace and those not on it. At best one can assign to each 3-geometry a probability amplitude $\psi = \psi \, (^{(3)}\mathcal{G})$, which is highest in a finite region surrounding the classically forecast leaf of history, and seek to approximate a classical slicing of superspace by means of superposition and phase cancellation of the various wave functions. Misner, Thorne, and Wheeler conclude,

[38] For more on this see Essay XI.
[39] For discussion, see C. Misner, K. S. Thorne, and J. A. Wheeler, *Gravitation* (San Francisco: W. H. Freeman, 1973), 1180–95. Cf. 'There is no such thing as spacetime in the real world of quantum physics. . . . superspace leaves us space but not spacetime and therefore not time. With time gone the very ideas of "before" and "after" also lose their meaning.' (J. A. Wheeler, 'From Relativity to Mutability', in J. Mehra (ed.), *The Physicist's Conception of Nature* (Dordrecht: D. Reidel, 1973), 227; see also J. A. Wheeler, 'Beyond the Black Hole', in Harry Wolf (ed.), *Some Strangeness in the Proportion* (Reading, Mass.: Addison-Wesley, 1980), 346–50 and the therein cited literature; Isham, 'Creation of the Universe as a Quantum Process', 392–400.)

The uncertainty principle thus deprives us of any way whatsoever to predict, or even to give meaning to, 'the deterministic classical history of space evolving in time.' *No prediction of spacetime, therefore no meaning for spacetime* is the verdict of the quantum principle. That object which is central to all of classical general relativity, the four-dimensional spacetime geometry, simply does not exist, except in a classical approximation.

These considerations reveal that the concepts of spacetime and time are not primary but secondary ideas in the structure of physical theory. These structures are valid in the classical approximation. However, they have neither meaning nor application under circumstances where quantum geometrodynamic effects become important. Then one has to forgo that view of nature in which every event, past, present, or future, occupies its preordained position in a grand catalog called 'spacetime'. . . . There is no spacetime, there is no time, there is no before, there is no after. The question of what happens 'next' is without meaning.[40]

This static model of reality, based on the ontological primacy of superspace and reinforced by quantum indeterminacy, is, however, in complete contradiction to Smith's own understanding of the nature of time. For Smith is an ardent A-theorist, who defends the objectivity of temporal becoming, the irreducibility of the 'now', and the ineliminability of tensed discourse; indeed, the objectivity of the transient present plays a central role in his metaphysic.[41] On Smith's view, the metaphysic of time which underlies quantum geometrodynamics and, therefore, wave-functional models of cosmogony can only be regarded as thoroughly unacceptable, a distortion of the nature of reality of the most serious kind.

But should we join Smith in his advocacy of an A-theory of time to the exclusion of static models of reality? Here I can only refer the reader to Smith's own excellent writings in defence of his thesis. I am persuaded that he is correct. Therefore, I can only regard cosmological models based on a metaphysic of superspace as mathematical constructs having no correspondence with reality.

It seems to me, therefore, that even if we concede that it would be unreasonable, all things being equal, to posit divine *creatio ex nihilo* when a plausible, empirical hypothesis for the origin of the universe is available or even likely to become available, Smith has

---

[40] Misner *et al.*, *Gravitation*, 1182–3.
[41] See Quentin Smith, *Language and Time* (New York: Oxford University Press, 1993) and the various articles by Quentin Smith cited in Essay IV.

8### 8### 8### 8#### 8#### 8### 8### 8#### 8### 8#### 8#### 8### 8### 8#### 8#### 8#### 8#### 8#### 8#### 8##### 8###

failed to show that the theistic hypothesis is unreasonable. Moreover, for the theist, it is not the case that all things are equal in this matter, for he has independent reasons from philosophy and revelation, apart from the scientific evidence, for accepting *creatio ex nihilo*. If these reasons are sound, then he would be rational in accepting the theistic hypothesis even if a plausible empirical account of the world's origin were available—which at present it most certainly is not—though he might not in such a case have a clue about the moment of creation.

4. CONCLUSION

In conclusion, then, I think it is clear that Smith has failed to carry the second prong of his argument, namely, that the universe began to exist without being caused to do so. In his attempt to show that there is no good reason to accept the theistic hypothesis, he misconstrued the causal proposition at issue, appealed to false analogies of *ex nihilo* creation, contradicted himself in holding the singularity to be the source of the universe, failed to show why the origin of the universe *ex nihilo* is reasonable on models adjusted or unadjusted for quantum effects, and, most importantly, trivialized his whole argument through the reduction of causation to predictability in principle, thus making his conclusion an actual entailment of theism. Nor has he been any more successful in proving that the theistic hypothesis is unreasonable in light of the evidence. For he ignores the important epistemological questions concerning the circumstances under which it would be rational to accept divine *creatio ex nihilo*; he has failed to show that vacuum fluctuation models are or are likely to become plausible empirical explanations of the universe's origin; on the contrary, such models are probably best regarded as naturalistic metaphysical alternatives to the theistic hypothesis, but as such are fraught with conceptual difficulties; and, most importantly, such models, on pain of ontological absurdity, do not in fact support Smith's contention (*b*), so that they do not render unreasonable the hypothesis that God created the universe, including whatever wider spatiotemporal realms of reality might be imagined to exist.

VI

A Criticism of A Posteriori and A Priori
Arguments for a Cause of the Big Bang
Singularity

QUENTIN SMITH

Introductory Note

This essay is a further elaboration of the argument in Essay IV. After defining classical Big Bang cosmology in Section 1, I develop (in Sections 2 and 3) my earlier criticism of arguments such as Newton-Smith's that the Big Bang singularity has a natural cause. Sections 1, 2, and 3 are rather scientifically technical and may be skipped by the philosopher who is mainly interested in the philosophical aspects of the debate between Craig and myself about theism and atheism. Sections 4 and 5 respond to Craig's Essays I and V, in which he argues that the causal principle, *everything that begins to exist has a cause*, is a necessary a priori truth. Sections 4 and 5 present an argument that the causal principle is not a necessary a priori truth. This supplements my argument in Essay IV against Craig's theory of causality, for Essay IV considered only Craig's argument for the empirical or a posteriori warrant we have for the causal principle.

I. BIG BANG COSMOLOGY AND PHILOSOPHIES OF
BIG BANG COSMOLOGY

Big Bang cosmology is based on four sets of equations. The most fundamental are the field equations of the General Theory of Relativity (GTR), which relate the metric tensor g_{ab} and its

Portions of this essay are pub. in 'Did the Big Bang Have a Cause?', *British Journal for the Philosophy of Science*, forthcoming.
I should like to thank Robert Weingard for helpful comments on an earlier version of this paper.

derivatives to the energy–momentum tensor T_{ab}. The formula summarizing the ten field equations is

$$R_{ab} - \tfrac{1}{2}Rg_{ab} + \lambda g_{ab} = (8\pi G/c^2)T_{ab}.$$

R_{ab} is the Ricci tensor of the metric g_{ab}, R is the Ricci scalar, λ is the cosmological constant (probably zero), c is the velocity of light, and G is Newton's constant of gravitation. The field equations are solved for the universe if the relevant observational values are introduced. Since our universe is (approximately) homogeneous and isotropic, it is described by the Robertson–Walker metric, which is determined by the radius of the universe at a given time and the curvature of spacetime. The metric is

$$ds^2 = dt^2 - (1/c^2)a^2 d\sigma^2,$$

where ds is the spacetime interval between two events, a the scale factor representing the radius of the universe at a given time, and $d\sigma$ is the line element of a space with constant curvature. The application of this metric to the field equations provides us with the Friedman solutions, which are the heart of Big Bang cosmology. With the cosmological constant omitted, these solutions read:

$$-3(d^2 a/dt)^2 = 4\pi G(p + 3P/c^2)a$$
$$3(da/dt)^2 = 8\pi G p a^2 - 3kc^2.$$

In these equations a as before is the scale factor representing the radius of the universe at a given time. da/dt is the first derivative of a with respect to time; it measures the rate of change of a with time (the rate at which the universe expands or contracts). $d^2 a/dt^2$ is the second derivative of a with respect to time; it measures the rate of change of da/dt (the acceleration of the expansion or the deceleration of the contraction). P is the pressure of matter and p its density. k is a constant which takes one of three values: 0 for a flat Euclidean space, -1 for a hyperbolic space, or $+1$ for a spherical space.

If Friedman's solutions are supplemented by the Hawking–Penrose singularity theorems, and the theorems are satisfied, it follows that our Friedman universe began to exist with a Big Bang singularity. The theorems state that a singularity is inevitable given the following five conditions:

(1) GTR holds true of the universe.

(2) There are no closed timelike curves (the principle of causality is not violated).

(3) Gravity is always attractive; that is, for any timelike vector V^a, the energy–momentum tensor of matter satisfies the inequality $(T_{ab} - \frac{1}{2}g_{ab}T)V^aV^b \geqslant 0$.

(4) The spacetime manifold is not too symmetric, such that every spacetime path of a particle or light ray encounters some matter or randomly oriented curvature. That is, any timelike or null geodesic contains some point at which $V_{[a}R_{b]cd[e}V_{f]}V^cV^d = 0$.

(5) There is some point p such that all the past-directed spacetime paths from p start converging again. This condition implies that there is enough matter present in the universe to focus every past-directed spacetime path from some point p, such that the matter causes a time-reversed closed trapped surface.

If these conditions hold in our universe, it follows that there is a Big Bang singularity which constitutes the beginning of the universe and the earliest time. (Actually, as Weingard emphasizes,[1] the theorems merely tell us that there is a non-spacelike geodesic that is only finitely extendible into the past, which is consistent with the singularity being a timelike singularity from which only some world lines originate. However, as Hawking argues, these are 'special cases and are unstable; in the general case there will be a curvature singularity that will intersect every world line. Thus general relativity predicts a beginning of time.'[2]

The above-mentioned four sets of equations (the field equations, the Robertson–Walker line element, the Friedman solutions, and the singularity theorems) entail that the universe began without a cause. This is true in a threefold sense. (*a*) Since there is nothing earlier than the Big Bang singularity, nothing earlier than the singularity can cause it. (*b*) Since nothing different from the singularity exists simultaneously with the singularity, nothing simultaneous with the singularity can cause it. (*c*) Since (by condition (?) of the singularity theorems) there are no closed timelike curves, there is nothing later than the singularity that can cause it.

Distinct from Big Bang cosmology are *philosophies of Big Bang cosmology*, which interpret the theses of Big Bang cosmology in

---

[1] R. Weingard, 'Some Philosophical Aspects of Black Holes', *Synthese*, 42 (1979), 199.
[2] S. W. Hawking, 'Theoretical Advances in General Relativity', in H. Woolf (ed.), *Some Strangeness in the Proportion* (Reading, Mass.: Addison-Wesley, 1980), 149.

the light of certain philosophical theories or principles. Now it is prima-facie possible that there is a sound philosophy of Big Bang cosmology that entails or renders it probable that the Big Bang singularity has a cause. This would be the case, for example, if there were a sound philosophical argument that it is a synthetic a priori truth that everything that begins to exist has cause. Alternatively, a philosopher might argue strictly on a posteriori grounds and claim that cosmologists misused inductive logic and drew false conclusions about the uncaused character of the singularity from the observational evidence.

In this essay I discuss whether there are sound philosophical arguments for the thesis that the singularity has a cause. My conclusion will be negative. In Sections 2 and 3, I discuss whether or not there is a sound a posteriori argument for such a cause. Weingard has discussed some of the analogies between vacuum black hole singularities and the Big Bang singularity,[3] and in Section 3 I use some of his ideas to show how interesting arguments from analogy can be constructed for the conclusion that the Big Bang has a cause. I shall conclude, however, consistently with Weingard's remarks, that there is no convincing observational evidence that there are vacuum black hole singularities and hence that there is no reason to think the analogical arguments I construct are sound. In Section 4, I discuss William Lane Craig's argument that there is an a priori argument for a cause of the Big Bang based on the synthetic a priori principle that everything that begins to exist has a cause.[4] I shall criticize Craig's Kant-based argument for this principle and conclude that this principle does not belong to the class of synthetic a priori truths.

My concern in this essay is with classical Big Bang cosmology or the so-called standard hot Big Bang model. I shall not discuss the new ideas introduced by string theory,[5] the Hartle–Hawking wave-

---

[3] Weingard, 'Some Philosophical Aspects of Black Holes'.

[4] William Lane Craig, *The* Kalām *Cosmological Argument* (New York: Harper & Row, 1979); 'God, Creation and Mr Davies', *British Journal for the Philosophy of Science*, 37 (1986), 163–75; Essay V.

[5] See e.g. R. Weingard, 'A Philosopher Looks at String Theory', in A. Fine and J. Leplin (eds.), *PSA 1988*, ii (East Lansing, Mich.: Philosophy of Science Association, 1989); and M. Green, J. Schwarz, and E. Witten, *Superstring Theory* (Cambridge: Cambridge University Press, 1987).

function theory,[6] Guth's original inflationary theory,[7] Linde's and Albrecht and Steinhardt's new inflationary theory,[8] Linde's chaotic inflationary theory,[9] Tryon's,[10] Gott's,[11] Zeldovich's,[12] and others' vacuum fluctuation theories, and DeWitt's grafting of Everett's many-worlds interpretation of quantum mechanics on to Big Bang cosmology.[13] These theories are certainly worth discussing on their own merits[14] but discussing them in the present essay would take us too far afield from our restricted topic. My concern is only with whether there is a sound *philosophy of classical Big Bang cosmology* that entails or renders probable that the singularity has a cause. But I should emphasize that classical Big Bang cosmology will probably be replaced some day by a mature version of one of the aforementioned theories, or by some other cosmology based on a quantum theory of gravity, and consequently the conclusions about the universe drawn in this essay that are based on the classical theory should be understood as having a correspondingly provisional character.

An assumption I am making here is that the initial singularity is real rather than a theoretical fiction. Although some[15] have claimed

[6] J. Hartle and S. W. Hawking, 'Wave Function of the Universe', *Physical Review*, D28 (1983), 2960–75.

[7] A. Guth, 'Inflationary Universe: A Possible Solution to the Horizon and Flatness Problems', *Physical Review*, D23 (1981), 347–56.

[8] A. D. Linde, 'A New Inflationary Universe Scenario', *Physical Review Letters*, 108B (1982), 389–93; A. Albrecht and P. Steinhardt, *Physical Review Letters*, 48 (1982), 1220 ff.

[9] A. D. Linde, 'The Inflationary Universe', *Reports on Progress in Physics*, 47 (1984), 925–86.

[10] E. Tryon, 'Is the Universe a Vacuum Fluctuation?', *Nature*, 246 (1973), 396–7.

[11] Guth, 'Inflationary Universe'.

[12] L. Grischak and Y. B. Zeldovich, 'Complete Cosmological Theories', in M. Duff and C. Isham (eds.), *Quantum Structure of Space and Time* (Cambridge: Cambridge University Press, 1982).

[13] B. S. DeWitt, 'The Many-Universes Interpretation of Quantum Mechanics', in B. S. DeWitt and N. Graham (eds.), *The Many-Worlds Interpretation of Quantum Mechanics* (Princeton, NJ: Princeton University Press, 1973).

[14] Some are discussed in Quentin Smith, 'The Anthropic Principle and Many-Worlds Cosmologies', *Australasian Journal of Philosophy*, 63 (1985), 336–48; Quentin Smith, 'World Ensemble Explanations', *Pacific Philosophical Quarterly*, 67 (1986), 73–86; Quentin Smith, 'A Natural Explanation of the Existence and Laws of Our Universe', *Australasian Journal of Philosophy*, 68 (1990), 22–43; and Essay IV.

[15] R. Swinburne, *Space and Time*, 2nd edn. (New York: St Martin's Press, 1981), 254; Essay I, Sect. 2.2.

the singularity is a 'physically impossible' state, I believe there are no good arguments for its physical impossibility and that there are plausible arguments for its reality.[16] If the initial singularity is treated as a theoretical fiction, then the earliest temporal interval of each length is open in the earlier direction. This entails there is no earliest instant and is compatible with the principle that every event has a cause. In this sense, it may be argued that classical Big Bang cosmology is not committed to the thesis that the universe has a beginning and that this beginning is uncaused. However, my question in this essay, 'Did the Big Bang have a cause?', is intended to mean 'Did the Big Bang singularity have a cause?', where the singularity is treated as real.[17]

2. SOME PRIMA-FACIE PROBLEMS WITH A POSTERIORI ARGUMENTS FOR A CAUSE OF THE BIG BANG SINGULARITY

It is by no means obvious that there is a logically coherent a posteriori philosophical argument for a cause of the Big Bang singularity, let alone a sound argument. After all, how could an a posteriori argument possibly reach a different conclusion from Big Bang cosmology, which is ('by definition') the theory that results from empirical investigation of the large-scale structure and the beginning and ending (or beginninglessness and endlessness) of the universe? Since the philosopher does not have in his study a more powerful telescope than the cosmologists, he cannot develop an argument from new or additional observational evidence than that possessed by the cosmologists. Rather, there seems only one route open to the philosopher, namely, an argument that the cosmologists

---

[16] See Essay VII and Essay IX.

[17] It should be added that some attempts have been made within the context of classical Big Bang cosmology to avoid the prediction of a singularity altogether (whether it be treated as real or fictitious). For example, J. D. Bekenstein ('Nonsingular General-Relativistic Cosmologies', *Physical Review*, D11 (1975), 2072–5) has argued that the stress–energy tensor of a classical massive scalar field can be negative and prevent the closed universe singularities. (In this regard, also see G. Smith and R. Weingard, 'Quantum Cosmology and the Beginning of the Universe', *Philosophy of Science*, 57 (1990), 663–7.) R. Ellis (*General Relativity and Gravitation*, 9 (1978), 87–94) has also developed a static universe model that does not have an initial singularity.

have not correctly reasoned from the available observational evidence, that they have not adequately followed the principles of inductive logic or the rules of application of this logic. It might be suggested that this is the line of thinking behind the a posteriori argument of W. H. Newton-Smith that the Big Bang singularity has a cause. He writes:

It appears that the prospects for ever having evidence of a genuine first event are remote. For, supposing that the Big Bang emerged from a singularity of infinite density, it is hard to see what would constitute a reason for denying that that singularity itself emerged from some prior cosmological goings-on. And as we have reasons for supposing that macroscopic events have causal origins, we have reason to suppose that some prior state of the universe led to the production of this particular singularity. So the prospects for ever being warranted in positing a beginning of time are dim.[18]

However, it is not obvious that this passage expresses a *philosophy of Big Bang cosmology* or a *philosophical reinterpretation of the implications of the cosmological data* as distinct from a *misunderstanding of Big Bang cosmology by a philosopher*. Specifically, it is not clear whether Newton-Smith simply misunderstands the concept of a singularity, as this concept is defined in Big Bang cosmology, or whether he is advancing an argument that is intended to qualify or override the conclusions reached in Big Bang cosmology. According to the standard cosmological definition, a state of the universe is a singularity if and only if it is a boundary of spacetime, such that spacetime cannot be extended beyond it. The quoted passage suggests that Newton-Smith thinks that infinite density is the defining property of a singularity, such that it is essential to a singularity to be infinitely dense but accidental to be a boundary of spacetime. If this is in fact Newton-Smith's understanding, it is mistaken, since infinite density is an accidental property of a singularity and being a boundary of spacetime is the essential property. (Of course *physical singularities*, not *coordinate singularities*, are here at issue.) Consider, for example, the following definition of a singularity by Misner, Thorne, and Wheeler (which is based on

[18] W. H. Newton-Smith, *The Structure of Time* (Boston: Routledge & Kegan Paul, 1980), 111.

Schmidt's now standard definition[19]). They ask us to suppose that a spacetime path terminates at a certain point. 'Suppose, further, that it is impossible to extend the spacetime manifold beyond that termination point—e.g. because of infinite curvature there. Then that termination point, together with all adjacent termination points, is called a "singularity".'[20] Notice that Misner *et al.* are not making the claim 'If the point has infinite curvature [or density], it (together with adjacent points of infinite curvature) is a singularity', but the claim 'If spacetime is inextendible beyond the point, for whatever reason, it (together with adjacent termination points) is a singularity'. It may be, as a matter of fact, that all singularities are infinitely dense, but this is not a defining condition of singularities but merely explains why (in the actual cases) that certain points are singular. If Newton-Smith's scenario were true, i.e. if the state of infinite density that obtained about 15 billion years ago was caused by an earlier state of the universe, the infinitely dense state would *not* be a singularity, for spacetime would be extended beyond the point(s) and the point(s) would thereby not comprise a boundary of spacetime. But if this were the case classical Big Bang cosmology would need to be revised or subsumed under some broader theory of the universe.

This last remark suggests a second reading of Newton-Smith's passage, namely, that Newton-Smith is suggesting that (1) the Big Bang singularity is the initial boundary of the universe *as the universe is represented in Big Bang cosmology*, but that (2) there are reasons (based on Newton-Smith's more careful use of inductive principles in reasoning from the observational data) to think that the singularity is not in fact an initial boundary. But it is not obvious that this second reading enables Newton-Smith's argument to be reconstructed in a logically coherent way. One difficulty centres on the sense of 'the universe'. Classical Big Bang cosmology uses a definition of 'the universe' that is based on the following definitions of a spacetime and cosmological model.

(D1) S is a spacetime = df. S is an inextendible, connected, differential manifold that has a smooth, non-degenerate pseudo-Riemannian metric of Lorentz signature $(+, -, \ldots,$

[19] B. G. Schmidt, 'A New Definition of Singular Points in General Relativity', *General Relativity and Gravitation*, 1 (1971), 269–80.
[20] C. Misner, K. Thorne, and J. Wheeler, *Gravitation* (San Francisco: W. H. Freeman, 1973), 934.

−) and that has a continuous, non-vanishing vector field which assigns a timelike vector to every point.

It is often added that S has a global time function all of whose time slices are Cauchy surfaces, but I shall not assume this since this assumption is inconsistent with the thesis that there are white holes and we cannot rule out white holes without begging the question against a certain argument that the Big Bang singularity has a cause, as will appear in Section 3. It perhaps goes without mentioning that GTR allows other sorts of spacetimes than that defined in (D1), but (D1) is the definition thought to be actually instantiated and is the one standardly employed in Big Bang cosmological discussions.

In order to define 'the universe', we also need a definition of a cosmological model:

(D2) C is a cosmological model = df. C is a pair whose first element is a spacetime S and whose second element is a symmetric tensor field of type (2,0).

The definition of the universe operative in Big Bang cosmology is

(D3) U is the universe = df. U is a cosmological model C that is described by Friedman's solutions to the field equations and that satisfies the Hawking–Penrose singularity theorems.

(Note that 'cosmological model' is here used to refer to the physical reality described by a theory rather than to the theory itself.) The Big Bang singularities and other singularities (e.g. black hole singularities) are attached to the spacetime S as its boundaries. The Big Bang singularity is defined as the boundary of the universe in the earlier direction, the beginning-point of the universe.

Given these definitions, if we read Newton Smith's statement 'that some prior state of the universe led to the production of this particular singularity' as involving a use of the word 'universe' in the sense of (D3), we get an implicit contradiction, since the Big Bang singularity is *defined as* the boundary of the universe in the earlier direction. Newton-Smith's statement would reduce, by substitution of synonyms for synonyms, to the explicit contradiction 'The boundary of the universe in its earlier direction is produced by a state of the universe earlier than this boundary'. Thus, if Newton-Smith's passage is to be given a coherent interpretation, he cannot be taken as using 'universe' in the sense of (D3). If there are physical processes earlier than the Big Bang singularity, then these processes are parts of 'the universe' in some novel sense. Now we need not insist that the believer in such earlier processes produce a specific physical theory of these processes or of the spatiotemporal structure they occupy, since all that is necessary for the causal argument to go through is a general definition that encompasses both what went before and what went after the Big Bang singularity. We may use the definition (capitalizing 'Universe' to distinguish it from (D3) and capitalizing 'Spacetime' to distinguish it from (D1)):

(D4) U' is a Universe = df. U' is a Spacetime T occupied by physical processes, such that there is no Spacetime T' that is both continuous with T and contains T as a proper part.

Here 'Spacetime' is taken as undefined, with the only conditions upon it being that spacetime in the sense defined in (D1) is a proper part of it and that its other proper part—the part not defined by (D1)—is earlier than its proper part that is defined by (D1). In order to have a distinct expression for the proper part not defined by (D1), I shall use the term 'spatiotemporal structure'.

It seems logically possible that there is a Universe U', such that the Friedman universe U that begins with the Big Bang singularity and the spatiotemporal structure that precedes the singularity are both parts of U'. About this earlier part of the Universe U', we need not say anything definite at all, except that it is a part of U', is a spatiotemporal structure, is earlier than the singularity, and contains physical processes that cause the singularity. But note that 'inextendible' in (D1) should be taken as implying merely that S is inextendible to a larger spacetime S' that obeys the equations of GTR, such that (D1) allows that it is logically possible that S is extendible to a Spacetime T that contains as its other proper part a spatiotemporal structure that does not obey GTR.

Given these definitions, we may ask if there is a possible reconstruction of Newton-Smith's argument that is a viable argument for the claim that there is some Universe U' containing both the Friedman universe U and a spatiotemporal structure earlier than the singularity that is occupied by physical processes that cause the singularity. If we interpret the passage earlier quoted from Newton-Smith in a sufficiently general fashion, we may extract an argument that avoids the above-mentioned contradiction and has a true observational premiss, namely (1). The argument goes

(1) We have reasons for believing macroscopic events have causes.

Therefore,

(2) We have reasons for believing the Big Bang singularity has a cause.

But this argument is sound only if the suppressed minor premiss 'The Big Bang singularity is a macroscopic event' is true, which it is not. So far from being macroscopic, the singularity is so 'small' it is less than four-dimensional. If the universe is finite, the singularity has zero spatial dimensions and zero temporal dimensions; it is a point of infinitely compressed matter that exists for an instant only. If the universe is infinite, then any given finite volume of space is compressed to a single point at the singularity; in this case, there are an infinite number of points constituting the singularity and it may be considered as one-dimensional, as a 'line' of sorts. Thus, Newton-Smith does not give us a sound a posteriori argument for a cause of the Big Bang singularity. These considerations suggest that a different and more cosmologically sophisticated approach must be taken if we are to have any hope of constructing a sound a posteriori philosophical argument for a cause of the singularity.

3. SOME SOPHISTICATED A POSTERIORI ARGUMENTS FOR A CAUSE OF THE BIG BANG SINGULARITY

A minimal condition of a sophisticated a posteriori philosophical argument will be that it refer the Big Bang singularity to a class to which it actually belongs and which is relevant to the singularity's possession or non-possession of the property of having a cause. This reference class is obviously not the class of macroscopic events since the singularity is not a macroscopic event, but it is not the class of microscopic events either, since the singularity is not a microscopic event. Instead, the singularity is a boundary of the manifold of events and as such the reference class should consist only of singularities. One way to construct an a posteriori argument based on the class of singularities is an argument from analogy, which involves at least four formal concepts:

(x)   The individual x that is the object of our investigation.

(F)   The reference property F, which is the property that is possessed by all and only members of a reference class to which *x* belongs.

(*R*)   The reference class *R* whose members possess *F*.

(*G*)   The sample property *G*, which is the property whose possession or non-possession by *x* we are endeavouring to discover.

(*S*)   The sample class *S*, which is the class of all and only those members of the reference class *R* of which we know whether or not they possess the sample property *G*.

(*R* − *S*) The remainder class, which is the class of the members of the reference class *R* that is left over when the sample class *S* is subtracted from *R*. The remainder class consists at least of *x* and possibly only of *x*.

Now for an argument from analogy to be sound it must not only have true premisses and be logically correct but also satisfy the rules of application of inductive logic, most notably the requirement of total relevant evidence. In arguments from analogy, the total relevant evidence includes all the relevant similarities and dissimilarities between the entities in the sample class *S* and the individual *x*. An argument from analogy is sound if it has true premisses, is logically correct, and satisfies this rule of application:

(3) The members of the sample class *S* are more relevantly similar to than different from the individual *x*.

Suppose we let the individual *x* be the Big Bang singularity, the reference class *R* the class of all singularities, the sample property *G* the property of having a cause, the sample class *S* the class consisting of all the singularities of which we know whether or not they have a cause, and the remainder class *R* − *S* the class consisting of all the singularities of which we do not know whether or not they have a cause. Consider this simplified argument from analogy.

(4) Every member of the class of singularities of which we know whether or not it has a cause has a cause.

(5) The Big Bang singularity is a member of the class of singularities.

Therefore, it is more probable than not that

(6) The Big Bang singularity has a cause.

The class of singularities, it might be argued, includes at least (*a*) Big Bang singularities, (*b*) big crunch singularities, (*c*) Schwarzschild gravitational collapse black hole singularities (*d*) Schwarzschild vacuum black hole singularities, (*e*) Reissner–Nordstrom gravitational collapse black hole singularities, (*f*) Reissner–Nordstrom vacuum black hole singularities, (*g*) Kerr gravitational collapse black hole singularities, and (*h*) Kerr vacuum black hole singularities. It might be argued that we know of each of these singularities, except for (*a*), that it has a cause, and since each of these singularities is more similar to than different from the Big Bang singularity, it follows that it is more probable than not that the Big Bang singularity possesses the sample property *having a cause*.

This argument, however, needs some serious modifications if it is to have any hope of getting off the ground. To begin with, we need to make explicit something I have been assuming, namely, that our talk of a singularity as having a cause requires that the concept of causality used in GTR be replaced by a broader conception of causality, for GTR recognizes as causes and effects only events that are parts of the four-dimensional spacetime manifold and singularities are boundaries rather than parts of the manifold.[21] Secondly, we need to narrow our reference class *R* since as it stands the rule of application (3) is not satisfied; that is, it is not the case that the members of the sample class *S* are more similar to than different from the individual *x*, the Big Bang singularity. Our sample class *S* includes big crunch and gravitational collapse black hole singularities and these singularities are more different from than similar to the Big Bang singularity, as will become implicitly clear in the following. We need to exclude these singularities from our reference class. The narrowing of our reference class requires a more detailed exploration of black hole singularities of the Schwarzschild, Reissner–Nordstrom, and Kerr types.

The geometry of a Schwarzschild black hole is

$$ds^2 = -(1 - 2M/r)dt^2 + \frac{dr^2}{1-2M/r} + r^2(d\theta^2 + \sin^2\theta d\phi^2),$$

where *M* is the mass of the body measured in geometric units, *r* the radial coordinate, and θ and ϕ the angular coordinates. This is the

---

[21] Accordingly, I am not using the definitions of 'cause' and 'effect' employed in Smith, 'A Natural Explanation of the Existence and Laws of Our Universe'. Furthermore, I depart from that paper by adopting a broader definition of 'time' that allows me to say that the singularities are temporally related to the events in the spacetime manifold. There is no inconsistency between the two papers: only a use of different definitions. On this point, see 'A Natural Explanation of the Existence and Laws of Our Universe', 38.

geometry of a spherical, non-rotating, and uncharged gravitating body, which collapses to a singularity when $r = 0$. But this Schwarzschild black hole singularity should not be included in our reference class, since this singularity is an *end-point* of spacetime curves in the Friedman universe U and therefore is not sufficiently similar to the Big Bang singularity to give us a sound argument from analogy. This black hole singularity possesses the properties

(I) being an end-point of timelike and null curves in U,
(J) being an effect of a body's B gravitational collapse, such that B is a part of U,

whereas the Big Bang singularity does not possess (I) or (J) but instead possesses

(K) being a beginning-point of timelike and null curves.

Accordingly, the Schwarzschild gravitational collapse singularity is not sufficiently similar to the Big Bang singularity to support a causal argument from analogy.

Our best hope is to follow the clue offered by Robert Weingard about the relation between the Schwarzschild black hole singularities and the initial Big Bang singularity and the final big crunch singularity:

Now it can be pointed out that while in the case of the Robertson–Walker final singularity, there is a corresponding Schwarzschild singularity that is connected with the collapse of a gravitating body, *the Schwarzschild singularity corresponding to the Robertson–Walker initial singularity is within a vacuum black hole—one that does not arise from gravitational collapse.*[22]

Thus, it is to the Schwarzschild vacuum black hole that we should look if we want to discover an analogy with the initial singularity. The vacuum black hole is described by taking the limit of the above-stated line element in which the radius of the gravitating body is set to 0. This black hole may be described in terms of two distinct asymptotically flat regions of a spacetime, each of which contains a singularity. These singularities connect and form a worm-hole that expands to maximum radius of $4\pi M$ and maximum surface area $16\pi M^2$ and then contracts and pinches off, leaving two disconnected singularities again. World lines enter the spacetime regions from the original singularities and world lines from these

[22] Weingard, 'Some Philosophical Aspects of Black Holes', 198, my italics.

regions crash into the singularities that exist after the pinching-off of the worm-hole. The singularities that exist prior to the pinching-off of the worm-hole are white hole singularities, singularities that (like the Big Bang singularity) emit rather than absorb particles; they have the property K but not I. These white hole singularities are further analogous to the Big Bang singularity in that the particle creation associated with them occurs in the highly curved spacetime in the region of the singularity.[23] These white hole singularities may be included in a reference class that is narrower than the one mentioned in (5) and thus may provide a stronger analogical argument for (6).

Suppose, then, that our reference class includes only these Schwarzschild white hole singularities and the Big Bang singularity. Our analogical argument will succeed if (*a*) the sample class includes all and only the Schwarzschild white hole singularities, (*b*) we know of each of these singularities that it has a cause, and (*c*) the set of the other relevant properties of the white hole singularities is more similar than different to the set of relevant properties of the Big Bang singularity. These three conditions would justify the conclusion that it is more probable than not that the Big Bang singularity has a cause.

But there is an apparent difficulty in the way of such an argument. The Schwarzschild white holes do not arise from a gravitational collapse but are built into the initial conditions of the universe. Misner, Thorne, and Wheeler observe that the Schwarzschild vacuum black hole singularities 'can exist only if the expanding universe . . . was "born" with the necessary initial conditions—with "$r = 0$" Schwarzschild singularities ready and waiting to blossom forth into wormholes'.[24] The Schwarzschild white hole singularities are retarded pieces of the Big Bang singularity and abide in a quiescent state until an unpredictable time when worm-holes arise from them. The problem this poses is that the Schwarzschild white hole singularities, *qua (retarded) parts of the Big Bang singularity, are no more effects of a causal process than the Big Bang singularity itself.* We will have reason to believe the white singularities have causes only if we already have reason to believe the Big Bang singularity has a cause, which is inimical to the

---

[23] For a development of this point, see ibid. 202–6.
[24] Misner *et al.*, *Gravitation*, 842–3.

prospects of a sound analogical argument. An independent reason to reject this argument is that the sample class is empty; there is no observational evidence whatsoever that there are Schwarzschild white hole singularities.

However, there are two other main classes of black holes, Reissner–Nordstrom black holes (which are electrically charged) and Kerr black holes (which have angular momentum), and these may provide us with a suitable sample class. Reissner–Nordstrom black holes have an electrically charged, non-rotating, and spherically symmetric gravitating body which occupies a spacetime structure with the geometry (where Q is the electric charge):

$$ds^2 = - (1 - 2M/r + Q^2/r^2)dt^2 + (1 - 2M/r + Q^2/r^2)^{-1}dr^2 + r^2(d\theta^2 + \sin^2\theta d\phi^2).$$

If this describes a gravitating body, then we will have only a singularity that is an end-point of curves and not a beginning-point. But let us take the limit of this solution in which the radius of the charged gravitating body is set to 0. This gives us a vacuum black hole, which has a worm-hole to another universe through which a particle can travel (unlike the case with the Schwarzschild worm-hole) and also a timelike singularity that connects the two universes. For an observer emerging into one of these universes from the worm-hole, the singularity will appear as a white hole singularity, a beginning-point of curves, and thus will possess the property K also possessed by the Big Bang singularity and the Schwarzschild white hole singularity.

But Reissner–Nordstrom white hole singularities are less analogous to the Big Bang singularity than the Schwarzschild white hole singularity since the Reissner–Nordstrom singularities are timelike singularities and the Big Bang and Schwarzschild singularities are spacelike singularities. This would provide a very weak causal argument from analogy. More importantly, there is simply no observational evidence that there are any Reissner–Nordstrom white hole singularities and so once again we are left with an empty sample class.

The only other type of black hole singularity besides the Schwarzschild and Reissner–Nordstrom singularities is the Kerr black hole singularity. The Kerr metric is

$$ds^2 = dt^2 - p^2/dr^2 - p^2 d\theta^2 - (r^2 + a^2)\sin^2\theta d\phi^2 - 2GMr/p^2(dt - a\sin^2\theta d\phi)^2,$$

where $p^2 = r^2 + a\cos^2\theta$, $\Delta = r^2 - 2GMr + a^2$, G = gravitational constant.[25] A Kerr black hole is rotating and has a singularity at $r = 0$. If we apply the metric to a vacuum black hole, we will have at $r = 0$ a disc singularity (alternatively, one can interpret the Kerr black hole in terms of closed timelike curves, as Weingard explains,[26] but we shall not take this route here). The edge of the disc is a singularity where curvature is infinite and the rest of the disc is an *extendible* edge of spacetime, i.e. an end of world lines that can be extended to a larger spacetime. This will violate cosmic completeness (the principle that a universe has an inextendible spacetime) but this violation will occur only with the horizon of the Kerr black hole (by the cosmic censorship principle). The white hole singularity will be the edge of the disc as considered from the viewpoint of the new region of spacetime into which the world lines can be extended. Thus, we again have a possible class of caused singularities that have the property K and are partially analogous to the Big Bang singularity.

But the problem once again is that there simply are no Kerr white hole singularities to give us a non-empty sample class. But the situation is not as dismal as with the Schwarzschild and Reissner–Nordstrom cases, since there are gravitational collapse Kerr black hole singularities (in fact, all actually extant gravitational collapse black hole singularities are of the Kerr type) and it is possible that the Kerr vacuum solution *partially describes* the Kerr gravitational collapse black holes. Specifically, it is possible that the vacuum solutions apply to the extent that they allow the edge of the disc formed from the collapsed star to be a white hole singularity as seen from the perspective of the spacetime into which the world lines could be extended. This would give us a non-empty sample class of white hole singularities. Unfortunately, however, this is not the actual situation. Whereas the Kerr gravitational collapse solutions describe the exterior geometry of all actual gravitational collapse black holes, the interior geometry (which I was suggesting might be partially described by the Kerr vacuum solutions) *is not* in fact described by the Kerr vacuum solutions but is more accurately

---

[25] For more details, see Weingard, 'Some Philosophical Aspects of Black Holes', 217, and Misner *et al.*, *Gravitation*, 877.

[26] Weingard, 'Some Philosophical Aspects of Black Holes', 207–12; R. Weingard, 'General Relativity and the Conceivability of Time Travel', *Philosophy of Science*, 46 (1979), 331–2.

described by the Schwarzschild gravitational collapse solutions, where the infalling matter hits a spacelike singularity and is there crushed out of existence (with no white holes formed).[27]

The upshot of these various considerations is that the prospects are exceedingly dim for an a posteriori argument for a cause of the Big Bang singularity. In order to obtain a sample class sufficiently analogous to the Big Bang singularity, we needed to exclude all singularities but white hole singularities. But whereas there are a variety of *types* of white holes that may be similar enough to the Big Bang singularity to warrant a causal argument from analogy, the sad fact is that there *are* no instances of these types (as far as current observational evidence suggests). Consequently, in direct opposition to Newton-Smith's conclusion ('it is hard to see what would constitute a reason for denying that that [the Big Bang] singularity itself emerged from some prior cosmological goings-on'), we must draw the inference that *it is hard to see what would constitute a reason for affirming that that Big Bang singularity itself emerged from some prior cosmological goings-on.*

4. A PRIORI ARGUMENTS FOR A CAUSE OF THE BIG BANG

The obvious candidate for an a priori argument for a cause of the Big Bang would be based on a premiss that a relevant principle of causality is a necessary truth, specifically, a *synthetically necessary a priori truth*. The leading proponent of such an argument is William Lane Craig;[28] he maintains that it is a 'synthetic a priori proposition'[29] that 'everything that begins to exist has a cause of its existence'.[30] This provides the premiss

(7) Everything that begins to exist is caused to begin to exist,

which, along with the premiss that

---

[27] For more details, see R. Wald, *Space, Time and Gravity* (Chicago: University of Chicago Press, 1977), 78–91.

[28] See Craig, *The* Kalām *Cosmological Argument*; Craig, 'God, Creation and Mr Davies'. Craig also argues for the further thesis that the cause is God, but I shall not address this theistic issue here. I have discussed the connection between theism and Big Bang cosmology in Essays VII and IX.

[29] Essay I, Sect. 3.

[30] Ibid.

(8) The Big Bang singularity is the beginning of the existence of the universe U,

entails

(9) The Big Bang singularity has a cause.

In Essay IV I argued that (7) is false since it is counter-exampled by virtual particles, which come into existence without a cause. I stated that this counter-argument against Craig is sound if 'cause' is defined in terms of predictability in principle, i.e. if causality is understood 'in terms of a law enabling single predictions to be deduced, precise predictions of individuals events or states'.[31] This definition of causality is a version of the 'sufficient condition' definition of causality, where something c is a cause of something e only if c is a sufficient condition of e. In his response,[32] Craig concedes that virtual particles are uncaused in this sense and correctly points out that they have causes in a different sense of 'cause', that associated with the notion of a physically necessary condition. A quantum vacuum is a physically necessary condition of a virtual particle coming into existence and, in this 'physically necessary' sense of causation, virtual particles may be said to have causes. A probabilistic definition of causality would also enable us to say that virtual particles have causes, for given a quantum vacuum there is a certain probability that virtual particles will be emitted by it.

But Craig denies that the sufficient condition definition of causality is applicable when dealing with the cause of the universe's coming into existence. He says 'since according to classical theism creation is a freely willed act of God, it follows necessarily that the beginning and structure of the universe were in principle unpredictable even though they were caused by God'.[33] It is not obvious that this is the case. Strictly speaking it is not God, but God's willing the universe to come into existence, that causes the universe to exist. (Events, not things, are causes.) But surely God's willing the universe to come into existence is a sufficient condition of it coming into existence. And surely the beginning and structure of the universe are in principle predictable in this case. Consider the argument

(10) God wills that the universe begin to exist with the structure S.

---

[31] Essay IV, Sect. 3.   [32] Essay V, Sect. 1.   [33] Ibid.

Therefore,

> (11) The probability is one that the universe will begin to exist with the structure S.

(10)–(11) is a paradigm of a valid argument and this shows that this conception of causality is well suited to a discussion of classical theism and the beginning of the universe. Note that the event of God's willing the universe to begin to exist is in principle unpredictable, but that does not entail that the universe's beginning to exist cannot be predicted with a probability of one on the basis of the premiss (10) about its cause. Analogously, the fact that God's willing does not have a sufficient condition of its occurrence does not entail that the universe's beginning of existence does not have a sufficient condition (namely, God's willing).

In any case, it is best to have as broad as possible a definition of causality in order not to rule out certain possibilities. Craig's reliance on the non-logical necessary condition definition of causality is inadequate since we could then not count a sufficient condition of something as a cause if it were not a non-logical necessary condition. (Surely, my kicking a ball is the cause of it moving; but my kicking is a sufficient condition of it moving and is not a non-logical necessary condition of its moving—the ball could move even if I did not kick it, e.g. if there were an earthquake that moved it.) The probabilistic definition of causality is the broadest and we may operate with it, since it covers both the sufficient causal conditions (which have a probability of one) and the relevant non-logical necessary conditions. The following is a probabilistic definition of causality that is adequate for our purposes.

In stating this definition, I will use the upper-case letters C and E to refer to event-types (the cause and effect) and let the lower-case letters c and e refer to an instance of C and E respectively. (The beginning to exist of a singularity counts as an event. This is true in the broad philosophical sense of 'event'. In the narrow GTR sense, an 'event' is an element in the four-dimensional manifold.) An event c is a probabilistic cause of an event e if and only if the following five conditions are met:

> (*a*) The events c and e both occur.
> (*b*) $P(E/C) = 0 < n \leqslant 1$; that is, given C, the probability that E occurs is greater than zero and either equal to one (determining causes are limiting cases of probabilistic causes) or less than one.

(c) $P(E/C) > P(E/-C)$; that is, that E is more likely to occur if C occurs than if C does not occur.

(d) A screening-off condition, which deals with such problems as 'the problem of the common cause'. Suppose that E and F are both effects of C and only C and that F invariably precedes E and succeeds C when E is caused by C. In this case, $P(E/F) > P(E/-F)$, even though F is not a cause of E. F needs to be screened off from E by a condition that includes the formula $P(E/CF) = P(E/C)$; if the probability that E occurs given C and F is the same as the probability that E occurs given only C, and F is later than C, then F is not a cause of E. Thus, a probabilistic cause must meet the fourth condition that it cannot be represented by F in the formula $P(E/CF) = P(E/C)$, where F is later than C.

(e) A condition is needed to deal with Simpson's paradox: A cause C of E ought to increase the probability of E but this increase in probability may not show up if there is another causal factor G such that G is correlated with C and interferes with C in such a way that in the presence of C the probability of E is decreased or not increased. The fifth condition is that C satisfies the formula $P(E/C) > P(E/-C)$ in all situations in which correlated interfering causes are absent.

We shall examine the question if there is any reason to think that (7) is a synthetic a priori truth, where 'cause' is interpreted probabilistically in correspondence with the above definition.

Some philosophers reject the category of synthetic a priori truths, but I shall assume here there are some instances of this category. It is arguable, for example, that the proposition *no body is simultaneously both red all over and green all over* is a synthetic a priori truth. This proposition is synthetic in that the concept of not-red is not a proper part of the concept of green and the concept of not-green is not a proper part of the concept of red. This proposition is a priori in that it can be known independently of experience. The concept of *knowable independently of experience* is defined as follows:

(KIE) p can be known independently of experience = df. For each possible finite mind x, and for each possible world W in which x exists, p is true and x has the concepts in p, x has enough experience to come to know p simply by virtue of understanding p or simply by virtue of reasoning about p.

If x knows p simply by virtue of understanding p, then p is self-evident; p cannot be conceived to be false by anybody who understands p. In this case, p is known immediately. If x knows p by virtue of reasoning about p, then p is known by virtue of being deduced from propositions that are known to be true or by being an ineliminable premiss in the proof of a proposition that is known to be true. In this case, p is known mediately.

I shall first consider whether our causal proposition

(7) Everything that begins to exist is (probabilistically) caused to begin to exist

is self-evident. Some synthetic a priori propositions are self-evident; one example is the proposition *no body is simultaneously both red all over and green all over*. But it seems clear that our causal proposition (7) is not self-evident. I can conceive of something beginning to exist without a cause. A case in point is the universe. I can conceive of the universe existing at a certain time t (say at the time of the Big Bang singularity), such that (*a*) there is no time earlier than t, (*b*) nothing else exists at t, (*c*) nothing timelessly exists that causes the universe to begin to exist, and (*d*) there are no closed timelike curves whereby 'later' parts of the universe cause the universe to exist at t.

Craig claims there is a 'metaphysical intuition that something cannot come out of nothing'.[34] He writes, 'if nothing existed—no matter, no energy, no space, no time, no deity—if there were absolutely nothing, then it seems unintelligible to say that something should spring into existence'.[35] But clearly this *is* intelligible. It is quite easy to understand the above proposition that the Big Bang singularity exists at t in conjunction with conditions (*a*)–(*d*). This is hardly like trying to understand the proposition *two plus two equals fifteen* or the nonsensical sequence *that orange below for monkey*. I think Craig's real point is better stated in his earlier book *The* Kalām *Cosmological Argument*, where he writes that the causal proposition 'is so intuitively obvious, especially when applied to the universe, that probably no one in his right mind *really* believes it to be false'.[36] His point here seems to be that this is self-evident, that it cannot be conceived to be false. But this claim itself seems obviously false; I find it quite easy to conceive of the universe beginning to exist without a cause. Given classical Big Bang

[34] Ibid.   [35] Ibid.   [36] Essay I, Sect. 3.

cosmology, it is easy to entertain as true the proposition that the Big Bang singularity exists in conjunction with conditions $(a)-(d)$. I find this uncaused beginning astonishing, amazing, 'mind-boggling', and utterly awesome, but that is different from saying I cannot conceive it to be the case.

Craig writes in Essay V that 'It does seem to me that only an aversion to the theism implied by the principle in the present context would lead the empiricist to think that the denial of the [causal] principle is more plausible than the principle itself.'[37] I have three responses to this.

First, the aversion to theism may be rational. For example, it may be that the atheist finds the argument against God's existence based on the occurrence of gratuitous evil to be more convincing than the causal principle. The atheist may contemplate some grotesque natural evil and consider which is more plausible, that God created a universe with this natural evil or that the causal principle is not true, and conclude that it is more plausible that the causal principle is not true.

Secondly, if Craig means an 'emotional aversion' to theism, then his method of argument can cut both ways. For the atheist may simply respond that Craig's adherence to the causal principle is based more on his emotional need to believe in a god than upon his rational faculties.

Thirdly, it is simply false that a rational or emotional aversion to theism is the central issue for a person who rejects the causal principle in the context of discussions of the Big Bang singularity. If I found the causal principle to be self-evident, I need not find myself required to accept theism. I could easily assume the Big Bang singularity has some other cause than a divine one. I could adopt on a priori grounds Newton-Smith's theory that it has some prior physical cause and avoid the obligation to construct some a posteriori argument for this thesis. Alternatively, I could adopt John Leslie's theory that the value of goodness created the singularity[38] or Thomas Nagel's suggestion that something humanly inconceivable caused the singularity.[39]

I think Craig's own remarks suggest that he himself finds an

---

[37] Essay V, n. 13.
[38] John Leslie, *Universes* (London: Routledge, 1989).
[39] Thomas Nagel, *The View from Nowhere* (New York: Oxford University Press, 1986).

uncaused beginning conceivable. He writes that 'We can in our mind's eye picture the universe springing into existence uncaused, but the fact that we can construct and label such a mental picture does not mean the origin of our universe could have really come about in this way ... Just because we can imagine something's beginning to exist without a cause it does not mean this could ever occur in reality.'[40] I believe that Craig is conceiving this possibility when he is constructing this mental picture and thus that his own remarks count against his claims of inconceivability.

But if we accept this is imaginable and conceivable, what right do we have to dismiss this as a reason for believing this could occur in reality? '*p* is imaginable or conceivable provides no reason to think that *p* is possibly true' is not in every case a valid argument. I can conceive and imagine that a disembodied mind exists. Is not my ability to conceive and imagine this a ground for believing that it is possible for a god or angel to exist?

Recent discussions of necessary a posteriori truths may seem to show that conceivability is no reason to think something is possible. For example, Putnam writes that 'it is conceivable that water isn't H_2O. It is conceivable but it isn't logically possible. Conceivability is no proof of logical possibility ... In [other] terminology, the point just made can be restated as: a statement can be (meta-physically) necessary and epistemically contingent.'[41] I believe these remarks are sound if they are interpreted in the following way (which does not pretend to be faithful to Putnam's own questionable theory). The remarks concern necessary a posteriori truths. A sentence (as used in the actual world) is a necessary a posteriori truth if and only if it actually expresses a necessarily true proposition but does not express a true proposition in some other possible world *W*, despite the fact that in *W* it obeys the same rule of use as it does in the actual world. The rule of use of 'Water is H_2O' is partly specified by the rule that 'water' refers directly to whatever exemplifies the microproperties that are causally responsible for the observable properties of being a colourless, odourless, tasteless, thirst-quenching liquid. Since in the actual world H_2O exemplifies these microproperties, 'Water is H_2O' actually expresses the necessarily true proposition *that H_2O is H_2O*. But in some other possible

[40] Essay I, Sect. 3.
[41] H. Putnam, *Philosophical Papers*, ii (Cambridge: Cambridge University Press, 1975), 233.

world *W*, something else exemplifies certain microproperties that are causally responsible for the observable properties of being a colourless, odourless, etc. liquid. Call this other substance *XYZ*. In *W*, the sentence 'Water is H_2O' obeys the same rule of use it actually obeys and yet expresses the necessarily false proposition *that XYZ is H_2O*. The sense in which I can conceive that water isn't H_2O is that I can conceive that 'Water is H_2O' expresses *that XYZ is H_2O* in the world *W*. The sense in which this sentence is an a posteriori necessary truth is that it requires empirical investigation of the actual world to determine whether the direct referent of 'water' is H_2O or some other substance.

Is there a parallel argument regarding the conceivability of the causal proposition? I think not, for the sentence 'Everything that begins to exist is caused to begin to exist' does not contain an indexical component (in the broad sense in which 'water' is indexical) that enables this sentence to express different propositions in different possible worlds (while obeying its actual rule of usage). This sentence expresses the same proposition in every world in which it obeys its actual rule of usage. Thus, there is no argument similar to Putnam's by which Craig can claim that the conceivability of an uncaused beginning provides no reason to believe this is metaphysically possible.

5. SOME KANTIAN A PRIORI ARGUMENTS FOR A CAUSE OF THE BIG BANG

These remarks suggest that if the only putative evidence that the causal proposition (7) is a necessary truth is self-evidence, then there is no evidence that it is a necessary truth and thus that there is no reason to believe that it is metaphysically impossible that the Big Bang singularity occurred uncaused. However, it is arguable that the causal proposition (7) is a *mediate synthetic a priori truth*, i.e. one that is not self-evident but which can be shown to be necessarily true by virtue of its deductive connections with other necessary truths. Craig wants his defence of the causal proposition to be thorough, so in his *The* Kalām *Cosmological Argument* he offers a Kantian-style argument for the causal proposition. According to Craig, the Kantian or neo-Kantian philosopher 'defends the validity of the causal proposition as the expression of the operation of a mental a priori category of causality which the mind brings to experience'.[42]

But if Kant's theory is to be introduced to defend the synthetic a priori status of the law of causality, Kant's theory must first be rectified in several respects, for it is unacceptable as it stands. For one thing, his theory implies the contradiction that the category of causality is inapplicable to the noumenal causes of phenomena. Secondly, Kant's causal principle is that every alteration of substance has a cause, whereas Craig's principle (7) is something Kant rejected outright. Kant claimed that it is impossible for anything to begin to exist and therefore that it is impossible for something to be caused to begin to exist; for Kant, every change is an alteration in the state of some permanent substance.[43] Thus, our second modification must be that the Kantian-type arguments be used to support (7) rather than Kant's own principle. Thirdly, Kant assumes that the a priori causal law is determinist and we are supposing the causal law is probabilistic. Fourthly, his idealism is inconsistent with scientific realism (i.e. with the view that there exist some imperceptible physical entities that are not dependent on the human mind), which is the better justified position (in my judgement[44]). I believe all these required modifications to Kant's theory would be acceptable to Craig (indeed, he makes some of them himself). But if Kant's theory is modified to deal with these problems, can it still be used to demonstrate that (7) is a synthetic a priori proposition? That is, can we soundly argue in a Kantian fashion that 'the causal principle (7) is a condition of the possibility of experience and therefore a synthetic a priori proposition'?

In order to answer these questions, we must first make more precise what is meant by 'experience' in a Kantian context. By 'experience' Kant means an objectifying interpretation of our representations in which 'we posit an object for our representations', such that our representations are apprehended 'not in so far as they are (as representations) objects, but only in so far as they stand for an object'.[45] Without this objectifying interpretation, 'We should

---

[42] Essay I, Sect. 3.

[43] I. Kant, *Critique of Pure Reason*, trans. N. Kemp-Smith (New York: Macmillan, 1929), A182/B225–A189/B232.

[44] Quentin Smith, *The Felt Meanings of the World: A Metaphysics of Feeling* (West Lafayette, Ind.: Purdue University Press, 1986).

[45] Kant, *Critique of Pure Reason*, A190/B235.

then have only a play of representations, relating to no object.'[46]
Let us call any synthetic proposition that is presupposed by our
objectifying interpretations of our representations a *transcendental
proposition*, such that the following definition holds:

(D5) *P* is a transcendental proposition = df. *P* is synthetic and
it is necessarily true that for any person *x*, if *x* objectively
interprets his representations, *x* is committed (at least
implicitly) to *P*.

(D5) is not stated by Kant but I think it captures a clear and
coherent part of his theory and is consistent with the above-
mentioned changes made to his theory. Let us suppose, for the
sake of argument, that the causal proposition (7) is a transcendental
proposition. Does it follow that (7) is synthetic and a priori? I think
it is clear that if synthetic a priori propositions are a species of
necessarily true propositions (propositions true in all possible worlds),
it follows that '*P* is a transcendental proposition' does not entail
'*P* is a synthetic a priori proposition'. For some transcendental
propositions are false in some possible worlds. For example,
the proposition that *there are objectifying interpretations* is a transcendental proposition, since no person can objectively interpret his
representations without being committed, at least implicitly, to this
proposition. Yet this proposition is not necessarily true, since it is
false in all the possible worlds that include only matter or only
representations that are not objectively interpreted. It follows,
therefore, that if (7) is transcendentally true, that does not show
that the Big Bang singularity cannot occur uncaused.

However, if (7) is merely a transcendental proposition, it still can
be used to block the argument that the Big Bang singularity is
actually uncaused. If (7) is transcendental, the proponent of an
uncaused Big Bang is committed to (7), since this proponent objectively interprets his representations. Accordingly, if somebody
maintains that the Big Bang is uncaused, he is contradicting
himself, since he is also committed to (7), which implies the Big
Bang has a cause. Consequently, it is encumbent upon a proponent
of the Big Bang cosmological theory (which entails the Big Bang
is actually uncaused) to show that (7) is not a transcendental
proposition.

[46] Ibid. A194/B239.

Kant's argument that his causal principle is transcendental is that it is required to distinguish the subjective succession of representations from the objective succession of events in the world. His basic assumption is that

> (12) Our representations must be objectively interpreted in such a way that every event is understood as having a definite and determinate position in an objective temporal order.

Kant believes this interpretation involves viewing events as following from one another in accordance with a rule. If we assume the rule is probabilist (an event of kind *y* follows with some degree of probability upon an event of kind *x*) and that it pertains to events that are beginnings of existence as well as to events that are changes of state (since both kinds of events make up the objective succession in the world), we may state the following general principle (which entails that (7) is a transcendental causal proposition):

> (13) Necessarily, for any person *x*, if *x* objectively interprets his representations, *x* is committed (at least implicitly) to the proposition that every event has a probabilist cause.

I believe, however, that (13) is false since the subjective succession of representations can be distinguished from the objective succession of events, consistently with assumption (12), if one allows there to be a first uncaused event. Consider this Kantian argument:

> (*a*) Let us suppose that there is nothing antecedent to an event, upon which it must follow according to a rule. (*b*) All succession of perception would then be only in the apprehension, that is, would be merely subjective, and would never enable us to determine objectively which perceptions are those that really precede and which are those that follow.[47]

We need not question the determinism implied by the 'must follow' or the idealism implied by the designation of objective events as 'perceptions' in order to see that this argument does not establish that there can be no uncaused first event. The inference from (*a*) to (*b*) is valid only if 'an event' in (*a*) means *any event*. If 'an event' in (*a*) means *some event*, the inference is invalid, for if there is an earliest uncaused event, and all later events are caused, then the succession of perceptions that relates to the later events would not

---

[47] *Ibid.* A194/B239. The lettering is mine.

be merely subjective. The later events would succeed one another in accordance with a causal rule and thereby their succession would be objective. The perception of the first event would also not be purely subjective, for this first event would be related by a causal rule to the second event and thereby would belong to the objective order. The first event x_1 might be related to the second event y_1 by the rule *an event of kind y follows from an event of kind x*. The fact that x_1 is a cause of a later event suffices to bring it into the objective succession; it need not in addition to this be an effect of something else. Therefore, (13) is false and (7) is not a transcendental proposition.

The above argument may be considered as a putative proof that (7) is a transcendental proposition, based on the premisses (12) and (13). However, if the causal proposition is not self-evident and has no proof, it does not follow that there is no a priori reason to believe it true. Some necessarily true a priori propositions are not self-evident and have no proof. One example is the Axiom of Choice or the Multiplication Axiom,

> (14) Given any class of mutually exclusive classes, of which none is null, there is at least one class which has exactly one term in common with each of the given classes.

(14) is commonly regarded as proposition that is not self-evident. Furthermore, (14) has no proof. But it can be known a priori since it is an ineliminable part of the proof of some self-evident proposition, namely,

> (15) The sum of v mutually exclusive classes, each having u terms, must have $v \times u$ terms.

The reason we introduce the condition of *being an ineliminable part of the proof of the proposition* is that not every premiss in a proof of some proposition is required if the proposition is to be proven; for example, Gödel's incompleteness theorem can be proven from ω-consistency but the premiss of ω-consistency is not a part of the simplest proof of this theorem and therefore is not required to prove it.

It is open to the defender of the transcendental necessity of the causal principle that it is an ineliminable part of a proof of some self-evident and transcendentally necessary proposition. If we resort to Kantian-style arguments, we may try and use a relevant transcendental causal proposition as a premiss in such a proof. There seems to be a proof suggested by Kant's First Analogy that uses the premiss

(16) Our representations must be objectively interpreted in such a way that no event is understood as an uncaused beginning of the universe.

It might be contended that Kant successfully uses (16) in a proof of the principle that

(17) It is a necessary condition of the possibility of experience that each time be perceptible and each time is perceptible only through the perceptibility of some occupant of that time.

(17) is not compelling but I will waive this objection and pretend, for the sake of argument, that it is true a priori and is self-evident. I think it can be shown that (16) is not a part of any proof of (17). If the universe began uncaused, its beginning would be an *absolute beginning* (i.e. the coming into existence of a substance and not merely an alteration of its state). Kant says of absolute beginnings that:

If we assume that something absolutely begins to be, we must have a point of time in which it was not. But to what are we to attach this point, if not to that which already exists? For a preceding empty time is not an object of perception. [Therefore, nothing can absolutely begin to be after a period of empty time.][48]

Let us grant Kant the following assumptions:

(18) An empty (unoccupied) time is not an object of perception.
(19) If we interpret something as absolutely beginning to be after a point of time at which it is not, this preceding time must be perceptible.
(20) If such a preceding time is perceptible, it must be 'attached to' (i.e. perceived in terms of) some object that occupies that time.

These three assumptions entail

(21) It is impossible that we interpret something as absolutely beginning to be after a period of empty time,

which in turn entails

[48] Ibid. A188/B231.

(22) It is impossible that we interpret the universe as beginning to be without any cause after a period of empty time.

However, even if (21) is thereby a part of a proof of (17), that hardly entails that (16) is part of such a proof, for the negation of (16) is consistent with (21). For it is the case that

(23) It is possible that we interpret the universe as beginning to be without any cause at t, such that t is the earliest time.

It might be objected that (23) is false since Kant demonstrated that we cannot interpret time as having a beginning. But I concede only that Kant argued this, since his argument is plainly unsound. His argument goes: 'The infinitude of time signifies nothing more than that every determinate magnitude of time is possible only through limitations of one single time that underlies it. The original representation, *time*, must therefore be given as unlimited.'[49] If this argument contains any true premiss, it is that for every determinate interval A of time, it is possible to conceive (represent) a longer interval B that includes A. But it does not follow from this that for every determinate interval A of time, there is a longer interval B inclusive of A that can be conceived (represented), for the conception (representation) of B mentioned in the true premiss may fail of reference.[50]

I conclude, then, that Kantian transcendental arguments offer little prospect for showing that a relevant causal proposition either has a proof or is a part of the proof of some self-evident proposition. In the absence of any other candidates for such proofs, and given the fact that no relevant causal proposition is self-evident, we should conclude that there is no reason to think that it is metaphysically impossible for the Big Bang singularity to occur uncaused. If classical Big Bang cosmology gives us reason to think it is uncaused, then we should believe it is uncaused (unless some other cosmological theory gives us empirical reason to think otherwise).

---

[49] Ibid. A32/B47–8.
[50] See Quentin Smith, 'On the Beginning of Time', *Noûs*, 19 (1985), 579–84.

2

The Atheistic Cosmological Argument

VII

Atheism, Theism, and Big Bang Cosmology
QUENTIN SMITH

Introductory Note

This essay begins a new debate about the relevance of Big Bang cosmology to the philosophy of religion. Part 1 concerned Craig's theistic cosmological argument, that is, Craig's argument that Big Bang cosmology and considerations about finitude and the past warrant the belief that God exists. Part 2 will concern an atheistic cosmological argument that I first lay out in the present essay. My argument is that classical Big Bang cosmology is inconsistent with theism due to the unpredictable nature of the Big Bang singularity. Since each of the subsequent essays in Part 2 are direct responses to the preceding essay, they will not require introductory notes.

1. INTRODUCTION

The idea that the Big Bang theory allows us to infer that the universe began to exist about 15 billion years ago has attracted the attention of many theists. This theory seemed to confirm or at least lend support to the theological doctrine of creation *ex nihilo*. Indeed, the suggestion of a divine creation seemed so compelling that the notion that 'God created the Big Bang' has taken a hold on popular consciousness and become a staple in the theistic component of 'educated common sense'. By contrast, the response of atheists and agnostics to this development has been comparatively lame. Whereas

This is excerpted from 'Atheism, Theism, and Big Bang Cosmology', first pub. in *Australasian Journal of Philosophy*, 69/1 (Mar. 1991), 48–65.

I am grateful to Richard Fallon and two anonymous referees for the *Australasian Journal of Philosophy* for helpful comments on an earlier version of this essay.

the theistic interpretation of the Big Bang has received both popular endorsement and serious philosophical defence (most notably by William Lane Craig and John Leslie[1]), the non-theistic interpretation remains largely undeveloped and unpromulgated. The task of this article is to fill this lacuna and develop a non-theistic interpretation of the Big Bang. I shall argue that the non-theistic interpretation is not merely an alternative candidate to the theistic interpretation, but is better justified than the theistic interpretation. In fact, I will argue for the strong claim that Big Bang cosmology is actually *inconsistent* with theism.

The cosmological theory that has been endowed with the theistic interpretation is the *classic Big Bang theory* (also known as the standard hot Big Bang theory), which is based on the Friedman models with their prediction of an original Big Bang singularity. In this paper I shall also work with this theory, as supplemented (as is now standard practice) with the singularity theorems and Hawking's principle of ignorance. But we must be careful about how we view the significance of this classical theory. We cannot say that it is 'the final truth' about the universe, since it is thought by many cosmologists that this classical theory will one day be replaced by a quantum cosmology that is based on a fully developed quantum theory of gravity. Accordingly, my argument in this essay cannot be 'If the classical Big Bang theory is true, God does not exist; the classical theory is true, therefore God does not exist'. Rather, my argument is simply that the existence of God is inconsistent with the classical Big Bang theory. I aim to produce a valid argument for God's non-existence, not a sound one.

There is also a second reason why the classical Big Bang theory cannot be viewed as the definitive theory of the universe. There are many other competing theories of the universe currently being considered, and some of these have at least as good a claim as the classical theory to be regarded as 'the best currently available theory' and 'the theory we should provisionally accept until the

---

[1] See William Lane Craig, *The* Kalām *Cosmological Argument* (New York: Harper & Row, 1979); 'God, Creation and Mr Davies', *British Journal for the Philosophy of Science*, 37 (1986), 163–75; 'Barrow and Tipler on the Anthropic Principle vs. Divine Design', *British Journal for the Philosophy of Science*, 39 (1988), 389–95; Essay V, Essay XI. Also see John Leslie, 'Anthropic Principle, World Ensemble, Design', *American Philosophical Quarterly*, 19 (1982), 141–51, 'Modern Cosmology and the Creation of Life', in E. McMullin (ed.), *Evolution and Creation* (South Bend, Ind.: University of Notre Dame Press, 1985), and numerous other articles.

complete quantum cosmology is developed'. These competitors[2] include (1) Guth's original inflationary theory, (2) Linde's and Albrecht and Steinhardt's new inflationary theory, (3) Linde's theory of chaotic inflation, (4) Tryon's, Gott's, and others' theories that there are many universes (one of which is ours) that emerged as 'vacuum fluctuations' from a background empty space, (5) Hartle and Hawking's theory that the universe's wave function is a function of three-dimensional spatial geometries but not of a fourth temporal dimension, (6) Everett's theory of branching universes, and many other theories of current interest. In order to keep this essay within manageable limits, I shall not consider these competing theories but shall confine myself to the classical Big Bang theory. This confinement is consistent with my limited aim of counteracting the theistic interpretation of this classical theory.

In Section 2 I set forth, in a relatively non-technical manner, the pertinent cosmological concepts. In Section 3 I offer an argument that these concepts are inconsistent with theism. In Sections 4–7 I state and respond to some objections to this argument.

2. THE BIG BANG COSMOLOGICAL THEORY

The Big Bang theory is largely based on Friedman's solutions to the so-called Einstein equation that lies at the heart of the General Theory of Relativity. The details may be found in Essays IV and VI and need only be mentioned in passing here. The ideas I wish to emphasize are the Hawking–Penrose singularity theorems and especially Hawking's principle of ignorance. The solutions for the Hawking–Penrose theorems in the general case show that there is a singularity that intersects every past-directed spacetime path and

[2] See (1) A. Guth, 'Inflationary Universe: A Possible Solution to the Horizon and Flatness Problems', *Physical Review*, D23 (1981), 347–56; (2) A. D. Linde, 'A New Inflationary Universe Scenario', *Physical Letters*, 108B (1982), 389–93, and A. Albrecht and P. I. Steinhardt, *Physical Review Letters*, 48 (1982), 1220 ff.; (3) A. D. Linde, 'The Inflationary Universe', *Reports on Progress in Physics*, 47 (1984), 925–86; (4) E. P. Tryon, 'Is the Universe a Vacuum Fluctuation?', *Nature*, 246 (1973), 396–7, and J. R. Gott, 'Creation of Open Universes from de Sitter Space', *Nature*, 295 (1982), 304–7; (5) J. B. Hartle and S. W. Hawking, 'Wave Function of the Universe', *Physical Review*, D28 (1983), 2960–75; (6) H. Everett, '"Relative State" Formulation of Quantum Mechanics', *Reviews of Modern Physics*, 29 (1957), 454–62. Some of these theories are discussed in Quentin Smith, 'World Ensemble Explanations', *Pacific Philosophical Quarterly*, 67 (1986), 73–86 and Essay IV.

constitutes the beginning of time. These solutions demonstrate, in Hawking's words, that even for imperfectly homogeneous universes 'general relativity predicts a beginning of time'.[3]

The singularity theorems are the part of Big Bang cosmology that support the claim that *there is* a Big Bang singularity. But the part of Big Bang cosmology that shall be crucial to my atheistic argument is the conception of the *nature* of this singularity. This conception is embodied in Hawking's principle of ignorance, which states that singularities are inherently chaotic and unpredictable. In Hawking's words,

A singularity is a place where the classical concepts of space and time break down as do all the known laws of physics because they are all formulated on a classical space-time background. In this paper it is claimed that this breakdown is not merely a result of our ignorance of the correct theory but that it represents a fundamental limitation to our ability to predict the future, a limitation that is analogous but additional to the limitation imposed by the normal quantum-mechanical uncertainty principle.[4]

One of the quantum-mechanical uncertainty relations is $\Delta p \cdot \Delta q \geq h/4\pi$, which implies that if the position q of a particle is definitely predictable then the momentum p of the particle is not, and vice versa. The principle of ignorance implies that one can definitely predict neither the position nor the momentum of any particle emitted from a singularity.[5] All possible values of the particle's position and momentum that are compatible with the limited information (if any) available about the interaction region are equally probable. But the principle of ignorance has further consequences. It implies that none of the physical values of the emitted particles are definitely predictable. The Big Bang singularity 'would thus emit all configurations of particles with equal probability'.[6]

If the singularity's emissions are completely unpredictable, then

[3] S. W. Hawking, 'Theoretical Advances in General Relativity', in H. Woolf (ed.), *Some Strangeness in the Proportion* (Reading, Mass.: Addison-Wesley, 1980), 149.

[4] S. W. Hawking, 'Breakdown of Predictability in Gravitational Collapse', *Physical Review*, D14 (1976), 2460.

[5] See S. W. Hawking, 'Is the End in Sight for Theoretical Physics?', in John Boslough, *Stephen Hawking's Universe* (New York: William Morrow, 1985), 145.

[6] Hawking, 'Breakdown of Predictability in Gravitational Collapse', 2460.

we should expect a totally chaotic outpouring from it. This expectation is consistent with Big Bang cosmologists' understanding of the early universe, for the early universe is thought to be in a state of maximal chaos (complete entropy). Particles were emitted in random microstates, which resulted in an overall macrostate of thermal equilibrium.[7]

It is important to understand the full significance of the principle of ignorance. If the Big Bang singularity behaves in a completely unpredictable manner, then no physical laws govern its behaviour. There is no law to place restrictions on what it can emit. As Paul Davies aptly comments, 'anything can come out of a naked singularity—in the case of the big bang the universe came out. Its creation represents the instantaneous suspension of physical laws, the sudden, abrupt flash of lawlessness that allowed something to come out of nothing.'[8] Here 'nothing' should be understood metaphorically as referring to something not a part of the four-dimensional spacetime continuum; the singularity is not a part of this continuum since it occupies less than three spatial dimensions. But Davies is literally correct in implying that the singularity entails an instantaneous state of lawlessness. The singularity exists for an instant and during this instant no physical law obtains that could connect the singularity to later instants. Given the initial conditions of the singularity, nothing can be predicted about the future state of the universe. Each possible configuration of particles has the same probability of being emitted by the singularity. (If there are uncountably infinite possible configurations, then we must speak instead of the probability density of each possible configuration and assign probabilities to each of the countable number of intervals of possible configurations, given an appropriate partition.) At any instant arbitrarily close to the instant at which the singularity exists, physical laws do obtain and they govern the particles actually emitted from the singularity. This means that for any physical configuration C that occupies an instant arbitrarily close to the instant occupied by the singularity from which C was emitted, there obtain laws connecting C to the configurations occupying later instants but there obtains no law connecting C to the earlier singularity. C adopts a lawful evolution but has its ultimate origin in primordial lawlessness.

[7] Ibid. 2463.
[8] P. Davies, *The Edge of Infinity* (New York: Simon & Schuster, 1981), 161.

3. THE ATHEISTIC ARGUMENT

I shall use the aspects of Big Bang cosmology explicated in the last section as the scientific premises of my atheistic argument. In this section I will add two theological premises and deduce the statement that God does not exist. Following the construction of this argument, I will state and respond to several objections to it (Sections 4–7). The real force of the argument will not become apparent until the responses to these objections are given.

The two theological premises I need are

(1) If God exists and there is an earliest state E of the universe, then God created E.
(2) If God created E, then E is ensured either to contain animate creatures or to lead to a subsequent state of the universe that contains animate creatures.

Premiss (2) is entailed by two more basic theological premises, namely,

(3) God is omniscient, omnipotent, and perfectly benevolent.
(4) An animate universe is better than an inanimate universe.

Given (4), if God created a universe that was not ensured to be animate, then he would have created a universe not ensured to be of the better sort and thereby would be limited in his benevolence, power, or wisdom. But this contradicts (3). Therefore, (2) is true.

Some of the scientific ideas articulated in the last section, mainly the Hawking–Penrose singularity theorems, provide us with the summary premiss

(5) There is an earliest state of the universe and it is the Big Bang singularity.

(5) requires a terminological clarification regarding 'the universe'. By this phrase I mean the four-dimensional spacetime continuum and any n-dimensional physical state that is earlier or later than the four-dimensional continuum. Since the universe has a zero radius at the singularity, it is not then four-dimensional, but since the singularity is a physical state earlier than the four-dimensional continuum it can be considered to be the first state of the universe (this is discussed further in Section 6).

The scientific ideas also give us the premiss

(6) The earliest state of the universe is inanimate since the singu-

larity involves the life-hostile conditions of infinite temperature, infinite curvature, and infinite density.

Another scientific idea enunciated in the last section, the principle of ignorance, gives us the summary premiss

(7) The Big Bang singularity is inherently unpredictable and lawless and consequently there is no guarantee that it will emit a maximal configuration of particles that will evolve into an animate state of the universe. (A maximal configuration of particles is a complete state of the universe, the universe as a whole at one time.)

(5) and (7) entail

(8) The earliest state of the universe is not ensured to lead to an animate state of the universe.

We now come to the crux of our argument. Given (2), (6), and (8), we can infer that God could not have created the earliest state of the universe. It then follows, by (1), that God does not exist.

I will now state and respond to four objections to this atheistic argument.

4. THE FIRST OBJECTION: ANIMATE UNIVERSES ARE NOT REQUIRED BY GOD

This objection is based on the principle that there is no universe that is the best of all possible universes. For each universe U_1 there is a better universe U_2. Consequently, the fact that there is some universe better than whatever universe is the actual one is not only compatible with divine creation but is entailed by it. Therefore, the objection goes, the fact that an animate universe is better than an inanimate one is compatible with God creating as the earliest state something that by chance leads to an inanimate universe. Premisses (3) and (4) do not entail (2) and the atheistic argument therefore fails.

In response, I note first that many theists claim that there is a best of all possible universes and that God ensures that the one he creates is the best one. My argument implies at least that these theologies are mistaken. But it also tells against theologies that entail there is no best possible universe. These theologies, if they are at all consistent with what is ordinarily meant by 'God' and what most philosophers and theologians mean by 'God', must impose some minimal constraint on the value of the universe God creates. I believe the overwhelming majority of theists explicitly or implicitly accept the minimal constraint that the universe contain living creatures. The idea that God has no more reason to create an animate universe than an inanimate one is inconsistent with the kind of person we normally conceive God to be. The God of the Judaeo-Christian-Islamic tradition is obviously a God who ensures that there be life in the universe he creates. This requirement conforms to the theism of Swinburne, Craig, Leslie, Plantinga, Adams, Morris, and all or virtually all other contemporary theists. Swinburne, for example, defines 'orderly universes' as the ones required by animate creatures and affirms that 'God has overriding reason to make an orderly universe if he makes a universe at all'.[9] According to this standard conception of God, premisses (3) and (4) come with the suppressed premiss

(4a) If God chooses to create a universe, he will choose to create an animate rather than an inanimate universe.

Given (4a), (3) and (4) do entail (2) and the atheistic argument is valid.

5. SECOND OBJECTION: GOD CAN INTERVENE TO ENSURE AN ANIMATE UNIVERSE

The second objection is that the lawlessness of the Big Bang singularity is not logically incompatible with its being ensured by God to emit a life-producing maximal configuration of particles. For God could intervene at the instant of the singularity and supernaturally constrain the singularity to emit a life-producing configuration.

I believe this objection is incompatible with the rationality of God. If God intends to create a universe that contains living beings at some stage in its history, then there is no reason for him to begin the universe with an inherently unpredictable singularity. Indeed, it is positively irrational. It is a sign of incompetent planning to create as the first natural state something that requires immediate

[9] R. Swinburne, *The Existence of God* (Oxford: Clarendon Press, 1979), 147. Swinburne's full definition is that orderly universes are those required by both natural beauty and life. Cf. p. 146.

supernatural intervention to ensure that it leads to the desired result. The rational thing to do is to create some state that *by its own lawful nature leads* to a life-producing universe.

This response to the second objection can be developed in the context of a discussion of John Leslie's interpretation of Big Bang cosmology. Leslie points to data or figures (the 'anthropic coincidences') that suggest it is *highly improbable* that an animate universe would result from a Big Bang singularity.[10] There are many possible maximal configurations of particles that might be emitted from the singularity and only an extremely small number of these, Leslie suggests, lead towards animate states. But Leslie argues that this improbability tells *for* rather than against the hypothesis of divine creation. (I should note that Leslie works with a 'Neoplatonic' conception of God[11] but that makes no substantive difference to the validity of the arguments I shall examine.) He implies that if we suppose that God constrained the singularity's explosion to be directed away from the more probable alternatives of lifelessness and towards the very narrow range of alternatives that lead to life, then we can 'explain away' the apparent improbability of an animate universe evolving from the singularity. The alleged simplicity of this explanation, the distinctive value of life, and other relevant premisses are regarded as making this explanation a credible one. But this fails to take into account the above-mentioned problem regarding God's rationality and competence, which appears here in an aggravated form. It seems to me that Leslie's premiss that it is highly improbable that the Big Bang singularity would (if left to evolve naturally) lead to an animate universe is *inconsistent* with the conclusion that God created the singularity. If God created the universe with the aim of making it animate, it is illogical that he would have created as its first state *something whose natural evolution would lead with high probability only to inanimate states*. It does not agree with the idea of an efficient creation of an animate universe

---

[10] See Leslie's articles mentioned in n. 1.

[11] For Leslie, 'God' means one of two things. God 'may be identified as the world's creative ethical requiredness [i.e. the ethical requiredness that created the universe] . . . Alternatively [God may be identified] as an existing person, a person creatively responsible for every other existence, who owed his existence to his ethical requiredness.' See his 'Efforts to Explain All Existence', *Mind*, 87 (1978), 93. On the second conception of God, God as a person, it is appropriate to refer to him with a personal pronoun ('he'). But on the first conception, the impersonal pronoun 'it' is more appropriate. For simplicity's sake, I use 'he' in the main body of the paper.

that life is brought about through the first state being created with a natural tendency towards *lifelessness* and through this tendency being *counteracted* and *overridden* by the very agency that endowed it with this tendency. The following two propositions appear to be logically incompatible:

(1) God is a rational and competent creator and he intends to create an animate universe.

(2) God creates as the first state of the universe a singularity whose natural tendency is towards lifelessness.

The problem involved here is essentially a problem of divine interference in or 'correction of' the divine creation. Leslie is 'opposed'[12] to the idea of 'divine interference' with natural processes and is unsympathetic to the idea that 'God occasionally intervenes [in the natural universe] with a helpful shove'[13] so as to ensure that life evolves. Leslie states that the hypothesis of such intervention involves an unsimple theory and for this reason is to be dispreferred. But such intervention is precisely what is required by his own account of the evolution of the early universe. His account supposes that God not only interferes with the singularity's explosion but also interferes with the subsequent evolution of the maximal configuration of particles that was emitted from the singularity. For example, Leslie mentions the theory that the early universe underwent a number of 'spontaneous symmetry breaking phases' during the first 10^{-4} seconds after the Big Bang singularity and that during these phases the four forces (gravitational, strong, weak, and electromagnetic) became separated. In the GUT era (from 10^{-43} seconds after the singularity to 10^{-35} seconds) the gravitational force is separated from the strong-electroweak force. During the electroweak era (from 10^{-35} seconds to 10^{-10} seconds) the strong force is separated from the electroweak force. During the free quark era (from 10^{-10} seconds to 10^{-4} seconds) the electromagnetic force is separated from the weak force. Each of these separations is a breaking of a symmetry (the unification of two or more forces) and each symmetry is broken in a random way. This means, in effect, that the strengths of the four forces are determined in random ways at the time they become separated. This is significant, Leslie indicates, since only a small range of the values these forces may

[12] Leslie, 'Modern Cosmology and the Creation of Life', 112.
[13] Ibid. 92.

possess are consistent with a life-supporting universe. For example, if the actual value of the weak fine structure constant ($a_w \sim 10^{-11}$) were slightly larger, supernovae would have been unable to eject the heavy materials that are necessary for organisms. If this value were slightly smaller, no hydrogen would have formed and consequently no stars and planets would have evolved. Similar considerations hold for the gravitational, electromagnetic, and strong forces. Given this, Leslie continues, it is 'exceedingly improbable'[14] that these symmetry-breaking phases would have resulted in the very narrow range of values required by a life-supporting universe. This improbability could be eliminated if we supposed that these values were not selected by natural random processes but were 'selected by God'. But this requires divine interference on a grand scale in the evolution of the universe. God would have to intervene in his creation at the Big Bang singularity to ensure that it emitted a maximal configuration of particles capable of undergoing the symmetry-breaking phases, *then again* during the GUT era to ensure that the separating gravitational force acquires the right value, *and then once again* during the electroweak era to ensure that the separating strong force acquires the right value, *and then once more* during the free quark era to ensure that the separating electromagnetic and weak forces acquire the right value. And these are only some of the interventions required (I have not even mentioned, for example, the interventions required to ensure that the elementary particles acquire the right masses). But why does Leslie think his theory avoids the implausibly complex theory of repeated divine interventions in natural processes? Because he *stipulates* that God's fixing of the values of the constants are not instances of such interventions. Interventions he defines as applying to less basic aspects of nature (such as creations of individual animal organisms).[15] But this stipulation seems arbitrary and implausible. If God's interference with the singularity's emission of particles and with the several symmetry-breaking phases are not examples of God interfering with natural states and processes, then I don't know what is.

Leslie suggests that the notion of divine interference with the processes of nature is implausible because it is less simple than the idea that God lets nature evolve on its own. But it seems to me

[14] Ibid. 95.   [15] Ibid. 91 and 112.

there is a more fundamental problem with this notion, at least as it applies to Leslie's scenario. This notion, in the context of Leslie's scenario, implies that the universe God created was so bungled that it needed his repeated intervention to steer it away from disaster and towards the desired life-producing states. God created a universe that time and again was probably headed towards *the very opposite result than the one he wanted* and only through interfering with its natural evolution could he ensure that it would lead to the result he desired. But this contradicts the principle that God is not a bungler ('a competent Creator does not create things he immediately or subsequently needs to set aright').

I should make explicit that the key idea in my argument is not that God is incompetent if he creates a universe whose laws he must *violate* if his intentions are to be realized, but that he is incompetent if he creates a universe requiring his *intervention* if his intentions are to be realized. A divine intervention in natural events is entailed by, but does not entail, a divine violation of natural laws, since God may intervene in an event (e.g. the explosion of the singularity) not governed by laws. Thus, the possible objection to my argument that 'if physical laws under-constrain the evolution of the universe, then God can constrain the universe to evolve into animate states without violating his physical laws' misses the point, that *intervention*, not violation, is the problem. However, if we assume Leslie's scenario, then we can say there are not only interventions but also violations, since in his scenario there are probabilistic laws governing the early evolution of the universe (which includes the symmetry-breaking phases) and God suspends (violates) these laws to ensure that the improbable life-producing outcomes result.

My conclusion is this. There are countless logically possible initial states of the universe that lead by a natural and lawlike evolution to animate states and if God had created the universe he would have selected one of these states. Given that the initial state posited by Big Bang cosmology is not one of these states, it follows that Big Bang cosmology is inconsistent with the hypothesis of divine creation.[16]

---

[16] I would add that my argument does not require that God create an animate universe in the most efficient way possible, since there may be no 'most efficient way possible', but merely that he create it in an efficient way (which minimally requires that no interventions be needed). Somewhat analogously, Keith Chrzan has soundly argued that 'there is no best possible world' does not entail 'there is no world

6. THIRD OBJECTION: THE SINGULARITY IS A THEORETICAL FICTION

The theist may attempt to avoid the difficulties of an unpredictable initial state and a divine intervention by supposing that the initial state of the universe is not an unpredictable singularity. The theist may continue to accept Big Bang cosmology except that she adopts rules for the interpretation of this theory that forbid reality to the singularity. These rules are based on a criterion of physical existence that the singularity fails to meet but which is met by the Big Bang explosion. These rules allow the theist to regard the Big Bang explosion, not the singularity, as the earliest state of the universe. (But now 'state' must be understood as a temporally extended state of a certain length rather than as an instantaneous one since the explosion is extended.) The Big Bang explosion is governed by physical laws and this explosion leads by a natural and lawful evolution to a state of the universe that contains animate creatures. The problem of God creating as the first state some totally unpredictable state is thereby avoided and the theist is able to ascribe a rational behaviour to God in creating as the first state something that naturally evolves into an animate universe.

In dealing with this third objection I shall ignore the problem of the unpredictable symmetry-breaking phases that Leslie introduces into his scenario and that would seem to vitiate the hypothesis that the Big Bang explosion predictably evolves into animate states.

without evil' and therefore that the 'no best possible world' theodicy fails to demonstrate that evil is a necessary implication of creation and thus fails to explain how God's existence is compatible with the actual world. See Keith Chrzan, 'The Irrelevance of the No Best Possible World Defence', *Philosophia*, 17 (1987), 161–7. The analogy can be seen if we substitute 'most efficient' for 'best possible' and 'without divine intervention' for 'without evil' in the above sentences. I also reject the supposition that the Hawking–Penrose theorems and the principle of ignorance are *metaphysically necessary* laws of nature and therefore that God had no alternative to creating a singularity that required his intervention. In his interesting article on 'Explaining Existence', *Canadian Journal of Philosophy*, 16 (1986), 713–22, Chris Mortensen entertains the supposition that the laws governing the beginning of the universe are necessary, but concludes, soundly I believe, that this supposition is not particularly credible. I would add that the Kripke–Putnam argument that some laws are necessary (e.g. that water is H_2O), even if sound, does not apply to the singularity theorems, for the Kripke–Putnam argument applies only to laws involving ostensively defined terms (e.g. 'water') and 'singularity' is not ostensively defined. See Jarrett Leplin, 'Is Essentialism Unscientific?' *Philosophy of Science*, 55 (1988), 493–510 and 'Reference and Scientific Realism', *Studies in History and Philosophy of Science*, 10 (1979), 265–85.

Although it is widely—but not universally—accepted today that such phases occur, these phases are not entailed by classical Big Bang cosmology and accordingly it is not appropriate to introduce them when criticizing theistic interpretations of this cosmology that do not themselves introduce the phases. Thus, in responding to the third objection I will not argue that there remain unpredictabilities even if the singularity is omitted but will argue instead that there is no justification for rejecting the singularity with its unpredictability.

Let me begin by noting that the description or definition of the Big Bang singularity as a mere idealization does *not* belong to Big Bang cosmology itself and thus that if this view of the singularity is to be justified some strong and independent philosophical arguments will be needed. Big Bang cosmology represents the singularity as a unique sort of reality, a *physical singularity*, but it is represented as real none the less. This is evinced by the fact that past-directed spacetime paths in the early universe are not modelled on half-open intervals that approach arbitrarily close to but never reach the ideal limit, but on closed intervals one of the end-points of which is the singularity. In the words of Penrose, 'the essential feature of a past spacelike singularity [the Big Bang singularity] is that it supplies a past singular end-point to the otherwise past-endless timelike curve'.[17] (A timelike curve is a spacetime path of a particle.) In the words of Geroch and Horowitz, converging past-directed spacetime paths are not commonly thought to merely approach with arbitrary closeness the same singular point but are thought to actually '*reach* the same singular point',[18] which requires the actual physical existence of the singular point. Furthermore, this point is thought by physicists to be earlier in time than the Big Bang explosion. Penrose articulates the common view that in the case of a finite universe 'we think of the initial singularity as a single point . . . [which] *gives rise*

---

[17] R. Penrose, 'Singularities in Cosmology', in M. S. Longair (ed.), *Confrontation of Cosmological Theories with Observational Data* (Dordrecht: D. Reidel, 1974), 264. Penrose shows how the zero dimensional singularity can be conformally rescaled as a three-dimensional singularity, which testifies further to the fact that the singularity is thought of as something real.

[18] R. Geroch and G. Horowitz, 'Global Structure of Spacetime', in S. W. Hawking and W. Israel (eds.), *General Relativity* (New York: Cambridge University Press, 1979), 267. Geroch and Horowitz go on to argue for the non-standard position that a study of the global properties of singular spacetimes is a more fruitful line of research than attempts to provide constructions of local singular points.

to an infinity of causally disconnected regions at the next instant',[19] a conception that clearly entails the physical and temporal reality of the initial singularity.

Given this realist representation of the singularity, the theists must have strong reasons indeed to support the interpretation of the singularity as a mere idealization. They must establish some convincing criterion of physical existence and show that the singularity fails to meet this criterion. This has been attempted by William Lane Craig. Craig argues that no infinitely complex object can be real and the singularity cannot be real since it has infinite values, such as infinite density; 'there can be no object in the real world that possesses infinite density, for if it had any size at all, it would not be *infinitely* dense'.[20] Craig's arguments against infinite realities in his book are aimed at showing that no reality can be mapped on to a Cantorian transfinite set. I have elsewhere[21] countered Craig's arguments but I would like to show here that even if his arguments were sound they would not count against the reality of the Big Bang singularity. When it is said that the Big Bang singularity has an infinite density, infinite temperature, and infinite curvature, it is not being said that the singularity has parts or properties that map on to a set with an aleph-zero or aleph-one cardinality. Rather, three things are implied and each of them is compatible with Craig's rejection of Cantorian realities:

The theory that there is an infinite singularity implies, first of all, that at any instant arbitrarily close to the Big Bang singularity the density, temperature, and curvature of the universe have arbitrarily high finite values. The values become higher and higher as we regress closer and closer to the singularity, such that for any arbitrarily high finite value there is an instant at which the density, temperature, and curvature of the universe possess that value.

The theory of the infinite singularity implies, secondly, that

---

[19] Penrose, 'Singularities in Cosmology', 264; the italics are mine. Penrose is best interpreted as speaking loosely in this passage, for strictly speaking there is no 'next instant' after the instant of the singularity (if time is dense or continuous) and the singular point does not topologically transform to an 'infinite' number of causally disconnected regions but to an arbitrarily large finite number.

[20] Essay I, Sect. 2.2.

[21] See Essay II and 'A New Typology of Temporal and Atemporal Permanence', *Noûs*, 23 (1989), 307–30, sect. 6. For a correction to one of my arguments in Essay II, see E. Eells, 'Quentin Smith on Infinity and the Past', *Philosophy of Science*, 55 (1988), 453–5.

when the singularity is reached the values become infinite. But this does *not* mean that the density, temperature, and curvature of the universe have values involving the numbers $\aleph_0$ or $\aleph_1$. Consider the phenomenon of density, which is the ratio of mass to unit volume (density = mass/volume). If the universe is finite and the big bang singularity a single point, then at the first instant the entire mass of the universe is compressed into a space with zero volume. The density of the point is $n/0$, where n is the extremely high but finite number of kilograms of mass in the universe. Since it is impermissible to divide by zero, the ratio of mass to unit volume has no meaningful and measurable value and *in this sense is infinite*. Although philosophers frequently misunderstand this use of the word 'infinite' by physicists, this usage has been clearly grasped by Milton Munitz in his recent discussion of the Big Bang theory. He notes that

the density of a homogeneous material is mass per unit volume—for example, grams per cubic centimeter. Given both a zero value and the conservation of the mass-energy of the universe [at the Big Bang singularity], no finite value can be given to the ratio of the latter to the former (it is forbidden to divide by zero). This is normally expressed by saying that the density becomes *infinite*. It would be more accurate to say the standard meaning of 'density' cannot be employed in this situation. The density cannot be assigned a finite measurable value, as is the case in all standard applications of the concept.[22]

The theory of the infinite singularity implies, thirdly, that the space of the singularity topologically transforms into the three-dimensional space of the universe at the Big Bang explosion. It is a familiar notion in the mathematical discipline of topology that a space with a topology of a point can assume the topology of a finite three-dimensional space. The topological transformation of the zero-dimensional space to the three-dimensional space is precisely the Big Bang explosion. But I am not saying here that the zero-dimensional space is homeomorphic to the three-dimensional space, where x is homeomorphic to y if there exists a continuous bijective map f of x on to y such that the inverse map f^{-1} is also continuous. Rather, I am saying that a space with the topology of a point assumes, at a subsequent time, the topology of a finite three-dimensional space. Such topological transformations are possible

[22] Milton Munitz, *Cosmic Understanding* (Princeton, NJ: Princeton University Press, 1986), 111.

but it is not possible, for instance, for a space with the topology of a point to assume, at a subsequent time, the topology of an infinite three-dimensional space (where 'infinite' is used in the Cantorian sense). If our universe is infinite, then the Big Bang singularity must have consisted of an infinite number of points and must have been at least one dimension, with each of the points 'topologically exploding' into a different finite three-dimensional region. Paul Davies comments that if the universe is finite

one can really suppose that the entire universe began compressed into one point. On the other hand, if space is infinite, we have the mathematically delicate issue of conflicting infinities, because infinitely extended space becomes infinitely compressed at the beginning of the big bang. This means that any given *finite* volume of the present universe, however large one chooses it to be, was compressed to a single point at the beginning. Nevertheless, it would not be correct to say *all* the universe was at one place then, for there is no way that a space with the topology of a point can suddenly assume the topology of a space with infinite extent.[23]

It might be conceded that the notion that the singularity is real escapes Craig's criticism, since it is not 'infinite' in a Cantorian sense, but argued that the concept of the singularity is defective for other reasons. For example, how can the entire mass of a finite universe be compressed into a point? The mass is three-dimensional and the point is zero-dimensional, which involves a contradiction. But this is a misunderstanding. The mass as compressed into the point is not ordinary mass, three-dimensional mass, but *infinitely compressed mass*, which means that it has lost its three-dimensionality and assumed the dimensionality of the point it occupies. The assertion that at the instant of the singularity n kilograms of mass is infinitely compressed in a zero volume implies in part that (1) at this instant there exists no three-dimensional mass, (2) at this instant there exists only one zero-dimensional point, that (3) this point subsequently assumes the topology of a three-dimensional space, and that (4) this subsequent three-dimensional space is occupied by n kilograms of mass. Of course this singular point can assume the topology of a three-dimensional space that contains *any* finite number of kilograms of mass—the actual number, n, is randomly 'selected' from the range of possibilities—and this is one of the reasons the singularity is wholly unpredictable.

[23] Davies, *The Edge of Infinity*, 159.

I believe, therefore, that there is no good reason for rejecting the reality of the Big Bang singularity and the attendant unpredictability. If Craig is to justify his claim that the assumption that it is real is an illegitimate 'ontologizing' of a mathematical construct, he must provide some reason to support this claim other than his arguments against Cantorian infinities. His recent and related claim that 'a physical state in which all spatial and temporal dimensions are zero is a mathematical idealization whose ontological counterpart is nothing'[24] is made with no effort to support it and should be rejected as an unjustified scepticism about a widely held scientific thesis.

7. FOURTH OBJECTION: UNPREDICTABILITY DOES NOT ENTAIL THERE IS NO DIVINE KNOWLEDGE

I have said the Big Bang singularity is unpredictable. It might be objected that the fact that *we* cannot predict what comes out of the singularity is consistent with God being able to predict what will emerge from it. God is omniscient, which implies he can know things that are unknowable by humans.

But this objection is based on several questionable assumptions, one of which concerns the meaning of the word 'unpredictable' as it is used in the formulation of Hawking's principle of ignorance. What is meant is *unpredictability in principle*, which entails but is different from *unpredictability by us*. The qualifier 'in principle' is added to indicate that the unpredictability is due to the fact that *no natural laws govern the state(s)*. If something is merely unpredictable by us, that is consistent with saying that it is governed by a natural law that is not knowable by humans. But if there is an 'in principle' unpredictability, then there is no natural law to be known, by God or any other knower. Since there is no natural law governing the singularity, God has no basis on which to compute what will emerge from the singularity. As Davies says, the instantaneous existence of the singularity and the subsequent explosion is an 'abrupt flash of lawlessness'.

Some might claim that 'unpredictability in principle' as used in quantum mechanics (and thus in Hawking's theory, which is partly

---

[24] Essay V, Sect. 2.

based on quantum mechanics) should be interpreted as meaning the same as 'unpredictability by us' since the most plausible interpretations of quantum mechanics (e.g. the Copenhagen interpretation) are anti-realist. But this claim, while perhaps justified on the old assumption that the Everett interpretation is the only realist one consistent with quantum mechanics, is not justified today, given that some plausible realist interpretations have been recently developed, such as, for example, Storrs McCall's 'branched model' interpretation.[25]

But this reference to a realist interpretation of the singularity's unpredictability does not do full justice to the objection that 'unpredictability does not entail there is no divine knowledge'. For the objector might claim that God can 'know in advance' the result of the singularity's explosion *even if there is no law on the basis of which he can form a prediction*. It might be said that just as God knows, logically prior to creation, the free decisions humans would make if they were in certain circumstances, so he knows, logically prior to creation, the way the singularity would explode if it were to be the first state of the universe. The theist may allege that in addition to the familiar sorts of counterfactuals, we may introduce a new sort, 'counterfactuals of singularities', one of which is the counterfactual

(1) If a Big Bang singularity were to be the earliest state of the universe, this singularity would emit a life-producing configuration of particles.

The theist may allege that (1) is true logically prior to creation and that God's pre-creation knowledge of (1) serves as his reason for his creation of a universe with a Big Bang singularity.

But this argument is unsound, since the supposition that (1) is true logically prior to creation is inconsistent with the semantic properties of counterfactuals. As Jonathan Bennett and Wayne Davies have argued,[26] counterfactuals are true iff the antecedent and consequent are both true in the possible world most similar to the actual world *before* the time specified in the antecedent. This

[25] Storrs McCall, 'Interpreting Quantum Mechanics via Quantum Probabilities', mimeograph (1989).

[26] Jonathan Bennett, 'Counterfactuals and Possible Worlds', *Canadian Journal of Philosophy*, 4 (1974), 381–402; Wayne Davies, 'Indicative and Subjunctive Conditionals', *Philosophical Review*, 88 (1979), 544–64.

entails that there are no possible conditions in which (1) is true, since the time specified in its antecedent is the earliest time.

But the theist need not accept the Bennett–Davies theory of counterfactuals. He may accept one of the theories of Robert Stalnaker and Richmond Thomason and Frank Jackson,[27] according to which a counterfactual is true iff its antecedent and consequent are both true in a possible world whose *total history* is most similar to that of the actual world. Or the theist may accept David Lewis's theory,[28] that counterfactuals are true iff some world in which the antecedent and consequent are both true is more similar in its overall history to the actual world than any world in which the antecedent is true and the consequent false.

But these theories of counterfactuals are of no avail since they one and all entail that a counterfactual is true only if *there is an actual world* that serves as a relatum of the similarity relation. According to the Bennett–Davies theories, the relatum is all the states of the actual world up to a certain time and according to the theories of Stalnaker, Lewis, and others, the relatum is all the states of the actual world. Since (1) is supposed to be true logically prior to creation, its truth conditions cannot include all the states (or all the states up to a time) of the actual world, which contradicts the truth condition requirements of counterfactuals.

But a theist familiar with the corpus of William Lane Craig might be able to come up with a response to this argument. Craig does not discuss 'counterfactuals of singularities' but he does discuss counterfactuals of freedom and some of his arguments may be borrowed by a defender of the truth of (1). In response to the objection that there is no actual world logically prior to creation in relation to which counterfactuals of freedom could be evaluated as true, Craig maintains that a part of our world is actual prior to creation, namely the part consisting of logically necessary states of affairs and counterfactual states of affairs concerning the free decisions of creatures.

---

[27] Robert Stalnaker, 'A Theory of Conditionals', in N. Rescher (ed.), *Studies in Logical Theory* (Oxford: Blackwell, 1968), 92–112; Richmond Thomason and Robert Stalnaker, 'A Semantic Analysis of Conditional Logic', *Theoria*, 36 (1970), 23–42; Frank Jackson, 'On Assertion and Indicative Conditionals', *Philosophical Review*, 88 (1979), 565 ff.
[28] David Lewis, *Counterfactuals* (Cambridge, Mass.: Harvard University Press, 1973).

Since the relevant states of affairs are actual, one can hold to both the doctrine of divine middle knowledge [i.e. that God knows counterfactuals of freedom prior to creation] and the current explanation of what it means for a counterfactual to be true: in those possible worlds which are most similar to the actual world (in so far as it exists at [this logical] moment [prior to creation]) and in which the antecedent is true, the consequent is also true.[29]

But this response is untenable, since the current explanation of counterfactuals is that their truth conditions include either *all the states of the actual world* or *all the states of the actual world earlier than a certain time*, and the counterfactuals that are allegedly objects of God's middle knowledge meet neither of these two requirements. They are supposed to be true logically prior to the creation of the earliest state and therefore cannot include in their truth conditions all the states of the actual world or all the states earlier than a certain time.

Of course, the theist may reject the current explanation of counterfactuals. He may hold that counterfactuals of freedom (or of singularities) are true iff their antecedents and consequents are both true in the possible world most similar to the actual world *in so far as the actual world exists at the moment logically prior to creation*. This seems to be Craig's position, although he mistakenly claims it is consistent with 'the current explanation of what it means for a counterfactual to be true'. Now Craig holds, as we have seen, that at this logically prior moment there obtain all logically necessary states of affairs and all counterfactual states of affairs concerning free decisions of creatures. In response to the objection that counterfactuals of freedom cannot be true at this logically prior moment, since the actual world is not then actual, he claims that it is partly actual, since it includes in part the counterfactual states of affairs, i.e. the 'states of affairs corresponding to true counterfactuals concerning creaturely freedom'.[30] But this argument is viciously circular. In order to demonstrate that counterfactuals of freedom are true logically prior to creation, it is assumed that counterfactuals of freedom are true logically prior to creation, i.e. that prior to creation there are 'states of affairs corresponding to true counterfactuals

[29] William Lane Craig, *The Only Wise God: The Compatibility of Divine Foreknowledge and Human Freedom* (Grand Rapids, Mich.: Baker Book House, 1987), 144.
[30] Ibid. 143.

concerning creaturely freedom'. To avoid this vicious circle, we must allow only the premiss that there obtain logically necessary states of affairs prior to creation. But this premiss is insufficient to establish the desired conclusion, since these states of affairs cannot ground the relations of transworld similarity required by logically contingent counterfactuals, the counterfactuals of freedom. It follows, then, that no sound argument can be constructed, in analogy to Craig's argument about counterfactuals of freedom, for the thesis that the 'counterfactual of singularity' (1) is true logically prior to creation. It is logically incoherent to suppose that (1) is true logically prior to creation and therefore the fact that God is omniscient does not entail that he knows, logically prior to creation, that the Big Bang singularity would evolve into an animate universe.

8. CONCLUSION

If the arguments in this paper are sound, then God does not exist if Big Bang cosmology, or some relevantly similar theory, is true. If this cosmology is true, our universe exists without cause and without explanation.[31] There are numerous possible universes, and there is possibly no universe at all, and there is no reason why this one is actual rather than some other one or none at all. Now the theistically inclined person might think this grounds for despair, in that the alleged human need for a reason for existence, and other alleged needs, are unsatisfied. But I suggest that humans do or can possess a deeper level of experience than such anthropocentric despairs. We can forget about ourselves for a moment and open ourselves up to the startling impingement of reality itself. We can let ourselves become profoundly astonished by the fact that this universe exists at all. It is arguably a truth of the 'metaphysics of feeling' that this fact is indeed 'stupefying' and is most fully appreciated in such experiences as the one evoked in the following passage:[32]

---

[31] Big Bang cosmology may be modified in certain fundamental respects so that our universe has an explanation in terms of other universes, but the set of all universes will none the less remain unexplained. See Quentin Smith, 'A Natural Explanation of the Existence and Laws of Our Universe', *Australasian Journal of Philosophy*, 68 (1990), 22–43.

[32] Quentin Smith, *The Felt Meanings of the World: A Metaphysics of Feeling* (West Lafayette, Ind.: Purdue University Press, 1986), 300–1. In an important study, Milton Munitz has plausibly argued that it is *possible* that there is a reason for the

[This world] exists nonnecessarily, improbably, and causelessly. It exists *for absolutely no reason at all*. It is *inexplicably* and *stunningly actual* . . . The impact of this captivated realisation upon me is overwhelming. I am completely stunned. I take a few dazed steps in the dark meadow, and fall among the flowers. I lie stupefied, whirling without comprehension in this world through numberless worlds other than this one.

existence of the universe, such that this reason is not a 'reason' in the sense of a purpose, cause, scientific explanation, or evidence (justification) for a belief or statement, but in some unique sense not fully comprehensible by us. This argument is consistent, of course, with the position that there actually is no reason for the existence of the Big Bang universe and that it is not possible that this universe has a cause or purpose. See his *The Mystery of Existence: An Essay in Philosophical Cosmology* (New York: Appleton-Century-Crofts, 1965), esp. pt. 4 and the conclusion.

VIII

Theism and Big Bang Cosmology
WILLIAM LANE CRAIG

In Essay VII Quentin Smith attempts to turn the tables on those who have argued that standard Big Bang cosmology, by providing scientific confirmation for the beginning of the universe, constitutes indirect evidence for a personal Creator of the universe, or God. Smith maintains that a non-theistic interpretation of the Big Bang is not only better justified than the theistic interpretation, but that Big Bang cosmology is actually *inconsistent* with theism.

I. THEISTIC INTERPRETATION

The theistic interpretation of standard Big Bang cosmology is presumably the classical doctrine of *creatio ex nihilo*: that a finite time ago God brought the universe into being without a material cause and fashioned the cosmos according to His design. The crux of Smith's argument against this interpretation lies in the fact that at the initial cosmological singularity all the laws of physics break down, so that it becomes impossible to predict what will emerge from it. Having explained this, Smith lays out the following ten-step argument (I have moved step (2) for clarity's sake):

(1) If God exists and there is an earliest state E of the universe, then God created E.

(3) God is omniscient, omnipotent, and perfectly benevolent.

(4) An animate universe is better than an inanimate one.

(2) Therefore, if God created E, then E is ensured either to contain animate creatures or lead to a subsequent state of the universe that contains animate creatures. (from (3), (4))

(5) There is an earliest state of the universe, and it is the Big Bang singularity.

First pub. in *Australasian Journal of Philosophy*, 69/4 (1991), 492–503.

(6) The earliest state of the universe is inanimate.

(7) The Big Bang singularity is inherently unpredictable and lawless, and consequently there is no guarantee that it will emit a maximal configuration of particles that will evolve into an animate state of the universe.

(8) Therefore, the earliest state of the universe is not ensured to lead to an animate state of the universe. (from (5), (7))

(9) Therefore, God could not have created the earliest state of the universe. (from (2), (6), (8))

(10) Therefore, God does not exist. (from (1), (9))

Smith considers four objections to this argument, which I should like to examine in reverse order.

1. *God could providentially ensure via His middle knowledge that the initial cosmological singularity should spawn an eventually animate universe.* Smith is to be commended for his perceptiveness in discerning the relevance of the recently revived Molinist doctrine of divine middle knowledge to his argument.[1] Smith suggests that logically prior to God's creative decree, perhaps God knew the conditional future contingent:

---

[1] See L. de Molina, *On Divine Foreknowledge: Part IV of the 'Concordia'*, trans. with an intro. and notes by Alfred J. Freddoso (Ithaca, NY: Cornell University Press, 1988). According to Molina, in addition to the contingency in the world that springs from God's free action as the primary cause, there is a natural contingency that exists in the world either directly as a result of indeterministic secondary causes or indirectly as the product of causal chains stemming from indeterministic secondary causes. Molina maintained that logically prior to God's eternal decision to create the world, He possessed an exhaustive knowledge, not only of all metaphysically necessary states of affairs, but also of all conditional future contingents. These latter are conditional states of affairs indicating which naturally contingent effects would be produced, directly or indirectly, by indeterministic secondary causes, granted the obtaining of some condition that specifies a possible arrangement of secondary causes. By actualizing the state of affairs specified in the condition, God can ensure that the contingent state of affairs which would ensue does in fact ensue. In this way God can providentially govern a world containing indeterministic secondary causes without removing from that world its contingency. Although Molina's doctrine is usually employed to reconcile God's sovereignty with human freedom, it is equally well suited to harmonize divine providence with indeterminacy in nature, as Freddoso points out (in de Molina, *On Divine Foreknowledge*, 29). For an exposition of Molina's and his defender Suarez's views, see William Lane Craig, *The Problem of Divine Foreknowledge and Future Contingents from Aristotle to Suarez* (Leiden: E. J. Brill, 1988), chs. 7 and 8. For a defence of the doctrine of middle knowledge, see William Lane Craig, *Divine Foreknowledge and Human Freedom* (Leiden: E. J. Brill, 1990), ch. 13.

(C) If a Big Bang singularity were to be the earliest state of the universe, this singularity would emit a life-producing configuration of particles.

Knowing the truth of (C), God, wishing to create an animate universe, decreed that the condition specified in the antecedent of (C) obtain, thereby ensuring that an animate universe would exist, without, however, causally determining that it should. This analysis makes it evident that by (2), Smith really means

(2′) Therefore, □ (if God created *E*, either *E* contains animate creatures or $\boxed{P}$ (*E* leads to a subsequent state of the universe that contains animate creatures)),

where '□' indicates broadly logical necessity and '$\boxed{P}$' indicates physical necessity. Given the doctrine of divine middle knowledge, the inference of (2′) is invalid, rendering the whole argument unsound.

But Smith charges that the theory of middle knowledge is incoherent because, according to possible worlds analyses of the truth conditions of counterfactual conditionals, such conditionals can be true only if the actual world already exists to serve as a relatum of the similarity relation among worlds which allows one to specify the truth conditions for the counterfactuals in question. But a counterfactual like (C) is supposed to be true logically prior to the obtaining of the actual world and, indeed, to furnish a partial basis for God's decree to actualize this possible world. But since (C) cannot be true logically prior to God's decree, it follows that the doctrine of middle knowledge is incoherent.

Confronted with such an objection, a defender of middle knowledge could simply point out that this consideration only serves to expose a shortcoming in the currently fashionable possible worlds semantics for counterfactual conditionals. Smith has to treat possible worlds semantics as the final word on the truth conditions of counterfactual conditionals in order for his anti-theistic argument to go through, a course which is hardly prudent in view of the recency and controverted character of these theories. Moreover, these theories were not developed with such an exotic case as divine middle knowledge in mind, so that it would be hardly surprising should such theories prove inadequate for analysing God's knowledge and decrees. One inadequacy of such theories, for example, appears to be their inability to deal with counterfactuals with impossible antecedents. The theist wishes to assert, for instance, the truth of

(A) If God did not exist, the world would not exist

and the falsity of

(B) If God did not exist, the world would exist,

but according to the predominant possible worlds analysis both (A) and (B) are vacuously true, since the antecedent does not obtain in any possible world. Clearly, some other analysis of the truth conditions of counterfactual conditionals needs yet to be developed in order to handle what are here evidently mutually exclusive and significant claims. Similarly, objections to divine middle knowledge based on semantical theories never intended to handle such a case are not very compelling.

In fact, however, I think that divine middle knowledge can be shown to be consistent with possible worlds semantics for counterfactual conditionals. In order to show this, we need to say a word about the criteria of similarity between possible worlds. Plantinga invites us to consider two worlds W and W^* in which the antecedent of some counterfactual is true and which have identical initial segments up to some time t, at which the worlds diverge in virtue of the fact that the consequent of the counterfactual is true in W but not in W^*.[2] In such a case, it could be argued that both worlds are equally close to the actual world up to t (what happens after t can hardly be relevant to what would have happened at t had the conditions in the antecedent obtained), so that neither $p\square \to q$ nor $p\square \to -q$ is true. The defect of this argument against the truth of the counterfactuals in question, says Plantinga, is that it neglects causal or natural laws. If in W^* not all of the natural laws operative in the actual world are operative there, whereas in W they are, then W^* is not so similar to the actual world as is W. But Plantinga notes an intriguing consequence of such an answer. A salient feature of causal laws is that they support counterfactuals. Hence, instead of claiming that W^* differs in its natural laws from the actual world, we could with equal justification have said that W^* lacks some of the actual world's counterfactuals. Thus, one measure of similarity between worlds is the degree to which they share their counterfactuals. So even if W and W^* share the same initial segment, it does not follow that W and W^* are equally similar to the actual world. For one measure of similarity will be the degree to which each shares the same counterfactuals with the actual world. If

---

[2] A. Plantinga, *The Nature of Necessity* (Oxford: Clarendon Press, 1974), 176–8.

$p\square \rightarrow q$ is true in the actual world, then whichever of W or W^* shares that counterfactual will, all things being equal, be most similar to the actual world.

Now logically prior to God's decree to create the universe, God knew, first, all broadly logically necessary propositions and, secondly, all true counterfactual conditionals relevant to conditional future contingents. Corresponding to the logical sequence in God's knowledge there is a logical sequence in the instantiation of the actual world as well. First, all broadly logically necessary states of affairs obtain. For this reason it is technically misleading to speak of God's actualizing a possible world, for there is a wide range of states of affairs which God does not actualize. So even though the actual world does not obtain at this logical moment, it is none the less true that aspects of the actual world already obtain at this moment, namely, broadly logically necessary states of affairs.

In the second moment corresponding to God's middle knowledge, the actual world is even more fully instantiated than at the first moment. For now all those conditional future contingent states of affairs corresponding to true counterfactuals obtain. For example, that state of affairs *If Peter were in* c, *he would deny Christ three times* obtains. Of course, neither Peter nor the circumstances at this point exist, and God could decide at this juncture not to create Peter at all or not to place him in those circumstances; but still the state of affairs obtains that if this individual essence were instantiated in the actualized state of affairs envisioned, then the exemplification of that essence would deny Christ three times. Hence, at the second logical moment additional states of affairs obtain which are not actualized by God.

Then comes the divine decree to create, and God freely actualizes all remaining states of affairs of the actual world. Only at this point does the actual world in its entirety obtain.

The upshot of this is that it is not wholly correct to say with Smith that prior to the divine decree the actual world does not obtain, for certain aspects of it do and other aspects do not. And those conditional future contingent states of affairs that do obtain are sufficient for the truth of their corresponding counterfactuals, since the latter correspond with reality as it thus far exists, and possible worlds can be ranked in their similarity to the actual world as thus far instantiated in terms of degree of shared counterfactuals, thus supplying the truth conditions for a possible worlds analysis of the truth of such counterfactuals.

But Smith impugns such an analysis as 'viciously circular'. For in order to demonstrate that the relevant counterfactuals are true logically prior to creation, it is assumed that logically prior to creation there obtain conditional future contingent states of affairs corresponding to the true counterfactuals.

But how is this circular? The counterfactuals are true because the relevant states of affairs obtain; but the states of affairs do not obtain because their corresponding conditionals are true. Smith's feeling of circularity stems, I think, rather from the fact that on a possible worlds semantics the truth conditions for the true counterfactuals may be specified in terms of the similarity of other possible worlds to the actual world as it thus far obtains and that the criterion of similarity to be employed in making the comparison is whether they share the same true counterfactuals. But this does not prove a *vicious* circularity, but only that, as Plantinga explains 'we cannot as a rule *discover* the truth value of a counterfactual by asking whether its consequent holds in those worlds most similar to the actual world in which its antecedent holds'.[3] Smith's difficulty may be that he is looking to the possible worlds account for a reason *why* certain counterfactuals are true or false, and that account fails to provide any. But that account is not to be construed as explaining why certain counterfactuals are true or false; it only means to provide an account of what it is for a counterfactual to be true or false.[4] Because the relevant states of the actual world already obtain prior to the divine decree to create, a possible worlds analysis of the truth conditions of the counterfactuals that are at that point true or false can be given, since other possible worlds can be ranked in terms of their similarity to the actual world as it thus far obtains. Such an analysis will no doubt be a rather uninspiring affair, but it is not for all that viciously circular.

Hence, it seems to me that the modern Molinist will be unphased by Smith's anti-theistic argument. He may contend that in virtue of His middle knowledge of conditional future contingents, God knew logically prior to His decree to create the universe the truth of (C) and thus was able to ensure the existence of an animate universe

---

[3] Plantinga, *The Nature of Necessity*, 178.

[4] Plantinga draws an instructive comparison in this regard between the possible worlds account of the truth conditions of counterfactual conditionals and the possible worlds account of the modal concepts of necessity and possibility in A. Plantinga, 'Reply to Robert Adams', in J. Tomberlin and P. van Inwagen (eds.), *Alvin Plantinga* (Dordrecht: D. Reidel, 1985), 378.

wholly apart from His intervention in the series of secondary causes. Indeed, the Molinist is apt to regard his account as an exciting and enlightening explanation of how a sovereign God providentially orders and guides an inherently indeterministic universe.

2. *The initial cosmological singularity is not an existent.* That is to say, the singularity has no positive ontological status: as one traces the cosmic expansion back in time, the singularity represents the point at which the universe ceases to exist. It is not part of the universe, but represents the point at which the time-reversed contracting universe vanishes into non-being. There was no first instant of the universe juxtaposed to the singularity. The temporal series is like a series of fractions converging toward 0 as its limit: $\frac{1}{2}, \frac{1}{4}, \frac{1}{8}, \ldots, 0$. Just as there is no first fraction, so there is no first state of the universe. The initial singularity is thus the ontological equivalent of nothing. The universe originates out of nothing. The breakdown of the laws of physics and the attendant unpredictability is perspicuous in light of the fact that nothingness possesses no physics.

If such a metaphysical interpretation of the initial singularity is even possible, then premiss (5) is unsubstantiated and Smith's antitheistic argument is undercut. For under this interpretation the singularity is not the first created state of the universe after all. Rather any initial temporal segment of the universe which one chooses to consider (such as the first three minutes) will be governed throughout by the laws of physics and so be amenable to prediction.

But Smith objects to such an interpretation, pointing out that in standard Big Bang cosmology the singularity is a past end-point of all converging past-directed spacetime paths. These paths are not commonly thought to approach arbitrarily this end-point but to actually reach it and terminate in it.

But does this justify Smith's inference that the singularity is an existent, a physical reality? This is far from clear.[5] For the mathe-

---

[5] It must be said that Smith misuses his sources at this point. According to him, 'In the words of Geroch and Horowitz, converging past-directed spacetime paths are not commonly thought to merely approach with arbitrary closeness the same singular point but are thought to actually "*reach* the same singular point", which requires the actual physical existence of the singular point.' (Essay VII, Sect. 6). But in fact all that Geroch and Horowitz said was, 'Consider now an example of some physical statement one may wish to make, for example, that curves γ_1 and γ_2 in figure 5.28 reach the same physical point.' R. Geroch and G. T. Horowitz, 'Global Structure of Spacetimes', in S. W. Hawking and W. Israel (eds.), *General Relativity* (Cambridge:

matical model which represents the singularity as the end-point of all converging past-directed spacetime paths does not carry with it an interpretation of the ontology involved. That is a question for philosophical metaphysics. Why could not the end-point in which all spacetime paths terminate be a state of ontological nothingness? The past-directed paths converge on a point at which everything disappears.

Certain statements by theoretical physicists lend themselves readily to such an interpretation. Imagining a space traveller C approaching a black hole, Geroch writes,

Individual C is overwhelmed with curiosity about this black hole. He goes directly towards it and then, at event u, crosses the horizon. He then remains in the internal region for a stretch and finally hits the singularity (that is, his world-line goes to this region on the diagram) . . . Recall that there is no event available to C on the singularity itself . . . What, then, does 'the world-line of C hits the singularity' mean physically? Mathematically, what happens is that his world-line just stops. Physically, this would mean that C is 'snuffed out of existence'; after some finite time according to himself, he ceases to exist in space-time. From the point of view of C, it is easy to say what this means. He exists at his watch-reading of 1, at his watch-reading of $1\frac{1}{2}$, at his watch-reading of $1\frac{3}{4}$, at his watch-reading of $1\frac{7}{8}$, and so on, but he simply does not exist for watch-readings of 2 or greater.[6]

Cambridge University Press, 1979), 267. In their discussion that follows, they proceed to speak exclusively in terms of 'approaching the same singular point' (six times). Of course, the inference that the singular point is real is entirely Smith's own.

[6] R. Geroch, *General Relativity from A to B* (Chicago: University of Chicago Press, 1978), 194. Cf. a similar remark by Hawking ('The Occurrence of Singularities in Cosmology', iii: 'Causality and Singularities', *Proceedings of the Royal Society of London*, A300 (1967), 189). 'Although we have omitted the singular points from the definition of space-time, we may still be able to recognize the "holes" left where they have been cut out by the existence of geodesics which cannot be extended to arbitrary values of the affine parameter . . . [Null and timelike geodesic incompleteness] imply that there could be particles or photons whose histories would not exist after (or before) a certain value of the time or of the affine parameter as measured by them. Thus they would apparently be annihilated (or created). It is this feature which some writers have found so objectionable and which has led them to suggest that, under realistic conditions, the General Theory of Relativity would not predict the occurrence of singularities.' Notice, however, that Hawking contradicts himself in stating that particles can be regarded as either created or annihilated at a singular point. For creation and annihilation at a point are *not* time-reversible. If a particle is annihilated at a singular point formed as a result of gravitational self-collapse, then in the time-reversed expansion of that collapse, the particle is not

But this appears to be precisely that interpretation of the ontological status of the singularity which Smith rejects above on Geroch's authority. Clearly, in Geroch's opinion, the fact that the world line of a particle reaches the singularity on a diagram does not carry with it the ontological interpretation that physical reality begins at the singularity. One could consistently maintain that the initial singularity is a state of non-being and that physical reality approaches arbitrarily close to it without having a first instant of existence.

Indeed, such an interpretation of the initial cosmological singularity is suggested by certain leading philosophers of science. Kanitscheider, for example, writes,

an analogy from mathematics can be employed, which softens the sharp alternatives [between an origin of the world and the eternity of the world]. If one follows the course of the world into the past, such a cosmogonic reconstruction does not land on a hard, first-event called the initial singularity, which alone has causal successors but no predecessors; but events in the neighbourhood of $t = 0$ 'run out' like the real numbers in the open interval $(0, 1)$, as we approach zero. This fits nicely with the fact that the singularity itself is not an element of the manifold, and thus at $t = 0$ no event (displayed as a point of four-dimensional spacetime) can have occurred which, as the first event, drew all others after it, although it was itself uncaused.[7]

Or again, Fitzgerald states, 'If one regards the initial singularity as not physically possible, then one can drop it from one's world picture, treating it as a kind of mathematical limit lacking physical reality.'[8]

created at the singular point, for at that point it does not exist. Rather at any moment arbitrarily close to the singularity, the particle exists. In this sense, discussions of creation and annihilation resemble the ancient *sorites*-type paradoxes concerning starting and stopping (for an excellent discussion, see R. Sorabji, *Time, Creation and the Continuum* (Ithaca, NY: Cornell University Press, 1983), 403–21). As Sorabji points out, the solution to these paradoxes involves the insight that there is 'no first *position* occupied away from the finishing point . . . Hence there can equally be no first *instant* of being away from the finishing point.' (*Time, Creation and the Continuum*, 405.) If we substitute existence for motion in these paradoxes, we arrive at the sort of interpretation of singular points which I am suggesting. All that is physically real is part of the spacetime manifold, and singular points lying on the boundary of that manifold are, like the boundary itself, mathematical idealizations having no physical counterparts. If such an interpretation is even possible—which it certainly seems to be—then Smith's anti-theistic argument fails.

[7] B. Kanitscheider, *Kosmologie: Geschichte und Systematik in philosophischer Perspektive* (Stuttgart: Philipp Reclam, Jr., 1984), 261–2.
[8] P. Fitzgerald, 'Swinburne's Space and Time,' *Philosophy of Science*, 43 (1976),

What justification might be offered for such an interpretation? Simply put, an object which has no spatial dimensions and no temporal duration hardly seems to qualify as a physical object at all, but is rather a mathematical conceptualization. The singularity has zero dimensionality and exists for no length of time; it is in fact a mathematical point. The state of infinite density illustrates this. (Though the existence of infinite quantities no longer belongs to the definition of a singularity, the initial cosmological singularity from which our universe emerged doubtless involved such quantities, like infinite density.) The problem with a physical object's being infinitely dense is not, as Smith surmises, that it would involve Cantorian infinites, but that in order for an object, say, a ball, having a finite mass, to be infinitely dense, two diametrically opposite points on its surface would have to approach so close to each other that they would finally converge into a single mathematical point having no volume or dimensions at all. But such a point could hardly be called a physical object and seems ontologically equivalent to nothing. Moreover, this 'object' does not exist for any period of time; its temporal duration is simply zero. But anything having positive ontological status would seem necessarily to exist for some temporal duration; to say it exists only at a durationless instant is to ascribe reality to a mathematical chimera. Similarly Smith's arguments about converting the topology of a point into the topology of finite three-dimensional space concern mathematical operations which do not entail a particular ontology.

In short, unless Smith is able to offer compelling reasons to the contrary, it seems to me that the metaphysician is rational in interpreting the ontological status of the singularity as nothingness rather than as the first state of physical reality. But if so, then Smith has failed to prove any inconsistency between theism and Big Bang cosmology.

3. *God can intervene to ensure an animate universe.* This objection serves to point out the most glaring weakness of Smith's argument. For while the doctrine of middle knowledge and the interpretation of the ontological status of the initial cosmological singularity are controversial, it seems perfectly perspicuous to me that God is free

636. Similarly J. S. Earman, 'Till the End of Time', in J. S. Earman, C. N. Glymour, and J. J. Stachel (eds.), *Foundations of Space-Time Theories* (Minneapolis: University of Minnesota Press, 1977), 109–33.

to bring about an animate universe as He wills and that if He wants to causally direct the evolution of nature toward this desired end rather than have it all pre-programmed in the first stage of the universe, then that is His prerogative as the sovereign Lord of creation. In other words, Smith's inference in (2′) seems to me invalid.

His arguments in support of this inference make little advance over the old objections of seventeenth-century Deism against miracles. Like Voltaire, Smith maintains that God would be irrational and incompetent to create a universe which repeatedly required His intervention in order to arrive at His desired end. According to Smith, God's bringing about an animate universe by creating a world that time and again from its inception was headed towards the very opposite end than His previsioned one, such that His intervention was necessary to counteract this natural tendency, which He Himself had given it, 'does not agree with the idea of an efficient creation'.[9] Smith asserts, 'my argument does not require that God create an animate universe in the most efficient way possible . . . but merely that he create it in an efficient way (which minimally requires that no interventions be needed)'.[10]

But why think that God is incompetent because He does not conform to our standards of efficiency? Thomas Morris makes two points in this connection:

First of all, efficiency is always relative to a goal or set of intentions. Before you know whether a person is efficient in what she is doing, you must know what it is she intends to be doing, what goals and values are governing the activity she is engaged in. In order to be able to derive . . . the conclusion that if there is a God in charge of the world, he is grossly inefficient, one would have to know of all the relevant divine goals and values which would be operative in the creation and governance of a world such as ours. Otherwise, it could well be that given what God's intentions are, he has been perfectly efficient in his control over our universe.[11]

To illustrate, on Smith's principles a boy engaged in building a model airplane is irrational to prefer actually assembling the plane, in defiance of the tendency of the several parts to disorder in accordance with the law of entropy, over simply having, if he

---

[9] Essay VII, Sect. 5.
[10] Essay VII, n. 20.
[11] T. V. Morris, *The Logic of God Incarnate* (Ithaca, NY: Cornell University Press, 1986), 77–8.

could, in an instant the finished product. Or again, a chef ought to prefer simply having instantaneously, if he could, his gourmet meal to eat over going to all the effort of preparing the meal. What Smith has overlooked is the fact that the final end of some activity may not exhaust an agent's reasons for engaging in that activity. The boy and the chef are not deemed to be bungling and irrational incompetents, since part of their intention is to enjoy the activity in which they are engaged. But how does Smith know that God does not also have reasons for His being causally engaged in the activity of creation? Perhaps God delights in the creative activity of fashioning a world. Or perhaps He wanted to leave a general revelation of Himself in nature by creating a world which would never in all probability have resulted from the natural tendencies of things alone. Perhaps He has reasons of which we have no idea.

But, secondly, Morris asks,

what reason do we have to hold that efficiency is a great-making property at all? . . . What is the property of being efficient, anyway? An efficient person is a person who husbands his energy and time, achieving his goals with as little energy and time as possible. Efficiency is a good property to have if one has limited power or limited time, or both. But apart from such limitations, it is not clear at all that efficiency is the sort of property it is better to have than to lack. On the Anselmian conception of God, he is both omnipotent and eternal, suffering limitations with respect to neither power nor time. So it looks as if there is no good reason to think that efficiency is the sort of property an Anselmian being would have to exemplify.[12]

Smith believes that it contradicts God's omniscience, or rather His wisdom, to create a world in need of correction by intervention because this is incompetent planning. But why is it incompetent? So far as I can see, the only reason given by Smith why God's intervening in the natural order of causes is incompetent is because it is inefficient—but that is to impose an inappropriate, anthropocentric standard upon God.

In sum, there is no reason to think that perfect-being theology entails a Deist account of creation. The Christian theist in particular believes that God not only created and conserves the world in being, but also that He is living and active within it. He will therefore tend to welcome Smith's case for the necessity of divine

---

[12] Morris, *The Logic of God Incarnate*, 78.

interventions as a confirmation of God's intimate involvement with creation at a very fundamental level.

4. *God is under no necessity to create an animate universe.* Since it is possible that there is no best possible world, it is not necessary that God create such a world. The most that His omnibenevolence would seem to require is that He create a world in which the evil is not on balance greater than the good. But for any world He creates, there may always be a better one. Now as premiss (4) states, an animate universe seems clearly better, *ceteris paribus*, than an inanimate one. But it does not follow from (3) and (4) that God is obliged to create an animate universe. Smith has failed to demonstrate any inconsistency between (3), (4), and

(11) God freely chooses to create an inanimate universe.

But if this is the case, then the inference in (2′) is once again invalid. Smith rejoins that the theist must impose some minimal constraint on the value of the universe which God creates. One such minimal constraint is surely that the universe be animate. For the idea that God has no more reason to create an animate universe than an inanimate is inconsistent with the traditional concept of God in Western theism.

But where does Smith get the idea that the compatibility of (3) and (11) implies that God has no reason for creating an animate universe rather than an inanimate one? God's having a reason for creating an animate universe does not imply that, necessarily, God creates an animate universe. Analogously, God no doubt had a reason for creating the world, but it does not follow that He created necessarily. Hence, Smith's later additional premiss

(4′) If God chooses to create a universe, He will choose to create an animate rather than an inanimate universe

is not necessarily true. All that Smith's reasoning proves is

(4″) If God chose to create the universe, He had a reason to create an animate rather than an inanimate one.

But it does not follow from (4″) that necessarily, if God chose to create the universe, then its first state either is animate or leads by physical necessity to an animate state. For there is no inconsistency in His choosing to create an inanimate universe. Hence, Smith has no good grounds for inferring (2′), without which his argument fails.

Of course, Smith could reformulate his argument, using

(2″) Therefore, (if God chose to create an animate universe, then either the earliest state E is animate or ℙ (E leads to an animate state)).

But then we are back to the inadequate responses to the objections discussed above. Like the inference of (2′), the inference of (2″) seems patently invalid.

2. NON-THEISTIC INTERPRETATION

The implausibility of Smith's argument and the superiority of the theistic interpretation of the Big Bang become painfully apparent when we consider what alternative the non-theistic interpretation offers us.[13] This interpretation is that 'our universe exists without cause and without explanation'.[14] The universe (including, in Smith's view, the initial singularity) simply sprang into being out of nothing without a cause, endowed with a set of complex initial conditions so fantastically improbable as to defy comprehension, and as it evolved through each symmetry-breaking phase cosmological constants and quantities continued to fall out wholly by accident so as to accord with this delicate balance of life-permitting conditions. Such an interpretation seems implausible, if not ridiculous. The metaphysician can hardly be regarded as irrational if he reposes more confidence in the principle *ex nihilo nihil fit* than in Smith's mootable premisses. I want to underline the fact that I in no wise denigrate Smith's profound astonishment, which he poetically expresses, that the universe exists at all—on the contrary, I feel it, too. But that astonishment should not end in a mute stupefaction, but lead us, as Leibniz saw, to the intelligible explanation of the universe, the God of classical theism.

---

[13] For more on these alternatives, see Essay X.
[14] Essay VII, Sect. 8.

IX

A Defence of the Cosmological Argument for God's Non-existence

QUENTIN SMITH

I. INTRODUCTION

The advent of Big Bang cosmology in this century was a watershed for theists. Since the times of Copernicus and Darwin, many theists regarded science as hostile to their world-view and as requiring defence and retrenchment on the part of theism. But Big Bang cosmology in effect reversed this situation. The central idea of this cosmology, that the universe exploded into existence in a 'Big Bang' about 15 billion years ago or so, seemed tailor-made to a theistic viewpoint. Big Bang cosmology seemed to offer empirical evidence for the religious doctrine of creation *ex nihilo*. The theistic implications seemed so clear and exciting that even Pope Pius XII was led to comment that 'True science to an ever increasing degree discovers God as though God were waiting behind each door opened by science.'[1] But the theistic interpretation of the Big Bang has received not only widespread dissemination in popular culture and official sanction but also a sophisticated philosophical articulation. Richard Swinburne, John Leslie, and especially William Lane Craig[2]

Not previously published.

[1] Pope Pius XII, *Bulletin of the Atomic Scientists*, 8 (1952), 143–6.

[2] See Richard Swinburne, *The Existence of God* (Oxford: Clarendon Press, 1979) and *Space and Time*, 2nd edn. (New York: St Martin's Press, 1981). Swinburne doubts that the prediction of a first event by Big Bang cosmology is probably true but none the less shows how this prediction can be theologically construed. Also see John Leslie, 'Anthropic Principle, World Ensemble, Design', *American Philosophical Quarterly*, 19 (1982), 141–51, 'Modern Cosmology and the Creation of Life', in E. McMullin (ed.), *Evolution and Creation* (South Bend, Ind.: University of Notre Dame Press, 1985) and numerous other articles. Leslie, of course, operates with a Neoplatonic conception of God, but his arguments are obviously relevant to classical theism. The most developed theistic interpretation of Big Bang cosmology is William Lane Craig's. See his *The* Kalām *Cosmological Argument* (New York: Barnes &

have developed powerful arguments for theism based on a well-grounded knowledge of the cosmological data and ideas.

The response of atheists and agnostics to this development has been comparatively weak, indeed, almost invisible. An uncomfortable silence seems to be the rule when the issue arises among non-believers or else the subject is briefly and epigrammatically dismissed with a comment to the effect that 'science has no relevance to religion'. The reason for the apparent embarrassment of non-theists is not hard to find. Anthony Kenny suggests it in this summary statement:

According to the big bang theory, the whole matter of the universe began to exist at a particular time in the remote past. A proponent of such a theory, at least if he is an atheist, must believe that the matter of the universe came from nothing and by nothing.[3]

This idea disturbs many for the reason it disturbs C. D. Broad:

I must confess that I have a very great difficulty in supposing that there was a first phase in the world's history, i.e. a phase immediately before which there existed neither matter, nor minds, nor anything else. . . . I suspect that my difficulty about a first event or phase in the world's history is due to the fact that, whatever I may *say* when I am trying to give Hume a run for his money, I can not really *believe* in anything beginning to exist without being *caused* (in the old-fashioned sense of *produced* or *generated*) by something else which existed before and up to the moment when the entity in question began to exist. . . . I . . . find it impossible to give up the principle; and with that confession of the intellectual impotence of old age I must leave this topic.[4]

Motivated by concerns such as Broad's, some of the few non-theists who have been vocal on this subject have gone so far as to deny, without due justification, central tenets of Big Bang cosmology. Among physicists, the most notorious example is Fred Hoyle, who vehemently rejected the suggestion of a Big Bang that seemed to imply a Creator and unsuccessfully attempted to construe the

Noble, 1979), 'God, Creation and Mr Davies', *British Journal for the Philosophy of Science*, 37 (1986), 163–75, 'Barrow and Tipler on the Anthropic Principle vs. Divine Design', *British Journal for the Philosophy of Science*, 39 (1988), 389–95; Essay XI and Essay V.

[3] Anthony Kenny, *The Five Ways* (New York: Schocken Books, 1969), 66.

[4] C. D. Broad, 'Kant's Mathematical Antinomies', *Proceedings of the Aristotelian Society*, 40 (1955), 1–22. This passage and the passage from Kenny are quoted in Essay I, Sect. 3.

evidence for a Big Bang as evidence for an evolving 'bubble' within a larger unchanging and infinitely old universe (I am referring to his 1970s post-steady state theory[5]). An example of this contrary approach among philosophers is evinced by W. H. Newton-Smith. Newton-Smith felt himself compelled to maintain, in flat contradiction to the singularity theorems of Big Bang cosmology (which entail that there can be no earlier state of the universe than the Big Bang singularity), that the evidence that macroscopic events have causal origins gives us 'reason to suppose that some prior state of the universe led to the production of this particular singularity'.[6]

It seems to me, however, that non-theists are not put in such dire straits by Big Bang cosmology. Non-theists are not faced only with the alternatives of embarrassed silence, confessions of impotence, epigrammatic dismissals, or 'denial' when confronted with the apparently radical implications of Big Bang cosmology. It will be my purpose in this paper to show this by further developing a coherent and plausible atheistic interpretation of the Big Bang,[7] an interpretation that is not only able to stand up to the theistic interpretation but is in fact *better justified* than the theistic interpretation. But my argument is intended to establish even more than this, namely, that Big Bang cosmology is actually *inconsistent* with theism. I will argue that if Big Bang cosmology is true, then God does not exist.

2. THE ARGUMENT

This argument was initially presented in Essay VII and I shall briefly restate it and then offer a more comprehensive defence of the argument than was offered in that essay.

The argument involves the idea that the Big Bang singularity is inherently unpredictable, as is implied by Hawking's principle of ignorance. The significance of this principle can be easily missed. It

[5] See Fred Hoyle, *Astrophysical Journal*, 196 (1975), 661.

[6] W. H. Newton-Smith, *The Structure of Time* (London: Routledge & Kegan Paul, 1980), 111.

[7] Quentin Smith, 'The Anthropic Principle and Many Worlds Cosmologies', *Australasian Journal of Philosophy*, 63 (1985), 336–48; 'World Ensemble Explanations', *Pacific Philosophical Quarterly*, 67 (1986), 73–86; 'A Natural Explanation of the Existence of the Laws of Our Universe', *Australasian Journal of Philosophy*, 68 (Mar. 1990), 22–43; Essay IV.

implies that the Big Bang singularity behaves in a completely unpredictable manner *in the sense that no physical laws govern its behaviour*. The unpredictability of the singularity is not simply an epistemic affair, meaning that 'we humans cannot predict what will emerge from it, even though there is a law governing the singularity which, if known, would enable precise predictions to be made'. William Lane Craig assumes unpredictability to be merely epistemic; he writes that 'unpredictability [is] an epistemic affair which may or may not result from an ontological indeterminism. For clearly, it would be entirely consistent to maintain determinism on the quantum level even if *we* could not, even in principle, predict precisely such events.'[8] Now I grant that there are legitimate uses of 'unpredictability' that are merely epistemic in import, but this is not how the word is used in Hawking's principle of ignorance. The unpredictability that pertains to Hawking's principle of ignorance is an unpredictability that is a consequence of lawlessness, not of human inability to know the laws. There is no law, not even a probabilistic law, governing the singularity that places restrictions on what it can emit. Hawking writes that

A singularity can be regarded as a place where there is a breakdown of the classical concept of space-time as a manifold with a pseudo-Riemannian metric. Because all known laws of physics are formulated on a classical space-time background, they will all break down at a singularity. This is a great crisis for physics because it means that one cannot predict the future. One does not know what will come out of a singularity.[9]

Deterministic or even probabilistic laws cannot obtain on the quantum level in the singularity, since *there is no quantum level* in the singularity; the spacetime manifold that quantum processes presuppose has broken down. The singularity is a violent, terrifying cauldron of lawlessness. As Paul Davies notes, 'anything can come out of a naked singularity—in the case of the Big Bang the universe came out. Its creation represents the instantaneous suspension of physical laws, the sudden, abrupt flash of lawlessness that allowed something to come out of nothing.'[10] The question I shall examine is whether this primordial lawlessness is consistent with the hypothesis of divine creation. I shall argue it is not.

[8] Essay V, n. 4.
[9] S. W. Hawking, 'Breakdown of Predictability in Gravitational Collapse', *Physical Review*, D14 (1976), 2460.
[10] P. Davies, *The Edge of Infinity* (New York: Simon & Schuster, 1981), 161.

The cosmological argument for God's non-existence has two premisses that are based on classical Big Bang cosmology, namely,

(1) The Big Bang singularity is the earliest state of the universe.
(2) The earliest state of the universe is inanimate.

(2) follows from (1) since the singularity involves the life-hostile conditions of infinite temperature, infinite curvature, and infinite density.

The principle of ignorance gives us the summary premiss

(3) No law governs the Big Bang singularity and consequently there is no guarantee that it will emit a configuration of particles that will evolve into an animate universe.

(1)–(3) entail

(4) The earliest state of the universe is not guaranteed to evolve into an animate state of the universe.

My argument is that (4) is inconsistent with the hypothesis that God created the earliest state of the universe, since it is true of God that *if he created the earliest state of the universe, then he would have ensured that this state is animate or evolves into animate states of the universe.* But this conditional may be doubted and the argument rejected.

3. THE QUESTION OF GOD'S INTENTION TO CREATE AN ANIMATE UNIVERSE

I believe it is essential to the idea of God as the perfect being that if he creates a universe, he creates an animate universe, and therefore that if he creates a first state of the universe, he creates a state that is, or is guaranteed to evolve into, an animate state. If somebody says, 'It does not matter to God whether the universe he creates is animate or inanimate', the person is operating with a concept of God different from that embodied in the definition of God as the perfect being, as omniscient, omnibenevolent, omnipotent, etc. Even if there is no best possible world, it does not follow that it is consistent with the divine nature to create just any world. There are different types of worlds and God is constrained by his nature to create a world of a certain type (even though there may be no best world within the class of possible worlds constituting the type in question). William Craig admits that the divine perfections impose some constraint upon which type of world God creates. 'The most that his omnibenevolence would seem to require is that he create a world in which the evil is not on balance greater than the good.'[11] However, I believe these perfections impose further constraints, for example, that the type of world God creates be one in which there is no gratuitous evil. If the evil is not on balance greater than the good, but the world contains some gratuitous evil, then this world cannot be created by an omnibenevolent being. I believe a further constraint imposed by God's omnibenevolence is that the type of world he creates be animate. Omnibenevolence requires living creatures in relation to which God can exercise his benevolence. How can God be benevolent to a cloud of hydrogen gas? Divine benevolence involves advancing the good of creatures and in order to have *a good* a creature must have some desires or goals that God can enable it to attain by creating it and placing it in the appropriate circumstances. A cloud of hydrogen gas, however, has no good that God can advance.

It might be responded that God's benevolence need not require him to advance the good of his creatures. Divine benevolence requires only that God do something that is good. Since a lifeless universe is better than no universe at all, God is doing something good by creating a lifeless universe.

I reject the idea that God's benevolence does not require that he advance the good of his creatures and thus I reject the idea that God would create only creatures that have no good. But even waiving this objection, the above response is unsound. I would grant that a lifeless universe is better than no universe at all, but deny that God is being good or benevolent if he creates such a universe rather than an animate universe. In assessing whether or not the creation of a universe containing only hydrogen gas is an act of benevolence we must take into account not only that an inanimate universe is better than no universe but also that an animate universe is better than an inanimate universe. Is it consistent (in the broadly logical sense) with omnibenevolence to bring into existence nothing but a cloud of hydrogen gas, when one could have instead brought into existence a world with animals and intelligent and moral organisms? I think it is not consistent, but is instead a sign of moral callousness or indifference or a failure to recognize the distinctive

[11] Essay VIII, Sect. 1.

value of intelligent and moral organisms; these signified characteristics are not consistent with a god that is all-knowing and supremely good. It would be a gratuitous evil of omission if God created merely an inanimate world.

Craig writes that:

All that Smith's reasoning proves is

(4″) If God chose to create the universe, he had a reason to create an animate rather than an inanimate one.

But it does not follow from (4″) that necessarily, if God chose to create the universe, then its first state is either animate or leads by physical necessity to an animate state. For there is no inconsistency in His choosing to create an inanimate universe.[12]

I agree that this 'does not follow', but for reasons not mentioned by Craig. It does not follow because God may choose to create a universe with no first state, but infinite in its past-direction. It also does not follow since God may create a first state that leads with high probability (rather than physical necessity) to an animate state. Craig's reason seems to be that if God actually had a reason to create an animate rather than an inanimate universe, this is consistent with his choosing to create an inanimate rather than animate universe in some merely possible world. I do not think so, since God does not contingently, but necessarily, have a reason to create an animate rather than an inanimate universe. It contradicts his omnibenevolence to suppose there is a possible world where God has a reason to create an inanimate rather than an animate universe, or has no reason to create an animate rather than inanimate universe. And if God has a reason (an overriding reason is the sort we have in mind) to create an animate rather than an inanimate universe, then it would be irrational to create an inanimate universe. (A person is irrational if he has an overriding reason to do A and chooses to do not-A instead.) Thus, I believe that necessarily, if God chose to create a universe that has a beginning in time, then its first state is either animate or leads by physical necessity or high probability to an animate state. In Essay VII I said that God would *ensure* the first state is or leads to an animate state, but this constraint can be interpreted broadly so as to allow for 'high probability' as well as 'physical necessity'.

---

[12] Ibid.

I shall now respond to further objections to the atheistic cosmological argument. Some of these objections have been made by Craig in Essay VIII, some are suggested by the writings of Swinburne, and still others are formulated independently.

4. THE QUESTION OF DIVINE INTERVENTION

One objection to the argument of Section 2 is that it does not take into account the possibility of divine intervention. If the Big Bang singularity is lawless, then it is feasible for God to intervene at the instant of the singularity and supernaturally constrain it to explode in a certain way, namely, to explode by emitting a life-producing maximal configuration of particles. In this way, God can guarantee that the earliest state of the universe will evolve into an animate state.

But it is not at all obvious that this objection is consistent with the classical theist conception of the divine nature. God is omniscient, omnipotent, and perfectly rational and it is not a sign of a being with these attributes to create as the first state of the universe some inherently unpredictable entity that requires immediate 'corrective' intervention in order that the universe may be set on the right course. It is a mark of inefficiency, incompetent planning, and poor design to create as the first natural state something that needs supernatural intervention 'right off the bat' to ensure that it leads to the desired outcome.

William Craig objects to this argument that it counts against deism, but not against classical theism.[13] The god of the deists creates the first state (if there is a first state) and allows the universe to evolve henceforth on its own, without requiring any further divine activity. The god of the classical theists, on the other hand, is ceaselessly active in the universe, for this god is *continuously creating* (*conserving*) the universe. Accordingly, Craig concludes, the classical theist 'will therefore tend to welcome Smith's case for the necessity of divine interventions as a confirmation of God's intimate involvement with creation at a very fundamental level'.[14]

But this response is based on a failure to distinguish in a relevant way *divine intervention* from *divine conservation* (*continuous creation*).

[13] Ibid.   [14] Ibid.

The two concepts are incompatible in a certain respect, i.e. that an intervention is possible only if conservation is not being realized in a rational way. God intervenes in a natural process if and only if God violates a natural law governing that process or alters a natural process or state that is not governed by any law. Divine conservation requires that God sustains the universe in existence throughout the time that the universe exists, such that each subsequent state of the universe is produced by God as following upon the prior state. If the universe is conserved by a perfectly rational being, then the sequence of states (the way in which they follow upon one another) will be lawlike and rationally explicable rather than haphazard. If a branch is falling to the ground at time t, then at $t + 1$ the universe is conserved in such a way that the branch hits the ground rather than vanishes into nothingness and is replaced by an instantaneously appearing three-headed bat. The materials for a definition of a rational conservation are provided by Swinburne's remarks in this passage:

An argument from the universe to God may start from the existence of the universe today, or from its existence for as long as it has existed—whether a finite or an infinite time. Leibniz considers the argument in the latter form, and I shall follow him. So let us consider the series of states of the universe starting from the present and going backwards in time, S_1, S_2, S_3, and so on. (We can suppose each to last a small finite time.) Now clearly there are laws of nature L which bring about the evolution of S_3 from S_4, S_2 from S_3, and so on. (I shall assume for the purpose of simplicity of exposition that this process is a deterministic process, viz., that L and S_5 together provide a full explanation of S_4, L and S_4 a full explanation of S_3, and so on; we can ignore any minor element of indeterminism—nothing will turn on it.) So we get the following picture:

$$> \ldots S_5 \overset{L}{\rightarrow} S_4 \overset{L}{\rightarrow} S_3 \overset{L}{\rightarrow} S_2 \overset{L}{\rightarrow} S_1.$$

The series of states may be finite or infinite—which, we do not know. Now God might come into the picture in one of two ways, as responsible for L, and so as providing a complete explanation of the occurrence of each state S; or at the beginning of the series (if it has one) as starting the process off.[15]

If the universe has a beginning and classical theism is true, then God will come into the picture in both ways.

---

[15] Swinburne, *The Existence of God*, 120.

These remarks suggest the following definition of conservation by a perfectly rational (powerful, good, etc.) being. God conserves the universe if and only if for any two states S and S' of the universe, where S' is the immediate successor of S, God creates S' and S' has the property of *being the nomological consequent of S*. (If time is dense or continuous, then S' must be understood as a temporally extended state, for there is no instantaneous state S' that immediately succeeds an instantaneous or extended state S if time is dense or continuous.) S' is the nomological consequent of S if and only if there is a set of laws L such that the premisses that L obtains and S exists wholly or partially entail that S is succeeded by S'. The premisses that L obtains and S exists wholly entail that S' is the successor if L is a set of deterministic laws and these premisses partially entail (i.e. render probable to some degree) that S' is the successor if L is a set of probabilistic laws.

If God interferes with the universe by violating or suspending his laws, or by altering the natural tendency of a lawless state, then he is not engaged in a rational continuous creation of the universe (i.e. a producing of a subsequent state as a nomological consequent of a prior state). Now the 'god' of popular imagination or various religious scriptures is thought to interfere in such ways with the universe (e.g. by causing a bush to combust spontaneously or by enabling a person to survive a plane crash in which two hundred other people are killed), but the god with the attributes assigned by classical theism (omniscience, etc.) should not be understood in this way. This god is perfectly rational and conserves the universe in a lawlike and rationally explicable way. It is not consistent with the concept of this god that he[16] should create a lawless first state (the Big Bang singularity) that naturally tends towards the opposite state he desires and thereby requires a divine correction.

I would note that this argument need not be made to hinge upon the reality of the Big Bang singularity. If the singularity is ontologically nothing, as Craig argues, then divine interference still shows up in the symmetry breaking phases, as I argued in Essay VII. It would make no clear sense for God to ordain that $S2$ follow

---

[16] At some juncture I should mention that by always referring to God as 'he' rather than 'she' I am calling attention to the fact that the god of classical theism has its roots in the patriarchal religions of the Judaeo-Christian-Islamic traditions. Distinct from these religions are the goddess religions of early Crete, Malta, Sumer, Canaan, etc. I would reserve 'she' for the deity of goddess religions.

upon S_1 by a probabilistic law L and at the same time for God to violate L and provide a force or particle mass with a value it probably would not have possessed if L were left unviolated. However, I believe the singularity exists (as I shall further argue in a later section) and shall continue to assume this.

I would emphasize that my argument does not presuppose that there is a 'most rational, competent or efficient way of creating an animate universe' and therefore does not succumb to an analogue of the 'no best possible world' theodicy, such as the one developed by George Schlesinger.[17] My argument presupposes only that there are efficient ways and inefficient ways, where an efficient way is one whereby animate states evolve in accordance with natural laws and an inefficient way one whereby animate states do not evolve in accordance with natural laws but require divine intervention.

Craig rejoins to this that efficiency is not one of the divine perfections. He refers to Thomas Morris's theory that an efficient person is one who husbands his energy and time, achieving his ends with as little energy and time as possible. It is good to be efficient if one has limited time and energy, but if one's power and time is unlimited, then there is no reason to think efficiency is better than inefficiency.

I believe there is such a reason, namely, that efficiency is of positive aesthetic value and inefficiency of negative aesthetic value (apart from any considerations of whether one's power and time is limited or unlimited). Efficiency, like gracefulness, is one of the positive aesthetic values that supervene upon personal activities and all else being equal it is irrational for a good person to realize a negative value when he could have realized a positive one instead.

But the premiss that 'God is inefficient' is not necessary to conclude that 'God is irrational' if he creates an initial state that is not naturally tended towards an animate universe. One needs only the principles of rational action. Consider this argument:

(1) God intends to create an animate universe.
(2) It is possible for God to create an initial state S that deterministically or probably evolves towards an animate universe and it is possible for God to create an initial state S' that does not deterministically or probably evolve towards an animate universe.

---

[17] Schlesinger, *Religion and the Scientific Method* (Dordrecht: D. Reidel, 1977).

(3) If it is within any person's power to do A or (exclusive disjunction) B, and A certainly or probably advances the person's goals and B does not, then (all other things being equal) the person is rational with respect to A and B if and only if the person does A rather than B.

(4) God creates S'.

Therefore,

(5) God is irrational.

Since (5) contradicts the definition of God, it follows that S' is uncreated by God and thus that God does not exist. (I respond to the objection that it is not possible for God to create S in a later section.)

Craig suggests that God may have reasons for creating S' 'of which we have no idea',[18] reasons which would show that it is one of God's goals to have a first state that does not deterministically or probably evolve towards an animate universe. Craig's suggestion at best may block a deductive 'Big Bang cosmological argument' against God's existence, but not a probabilistic argument. I may grant that it is logically possible that God has reasons for creating S' of which we have no idea, but I would affirm there is not the slightest evidence of any such reason and therefore that it is rational to believe that this is improbable. (The difference between a deductive and probabilistic 'Big Bang cosmological argument against God's existence' is not material to the main thrust of my argument, since 'Big Bang cosmology is inconsistent with theism' may be read as 'Big Bang cosmology is certainly or probably inconsistent with theism'.)

However, Craig does introduce a possible reason for God's creation of S', namely, that God 'wanted to leave a general revelation of himself in nature by creating a world which would never in all probability have resulted from the natural tendencies of things alone'.[19] But this gets things backwards; the fact that the evolution of an animate universe appears to be due to random chance and improbable occurrences suggests that the universe is *not* designed for humans or other living creatures and therefore that God does not exist. If the universe is to manifest a 'general revelation' of God, then it must appear designed by God and it does not appear

---

[18] Essay VIII, Sect. 1.   [19] Ibid.

designed if the evolution of living creatures appears to be rendered improbable by the very nature of things.

The further possible reason Craig offers for God's creation of S, that 'Perhaps God delights in the creative activity of fashioning a world'[20] runs foul of the above-mentioned points, namely, that the 'creative activity' alluded to involves interfering with the natural tendency with which God endowed things and that this entails inefficiency, the performance of an aesthetically disvaluable action, and the performance of an action that conflicts with the principles of rational activity. God would not 'delight' in doing something inefficient, irrational, or aesthetically disvaluable.

Accordingly, while my argument against the hypothesis of divine intervention in the singularity and the symmetry-breaking phases may be a 'glaring weakness'[21] in an argument against the god of popular imagination or religious scriptures, it is not a weakness in an argument against a god conceived as possessing the divine perfections of perfect rationality, omniscience, omnipotence, and omnibenevolence.

5. THE QUESTION OF THE REALITY OF THE SINGULARITY

Craig argues that a crucial premiss of the atheistic argument, premiss (1) that 'The Big Bang singularity is the earliest state of the universe', is false since it is based on a reification of the singularity. Although I argued that physicists represent the singularity as real, Craig thinks certain statements by physicists lend themselves to the interpretation that the singularity is unreal. But the statements he quotes imply no such thing. He quotes Geroch to the effect that a space traveller C is 'snuffed out of existence' when C hits a black hole singularity.[22] This is indeed true, but not because the singularity does not exist. Rather, it is because the space traveller C becomes crushed to a dimensionless point and no person can exist if she is crushed to a point. But the singular point still exists. Craig also quotes Kanitscheider to the effect that the singularity is not an element of the spacetime manifold and is not an event, but this is something I have insisted myself and it certainly does not entail

---

[20] Ibid.   [21] Ibid.   [22] Ibid. 497–8.

that the singularity does not exist. The singularity is not an event or an element in the manifold since all such events belong to a four-dimensional continuum, whereas the singularity does not. In the case of a finite universe, it is a pointlike object that exists by itself for an instant before exploding and emitting a four-dimensional continuum.

Craig proceeds to assert that something with no duration and no spatial dimensions does not exist. 'Simply put, an object which has no spatial dimensions and no temporal duration hardly seems to qualify as a physical object at all, but is rather a mathematical conceptualization. . . . [S]uch a point could hardly be called a physical object and seems ontologically equivalent to nothing.'[23] But this assertion flies in the face of Special and General Relativity, which represents spacetime as a continuum of instantaneous spatial points. Each 'event' in the spacetime is a spatial point (something with zero spatial dimensions) and is instantaneous (something with zero temporal duration). The difference between the Big Bang singularity and elements of spacetime is that the elements have four coordinates and the singularity has no coordinates; this difference is reflected in the fact that the singularity is a boundary or edge of spacetime and an event or element is a part of spacetime. Each event is a spatial point that is connected to other spatial points along the dimensions of height, width, and depth, but the initial singularity of a finite universe is a spatial point that exists in isolation. But each element of the manifold and the singularity are similar in that each is spatially and temporally pointlike. Thus, if Craig's assertion is true, not only would the singularity be ontologically equivalent to nothing but everything that exists (every event) would be equivalent to nothing. This shows his statement is false. I conclude that there is reason to believe that the singularity is real (namely, that it follows from the Hawking–Penrose singularity theorems and is treated as real by physicists) and no reason to believe it is unreal. (Recall that I am working with classical Big Bang cosmology in this essay and am not taking into account quantum models, such as the Hartle–Hawking model, which result in the singularity being 'smeared out'.)

Richard Swinburne also believes that the singular point is a mathematical idealization. He provides an argument for this, namely,

---

[23] Ibid. 499.

that it is logically necessary that space be three-dimensional. Swinburne presents an argument against the logical possibility of two-dimensional objects and suggests that analogous arguments can be constructed against one-dimensional and zero-dimensional objects. He asks us to consider a two-dimensional surface that contains two-dimensional objects:

> it is clearly logically possible that the two-dimensional 'material objects' should be elevated above the surface or depressed below it . . . the logical possibility exists even if the physical possibility does not. Since it is logically possible that the 'material objects' be moved out of the surface, there must be places, and so points, outside the surface, since a place is wherever, it is logically possible, a material object could be.[24]

Therefore, Swinburne concludes, if there exists a two-dimensional object or surface there must also exist a third spatial dimension. Swinburne's argument instantiates the following invalid argument form:

(1) Fx is logically possible (i.e. it is logically possible for x to possess the property F).

(2) C is a necessary condition of Fx.

(3) x exists.

(4) Therefore, C exists.

The fact that Swinburne's argument has this form becomes clear if we state his argument as follows:

(1′) It is logically possible for any object on a two-dimensional surface to possess the property of *moving above or below the surface*.

(2′) A third spatial dimension is a necessary condition of any object on a two-dimensional surface moving above or below the surface.

(3′) There exists an object on a two-dimensional surface.

(4′) Therefore, there exists a third spatial dimension.

If (1′)–(4′) proves that objects on two-dimensional surfaces require a third spatial dimension, then the following argument proves that there is a heaven:

(1″) It is logically possible for any human body to be resurrected after death and occupy a heavenly space.

---

[24] Swinburne, *Space and Time* (New York: St Martin's Press, 1981), 125.

(2″) Heaven is a necessary condition of any human body being resurrected.

(3″) There are human bodies.

(4″) Therefore, there is a heaven.

The fallacy, if the reader has not already grasped it, is the assumption that a necessary condition of an object possessing a certain property must be actual if the object is actual. This of course is not so; the necessary condition need be actual only if the object's possession of the property is actual. I conclude that Swinburne has given us no reason to believe that it is impossible for there to be a Big Bang singularity that occupies less than three spatial dimensions. Given that Swinburne's argument fails, and that no other arguments against the coherency of the Big Bang singularity have been presented (at least of which I am aware), the above considerations warrant the conclusion that there is no reason to deny reality to the Big Bang singularity. Thus, the problem of unpredictability remains.

6. COUNTERFACTUALS OF SINGULARITIES

The problem of unpredictability would be solved if counterfactuals of singularities were true logically prior to creation. I argued in Essay VII that the supposition that these propositions are true logically prior to creation is inconsistent with possible world semantics (e.g. those of Stalnaker, Lewis, Davies, and others). There is an inconsistency since there are no relevant grounds of similarity among possible worlds that could render the counter-factuals of singularities true logically prior to creation.

I would agree with Craig's remark that 'Smith has to treat possible world semantics as the final word on the truth conditions of counter-factual conditionals in order for his anti theistic argument to go through.'[25] Craig rejects my argument by rejecting these semantics, but I am willing to rest my case on the validity of these semantics.

However, Craig attempts to show that the thesis that God knows (logically prior to creation) counterfactuals of singularities is 'con-sistent'[26] with possible world semantics. The structure of Craig's argument for the consistency is unclear to me but I will indicate the various remarks he makes cannot add up to a sound argument for the alleged consistency.

[25] Essay VIII, Sect. 1.   [26] Ibid.

Craig responds to the claim that there are no truth-makers for counterfactuals of singularities (logically prior to creation) by stating that counterfactual states of affairs concerning singularities serve as the truth-makers of the counterfactual propositions. 'The counterfactuals are true because the relevant states of affairs obtain.'[27] The counterfactual proposition *If a Big Bang singularity were to be the earliest state of the universe, this singularity would emit a life-producing configuration of particles* is made true by virtue of corresponding to the state of affairs *If a Big Bang singularity were to be the earliest state of the universe, this singularity would emit a life-producing configuration of particles*. According to Craig, similarity relations among worlds do not constitute 'a reason *why* certain counterfactuals are true or false'.[28] The reason why a counterfactual is true is that the corresponding counterfactual state of affairs obtains. However, I would object that this position is *not* consistent with the standard possible worlds semantics. According to these semantics, 'counterfactual states of affairs' are not the truth-makers of counterfactual propositions. There are no such states of affairs. Rather, relations of similarity among possible worlds are the truth-makers. The similarity relations have for their relata the histories of the worlds and the natural laws. These relations are the reason why the relevant counterfactual propositions are true.

But Craig (following Plantinga) seems to see an opening here since natural laws entail counterfactuals. Given this entailment, it seems possible to argue that the counterfactual states of affairs grounded in the natural laws help determine the similarity among worlds and function as truth-makers of the counterfactual propositions entailed by natural laws. The move is then made to the claim that counterfactuals of singularities also have counterfactual states of affairs for their truth-makers, states of affairs that obtain logically prior to creation.

Although I have difficulty with each of the moves in this argument, I will confine myself to pointing out that its initial move is untenable; the fact that natural laws are among the relata of the similarity relations and entail counterfactual propositions does not imply that there are counterfactual states of affairs that serve as the truth-makers of the entailed propositions. The truth-makers are rather the similarity relations among worlds that are grounded upon

----

[27] Ibid. 496.   [28] Ibid.

the world histories and the laws themselves. The truth-maker of *If a light ray were emitted from this flashlight, it would travel at 186 000 m.p.s.* is not a counterfactual state of affairs (it is not the putative state of affairs *If a light ray were emitted from this flashlight, it would travel at 186 000 m.p.s.*); rather, it is a certain similarity relation between the actual world and a world in which there obtains the natural law that light travels at 186 000 m.p.s. and in which a light ray is emitted from this flashlight. Craig says that 'one measure of similarity between worlds is the degree to which they share their counterfactuals.'[29] This is true in the sense that one measure of similarity between worlds is the degree to which they share their natural laws, which entail counterfactual propositions. But Craig does not show us how to get from this claim to the thesis that there are counterfactual states of affairs. Nor does he show us how to get from the claim that there are counterfactual states of affairs corresponding to the propositions entailed by natural laws to the further claim that there obtain (logically prior to creation) counterfactual states of affairs concerning singularities. He simply asserts that there are such states of affairs and that they obtain logically prior to creation. Thus, I do not think Craig has established his claims or demonstrated that his assertions are consistent with the possible worlds semantics of counterfactuals.

7. THE QUESTION OF THE RELATIVE SIMPLICITY OF THE THEISTIC AND ATHEISTIC HYPOTHESES

I shall turn now to an issue not previously discussed, namely, whether considerations of simplicity support the theistic or atheistic interpretation of Big Bang cosmology. Swinburne has advanced the argument that the hypothesis of divine creation is simpler than the atheistic hypothesis and more likely to be true. He claims that God is simpler than the physical universe and therefore is more likely than it to exist unexplained. 'If something has to occur unexplained, a complex physical universe is less to be expected than other things (e.g. God).'[30] If the physical universe is created by God then it has its explanation in God and consequently does not exist unexplained; in this case, only God exists unexplained. Since the hypothesis that

---

[29] Ibid. 495.   [30] Swinburne, *The Existence of God*, 130.

only God exists unexplained is simpler than the atheistic hypothesis, it is more likely to be true.

The principle Swinburne is appealing to is

(1) The simpler an existent is, the more likely it is to exist unexplained.

I believe, however, that even if we grant Swinburne this and other of his premises it can be shown that considerations of simplicity support atheism rather than theism. Swinburne's criterion of simplicity is that there is a simplicity 'about zero and infinity which particular finite numbers lack'.[31] For example, 'the hypothesis that some particle has zero mass, or infinite velocity is simpler than the hypothesis that it has a mass of 0.34127 of some unit, or a velocity of 301 000 km/sec.'[32] Likewise, a person with infinite power, knowledge, and goodness is simpler than a person with a certain finite degree of power, knowledge, and goodness. Furthermore, a person with infinite power, knowledge, etc. is simpler than a physical object that has particular finite values for its size, duration, velocity, density, etc. Assuming these premises, let us examine the hypothesis that a finite universe begins with an uncaused singularity. The singularity in question has *zero* spatial volume and *zero* temporal duration and *does not have particular finite values* for its density, temperature, or curvature. It seems reasonable to suppose that by virtue of these zero and non-finite values this instantaneous point is the simplest possible physical object. If we grant to Swinburne that God is the simplest possible person and hold that God and the uncaused singularity cannot both exist (for reasons stated in the atheistic argument in Section 2), then our alternatives are to suppose that either the simplest person exists and creates the four-dimensional spatiotemporal universe or the simplest physical object exists and emits the four-dimensional spatiotemporal universe. If we use criteria of simplicity, are there any reasons to prefer one of these hypotheses over the other? It seems reasonable to suppose that the simplest possible physical object is equally as simple as the simplest possible person, such that there is no basis to prefer one over the other on grounds of intrinsic simplicity. Swinburne holds that God exists unexplained and so God and the simplest physical object are also on a par in this respect. But the hypothesis that the four-dimensional spatiotemporal universe began from the simplest physical object is

[31] Ibid. 94.   [32] Ibid.

in one crucial respect simpler than the theistic hypothesis. It is simpler to suppose that the four-dimensional physical universe began from the simplest instance of the same basic kind as itself, namely, something physical, than it is to suppose that this universe began from the simplest instance of a different basic kind, namely, something non-physical and personal. The atheistic account of the origin of the four-dimensional universe posits phenomena of only one basic kind (physical phenomena), whereas the theistic account of its origin posits phenomena of two basic kinds (physical phenomena and disembodied personal phenomena). Thus on grounds of simplicity the postulation of a singularity that explodes in a Big Bang wins out over the postulation of a deity that creates the Big Bang explosion *ex nihilo*.

8. THE QUESTION OF THE METAPHYSICAL NECESSITY OF A BIG BANG UNIVERSE

According to essentialism, natural laws, such as the law that water is H_2O, are metaphysically necessary; they hold in all possible worlds such that God could not have created a universe in which they are violated. Consequently, if it is a natural law that a universe obeying the Friedman solutions to Einstein's equation and the Hawking–Penrose singularity theorems begins in a singularity, then God could not have created a Friedman–Hawking–Penrose (FHP) universe otherwise than by first creating an unpredictable singularity. Given this, and given his desire that the universe be animate, he would then have to intervene to ensure that the universe be animate. This would not be a sign of inefficiency or bungling since this would be the only possible way in which an animate universe could be guaranteed.

My response to this objection is that even if its essentialist assumption is sound, it does not follow that God must create a Big Bang singularity if he intends to create an animate universe. For the fact that certain natural laws are metaphysically necessary does not entail that they are necessarily instantiated. If we borrow the symbolism, if not the position, of D. M. Armstrong,[33] we may say that a metaphysically necessary natural law is of a form such as

[33] D. M. Armstrong, *What Is a Law of Nature?* (Cambridge: Cambridge University Press: 1983), 163. Armstrong rejects the idea that laws of nature are metaphysically

(L) $\Box(N(F,G))$,

where F and G are universals and N a relation between them. N is the relation of nomic necessitation. Armstrong takes N to be primitive, but I think we can define N in terms of coexemplification. (L) means that in every possible world in which F is exemplified, G is coexemplified. If F is water and G H_2O, then (L) says that in each world in which *being water* is exemplified, *being H_2O* is exemplified by whatever exemplifies *being water*. But (L) does not entail that F or G is *exemplified*. The fact that water is H_2O in every world in which there is water does not entail that there is water in every world. Analogously, the fact that a universe that satisfies the FHP laws begins in a Big Bang singularity in every world in which such a universe exists does not entail that there is a FHP universe in every world. For other sorts of universe are also possible, ones that satisfy other sets of laws, including sets of laws that enable an earliest state to be, or evolve predictably into, an animate state. If God exists and intended there to be an animate universe, he would have created one of these universes (or a beginningless animate universe).

This response to the essentialist objection might be rejected on the grounds that essentialism and the FHP theory jointly entail that the only metaphysically possible universes are FHP universes. Let F be the property *being a universe* and G the property *being a FHP universe*. According to (L), *being a universe* cannot be exemplified unless *being a FHP universe* is coexemplified.

I believe, however, that we can concede even this objection consistently with the soundness of the atheological argument. To see this, we must reflect on the evidence adduced for the metaphysical necessity of natural laws. Kripke, Putnam, and other originators of essentialism have recognized that some *reason* must be given for holding a natural law to be necessary that defeats the standard reason for regarding them to be contingent, namely, that they can be coherently conceived not to obtain. The reason for holding some principles to be necessary, such as tautologies (all unmarried men are men), analytic principles (all unmarried men are bachelors), and synthetic a priori principles (all completely green objects are

necessary. Alfred J. Freddoso, on the other hand, argues that natural laws are correctly represented by (L). See his 'The Necessity of Nature', in P. French, T. Uehling, and H. Wettstein (eds.), *Midwest Studies in Philosophy*, xi (Minneapolis: University of Minnesota Press, 1986), 215–42.

not simultaneously completely red), is that they cannot coherently be conceived to be false. But this is not the case for natural laws. As Putnam remarks, 'we can perfectly well imagine having experiences that would convince us (and that would make it rational to believe that) water *isn't* H_2O. In that sense, it is conceivable that water isn't H_2O.'[34] But in this case, conceivability of being otherwise is a defeated guide to contingency, for considerations of how the reference of 'water' is established, in conjunction with scientific observations, show that water is necessarily H_2O. But I will not strictly follow Putnam in presenting 'the argument from the rigidity of "water"' since subsequent formulations have provided improved versions. Keith Donnellan[35] offered a version improving on Putnam's, and Nathan Salmon[36] has improved upon Donnellan's version. But Paul Copeck[37] has recently improved upon Salmon's version and I shall partly borrow from Copeck's version in the following summary statement of this argument. The first premiss is a formalization of the rigid meaning of 'water' in terms of the word's ostensive definition and the second premiss is borrowed from current scientific theory:

(1) It is necessarily the case that: something is a sample of water iff it exemplifies dthat (the properties $P_1, \ldots, P_n$, such that $P_1, \ldots, P_n$ are causally responsible for the observable properties (e.g. being tasteless, odourless, and clear) of the substance of which *that* is a sample).

(2) This (liquid sample) has the chemical structure H_2O, such that *being H_2O* is the property causally responsible for the observable properties of being tasteless, odourless, clear, etc.

Therefore,

(3) It is necessarily the case that: every sample of water has the chemical structure H_2O.

The word 'dthat' in premiss (1) is Kaplan's rigidifying functor, which operates on 'that' to produce a demonstrative reference that is rigid. Now if we construct an analogous argument for the necessity of a universe being FHP, it would appear as

[34] Hilary Putnam, *Philosophical Papers*, ii (Cambridge: Cambridge University Press, 1975), 233.
[35] Keith Donnellan, 'Substance and Individuals', APA address, 1973.
[36] Nathan Salmon, *Reference and Essence* (Princeton, NJ: University Press, 1981).
[37] Paul Copeck, 'Review of Nathan Salmon's *Reference and Essence*', in *Journal of Philosophy*, 81 (1984), 261–70.

(4) It is necessarily the case that: something is an instance of a universe iff it exemplifies dthat (the properties $P_1, \ldots, P_n$, such that $P_1, \ldots, P_n$ are causally responsible for the observable properties (e.g. receding galactic clusters, the background microwave radiation of 2.7 K) of the kind of which that is an instance).

(5) This instance of a universe has a FHP structure, such that *being an FHP universe* is the property causally responsible for the observable properties of receding clusters, background radiation, etc.

Therefore,

(6) It is necessarily the case that: every instance of a universe has the property of *being an FHP universe*.

I shall not challenge the soundness of (4)–(6) but merely show its soundness is consistent with the soundness of the Big Bang cosmological argument for God's non-existence. It will be helpful if a parallel with the example of water is drawn. As Putnam has pointed out, there is another possible world W in which a substance has a certain chemical structure XYZ, such that XYZ is causally responsible for the substance's observable properties of being a clear, odourless, tasteless liquid. This substance is not water but something whose observational properties are indistinguishable from those of water. This substance may be called $water_1$, such that it is metaphysically necessary that $water_1$ is XYZ. Analogously, there is another possible world W in which the cosmic structure responsible for the observable properties of receding clusters, background radiation, etc. is not a FHP structure but some other structure, say ABC. That which has this structure is not a universe, since 'universe' rigidly refers to something with a FHP structure. But we can call it a $universe_1$, just as we can call XYZ $water_1$. There are still other worlds in which the relevant observational properties do not include receding clusters and background radiation but such properties as the systems of Ptolemy, Copernicus, or Newton were thought to exemplify. What is causally responsible for these properties may be called a $universe_2$, a $universe_3$, etc. Accordingly, the proponent of the atheological argument may grant that God could not have created an animate universe without creating a Big Bang singularity, but he will point out that it would be irrational and incompetent on the part of God to create an animate universe; the rational thing to do is to create an animate universe$_1$ or an animate universe$_2$, etc., such that these systems do not require divine interventions for animate states to be ensured.

For the reasons adduced in this section and earlier sections, I believe it rational to hold that the Big Bang cosmological argument for God's non-existence is defensible.

X

A Criticism of the Cosmological Argument for God's Non-existence

WILLIAM LANE CRAIG

I. INTRODUCTION

'The most efficacious way to prove that God exists is on the supposition that the world is eternal,' advised Thomas Aquinas. 'For, if the world and motion have a first beginning, some cause must clearly be posited to account for this origin of the world and of motion . . . since nothing brings itself from potency to act, or from non-being to being.'[1] In Thomas's thinking, once it is conceded that the world began to exist, the argument is for all practical purposes over: it is obvious that a First Cause must exist. He therefore sought to prove God's existence on the more neutral presupposition of the eternity of the world; besides, the temporal finitude of the world could be known only by revelation, since the philosophical arguments for a beginning of the universe were, in his opinion, unsound.

The discovery during this century that the universe is in a state of isotropic expansion has led, via a time-reversed extrapolation of the expansion, to the startling conclusion that at a point in the finite past the entire universe was contracted down to a state of infinite density, prior to which it did not exist. The standard Big Bang model, which has become the controlling paradigm for contemporary cosmology, thus drops into the theologian's lap just that crucial premiss which, according to Aquinas, makes God's existence practically undeniable.

Quentin Smith disagrees. He argues that the standard model 'is

---

This essay is an adaptation and expansion of 'God and the Initial Cosmological Singularity: A Reply to Quentin Smith', *Faith and Philosophy*, 9 (1992), 238–48.

[1] Thomas Aquinas, *Summa contra Gentiles*, 1.13.30 (Pegis translation).

actually *inconsistent* with theism' and that, therefore, an atheistic interpretation of the Big Bang 'is in fact *better justified* than the theistic interpretation'.[2] He claims, indeed, to have established 'a coherent and plausible atheistic interpretation' of the origin of the universe.

In support of this remarkable position, Smith presents the following argument:

(1) The Big Bang singularity is the earliest state of the universe.

(2) The earliest state of the universe is inanimate.

(3) No law governs the Big Bang singularity, and consequently there is no guarantee that it will emit a configuration of particles that will evolve into an animate universe.

(4) Therefore, the earliest state of the universe is not guaranteed to evolve into an animate state of the universe.

(5) If God creates a universe, He creates an animate universe.

(6) Therefore, if God created the earliest state of the universe, then He would have ensured that this state is animate or evolves into animate states of the universe.

(7) Therefore, God did not create the earliest state of the universe.

Smith takes this argument to be a Big Bang cosmological argument for the non-existence of God.

2. CRITIQUE OF SMITH'S ARGUMENT

Smith's argument seems multiply flawed. Consider, for example, premiss (1). The premiss is patient of two very different interpretations. This fact emerges in the argument's conclusion. From (1) and (7) it follows that

(8) God did not create the Big Bang singularity.

This Smith takes to mean

(8') The Big Bang singularity was an actual state uncreated by God,

which is alleged to be inconsistent with classical theism's doctrine of creation. But (8) could be taken to mean

(8″) God refrained from creating the Big Bang singularity,

[2] Essay IX, Sect. 1.

that is to say, He, on the pattern of certain contemporary cosmologists, chose to 'cut out' the singularity from the spacetime manifold and create that manifold without that initial singular point. If this is all that Smith's argument proves, then it is not inconsistent with classical theism. If we take his argument to imply (8″), then by (1) we understand

(1″) The Big Bang singularity is the earliest state of the universe in the standard model,

whereas Smith takes it to mean

(1′) The Big Bang singularity described by the standard model was the actual earliest state of the universe.

The theist who finds himself convinced by Smith's line of argument could escape inconsistency by denying (1′). Such a move would raise interesting epistemological questions concerning the rationality of belief in *creatio ex nihilo* to which Smith has yet to give attention.[3]

But (1′) is vulnerable on other, more plausible grounds than this. For the question arises of the ontological status of the singularity. It needs to be emphasized that this is *not* the same question as the reality of the singularity, as that expression is usually employed in contemporary cosmological theory. Certain singularities in physical theory are merely apparent, resulting from the coordinate system being used. For example, the Schwarzschild solution to Einstein's field equations in the General Theory of Relativity involves a coordinate singularity when the radius of the body in question equals twice its mass. This singularity results merely from the fact that Schwarzschild chose coordinates for his solution which are not applicable on this surface. By contrast, when the body's radius equals zero, a real, and not merely coordinate, singularity occurs. Now the initial cosmological singularity was certainly a real singularity. But that does not settle the question of its ontological status.

The ontological status of the Big Bang singularity is a *metaphysical* question concerning which one will be hard-pressed to find a discussion in scientific literature. The singularity does not exist in space and time; therefore it is not an event. Typically it is cryptically

---

[3] See Thomas V. Morris, *'Creatio ex nihilo'*, in *Anselmian Investigations* (Notre Dame, Ind.: University of Notre Dame Press, 1987), 151–60. Morris argues that belief in *creatio ex nihilo* gains in rationality as the number of empirical beliefs it forces us to abandon decreases. Cutting out the singularity would sacrifice a minimal number of such beliefs.

said to lie on the boundary of spacetime. But the ontological status of this boundary point is virtually never discussed.

For that reason I am not terribly impressed with Smith's statement that 'Cosmologists find no difficulty in the concept of a space that has zero dimensions (a spatial point) and that exists for an instant.'[4] My own experience is that a question concerning the ontological status of the initial cosmological singularity is likely to be met with bewilderment or disclaimers about not being a philosopher. Mathematical models containing singular points do not carry their metaphysical interpretation on their faces.

Now to my mind, at least, a good case can be made for the assertion that this singular point is ontologically equivalent to nothing.[5] Smith attempts a *reductio* of my argument by claiming that a continuous spacetime manifold could then not exist, since it is composed of point-events.[6] By now I think it is evident that I am dubious whether an ontological continuum does exist; instants and points seem to me to be mathematical fictions. But let that pass, for Smith's *reductio* fails on less controversial grounds than these. For instants of time and points of space are not typically conceived to be themselves intervals of time and of space, but mere *boundaries* of intervals. And it is consistent to hold that boundary points cannot exist independently of the intervals which they bound. If instants and points exist only as boundaries of intervals, then they have no independent ontological status and so cannot subsist alone. But in the case of the initial cosmological singularity, this point-instant is said to exist independently. Therefore, point-instants of the manifold can exist (as boundaries of intervals), while the singularity cannot.

The B-theorist would deny this distinction, since the singularity bounds the spacetime manifold. But this response is not open to the A-theorist because on his view temporal becoming is real and

---

[4] Quentin Smith, 'A Big Bang Cosmological Argument for God's Non-existence', *Faith and Philosophy*, 9 (1992), 225.

[5] See Essay VIII. Analogous equivalencies elsewhere in science may help to drive home the point. For example, in discussions of the conventionality of simultaneity in relativity theory, one speaks of synchronization of spatially separated clocks by means of the slow transport of clocks from one place to another. It is claimed that by transporting clocks at progressively slower velocities, one can approach absolute synchronization, which would result from a clock transported from one place to another at infinitely slow velocity. But no one takes infinitely slow transport of clocks as describing an actual procedure, since infinitely slow velocity is ontologically equivalent to rest, that is, to no transport at all!

[6] Essay IX, Sect. 5.

objective, and so, if temporal becoming is instantaneous, at the instant the singularity comes to exist, all other instants are non-existent, mere potentialities. Therefore, it would exist alone.[7] Indeed, it seems to me in general very difficult to reconcile the A-theory of time with the view that instants are not mere boundary points, but subsist as independent, degenerate intervals of zero duration. Not only does this raise the ancient puzzle of how the present moment can be an interval of zero temporal duration, given that past and future are ontologically unreal,[8] but the notion that the present is a solitary instant also seems to pose insuperable problems for the reality of temporal becoming, since instants have no immediate successors, so that one after another cannot elapse.[9]

Be that as it may, so long as it is consistent to hold that points and instants have reality only in so far as they bound intervals, Smith's *reductio* argument fails. He offers no direct refutation of the claim that a physical object existing for no time and having no extension is not a physical object at all. If the initial cosmological singularity is a mere conceptualization ontologically equivalent to nothing, Smith's premiss (1) is false and his argument fallacious, since the universe did not begin at the singularity. Rather the universe, the spacetime manifold, does not possess a first temporal instant, but exists at any moment arbitrarily close to the initial cosmological singularity. It is therefore governed throughout its existence by natural laws so that its becoming animate could be physically guaranteed from any arbitrarily designated initial temporal segment.

[7] If the A-theorist adopts an atomistic view of time, then he could maintain that the singularity is the boundary of the first chronon of time. But this would undermine Smith's argument because then the first state of reality would not be lawless and unordered.

[8] Aristotle, *Physics*, 4.10.217^{b}33–218^{a}9. For an excellent discussion of the early history of this conundrum, see Richard Sorabji, *Time, Creation and the Continuum* (Ithaca, NY: Cornell University Press, 1983), 7–63. Augustine in particular agonized eloquently over this problem (Augustine, *Confessions*, 9.15–28).

[9] See discussion in Adolf Grünbaum, 'Relativity and the Atomicity of Becoming', *Review of Metaphysics*, 4 (1950–1), 143–86. Recall that Grünbaum solves the problem only by denying the reality of temporal becoming. My own solution is not to adopt an atomistic view of time, but to maintain that only intervals of time are real or present and that the present interval (of arbitrarily designated length) may be subdivided into subintervals which are past, present, and future respectively. Thus, there is no such time as 'the present' *simpliciter*; it is always 'the present hour', 'the present second', etc. The process of division is potentially infinite and never arrives at instants. For a fine treatment see Andros Loizou, *The Reality of Time* (Brookfield, Vt.: Gower, 1986), 44–5.

But the theist need not prove even so much in order to remove the teeth from Smith's argument. Plantinga has reminded us that in dealing with defeaters of theism, it is not necessary to supply a rebutting defeater-defeater: an undercutting defeater-defeater may do.[10] So long as my interpretation of the ontological status of the singularity has even equal, if not superior, plausibility to Smith's, his argument for God's non-existence is undercut. At the very least, I think, Smith must in all honesty admit that the ontological status of the singularity is so poorly understood today that such an interpretation is as valid as his own. But if that is so, then premiss (1') is at best unsubstantiated, and therefore his argument fails to prove that the theistic interpretation is inconsistent and therefore that the atheistic interpretation is better justified, since the latter claim rests solely on the alleged inconsistency of the theistic interpretation.

Premisses (2) and (5) are also problematic. Smith's argument seems tacitly to assume that the only finite, animate beings that exist are those that exist in the physical universe, for he equates God's intending 'his creation to be animate' with God's intention to create an animate universe. But the problem is that according to Christian theism the physical universe does not exhaust the created order. There are also realms of spiritual substances, or angels, which are part of the created order. Suppose God created the angelic realms prior to creating the physical universe. In such a case, creation is already animate before the work of physical creation has begun. So why is God obliged to guarantee *ab initio* that the physical order is animate? Indeed, why must the physical order ever become animate in such a case? What these considerations suggest is that even if Smith's argument were effective against some bare-boned theism, it still might not have any relevance to Christian theism.

But premiss (5) has more serious shortcomings than this. For, we may ask, is (5) necessarily true? Are there no possible worlds in which God creates an inanimate universe? Smith thinks that 'It is essential to the idea of God in the Judaeo-Christian-Islamic tradition that if he creates a universe he creates an animate universe' and that God's creating an inanimate universe is therefore 'at odds' with classical theism.[11] But if we take Aquinas as our guide, that does

---

[10] Alvin Plantinga, 'Foundations of Theism', *Faith and Philosophy*, 3 (1986), 298–313.

[11] Smith, 'A Big Bang Cosmological Argument for God's Non-existence', 223.

not seem to be the case. On his view, rational creatures enhance the goodness of the universe, but there is no necessity that God create them. He writes, 'God wills man to have a reason in order that man may be; He wills man to be so that the universe may be complete; and He wills that the good of the universe be because it befits His goodness.'[12] Thomas goes on to explain that some things are willed by God with a necessity of supposition (for example, that man be endowed with reason, if God wills that man exist), others as useful but not necessary to some end, and still others as merely befitting His goodness. This last relation is conceived by him to be extremely weak; something so willed is willed by God's good pleasure as appropriate to, but not required by, His goodness. Hence, even if it is necessary that God will man's existence in order for the goodness of the universe to be complete, there is no necessity that God will that the goodness of the universe be complete. God could have willed that a universe without intelligent life—or without life at all—exist. This does not imply that God therefore has no reason for willing that animate beings exist. On the contrary, Aquinas affirms that a reason can be assigned for the divine will, but that this reason is *contingent*. Smith is therefore mistaken in thinking that willing an inanimate universe is impossible for God according to classical theism.

But Smith also argues that God cannot have a contingent reason for creating an animate universe since this 'contradicts his omnibenevolence'.[13] It is impossible that God have a reason for creating an inanimate universe because 'omnibenevolence requires living creatures in relation to which God can exercise his benevolence'.[14] But this point precisely supplies the thread for the unravelling of Smith's argument: benevolence is a relational property connoting *willing the good of others*. Since God is not morally obligated to create any world at all, the theist may hold that omnibenevolence is therefore, like *sovereignty* and *providence*, a contingent property of God. Smith does not deny that it is not immoral of God to refrain from creating; but if that is the case, it follows that omnibenevolence is not essential to God's nature. Rather goodness is; the property of being disposed to will the good of any others that exist. Such a dispositional property does not entail the existence of others to

---

[12] Thomas Aquinas, *Summa contra Gentiles*, 1.86.5.
[13] Essay IX, Sect. 3.   [14] Ibid.

whom benevolence would be shown. Smith denies that God is good
if He creates an inanimate universe, when He could have brought
into existence a world with animals and persons. But this is just the
old 'best of all possible worlds' argument in new guise; if there is
no best possible world, then a similar complaint could be voiced
about any world that God creates, so the objection is vacuous.
Smith would perhaps deny this, claiming that within the inanimate
type of world there is no best possible inanimate world and within
the animate type of world there is no best possible animate world,
but that God is morally obligated to choose a world from the latter
type over the former type. But it is not obvious why this is so, since
we can imagine innumerably many worlds of the former type which
would exceed in goodness worlds of the latter type (for example,
inanimate worlds of great beauty compared with animate worlds
filled with unredeemed and gratuitous evil). To say that God must
choose one of the latter type which exceeds in goodness all of the
former type immediately starts one down the infinite regress, since
the lines of one's typology are arbitrarily drawn by certain chosen
standards and one can always find better and better types of world,
just as one can find better and better worlds.

Besides all this, the Christian theist will deny Smith's assumption
that omnibenevolence requires living *creatures* as the objects of
God's benevolence. One of the beauties of the Christian doctrine of
the Trinity is that God is not a lonely monad, but a triad of persons
united in one nature. In the absence of creation, God enjoys the
fullness of the love and joy of the inner-Trinitarian fellowship; each
of the divine persons wills the good of the others. In the tri-unity of
His own being God's benevolence is fully expressed, and the wonder
of creation is that God should voluntarily and out of no necessity
of His own nature graciously choose to create finite persons and
invite them into this inner fellowship of the Godhead. God's omnibenevolence, whether taken to be a contingent or essential property
of God, does not therefore constrain Him to create an animate
universe, any more than two artists are morally obligated to beget
children.

Now consider the inference drawn in premiss (6), which seems
clearly invalid. Smith understands (6) to mean that if God 'creates a
first state of the universe, he creates a state that is, or is guaranteed
to evolve into, an animate state'.[15] But even if we concede the truth

---

[15] Ibid.

of (5), how does it follow that (6) is true? There are two ways in which a provident God could create an animate universe out of a necessarily inanimate initial singularity: (*a*) by His middle knowledge, God could have known that had He actualized the Big Bang singularity, an animate universe would have evolved from it, or (*b*) by His miraculous intervention, God could causally bring about an animate universe.

With respect to God's ensuring an animate universe by means of His middle knowledge, Smith is content to rest his case on the final validity of the possible worlds semantics for counterfactual conditionals.[16] But until such semantics show us how to deal with intuitively true or false counterfactuals with impossible antecedents, their adequacy must remain in doubt and with them Smith's argument.

Smith's original charge against the middle knowledge position was that it is viciously circular.[17] I attempted to answer this charge by explaining that those states of affairs which make counterfactuals of freedom true or false are actual logically prior to God's decree to create and therefore serve as one measure of similarity among worlds, an account which is not *viciously* circular.[18] Smith's rejoinder to this is curious. He asserts, 'According to these semantics, "counterfactual states of affairs" are not the truth-makers of counterfactual propositions. There are no such states of affairs . . . The truth-makers are rather the similarity relations among worlds that are grounded upon the world histories and the laws themselves.'[19] There are at least two things wrong with this response: (1) It confuses truth conditions with grounds of truth of a proposition. Possible worlds semantics does not even aspire to tell us *why* certain counterfactuals are true/false or the *grounds* of their truth. As a semantical theory it merely lays out the semantical conditions for a certain class of propositions' taking the values T or F respectively. It is a sort of calculus, if you will, that tells us what it means to say that a counterfactual proposition is true/false, but it neither tells us what makes it true/false nor makes any ontological pronouncement on whether counterfactual states of affairs exist.[20] (2) More importantly,

---

[16] Ibid., Sect. 6.   [17] Essay VII, Sect. 7.
[18] Essay VIII, Sect. 1.   [19] Essay IX, Sect. 6.
[20] This is especially evident if there are bivalent counterfactuals of freedom, for these cannot be made true or false on the basis of similarity relations alone, since countercausal freedom requires that one be able to choose differently in worlds

it is irrelevant. Suppose that the grounds of the truth of counterfactuals are just the similarity relations among worlds, as Smith maintains. Plantinga's salient point remains that *included* in these similarity relations is the worlds' degree of shared counterfactuals. The counterfactual propositions true at a world are true logically prior to the truth or falsity of contingent categorical propositions at that world and so can be known by God logically prior to His creative decree. It matters not whether we order logically prior to the full instantiation of a world either the relevant states of affairs or else the relevant similarity relations. So long as some such ordering is coherent—and the burden of proof is on Smith to show otherwise—the middle knowledge solution to God's ensuring an animate creation is viable.

Turning, then, to the second alternative of divine miraculous intervention, Smith claims that it is irrational and inefficient for God to create a first state of the universe which does not tend to the end for which the universe is created.[21] I argued that perfect-being theology does not, *pace* Smith, entail a Deist account of creation.[22] Surprisingly, Smith erroneously interprets me to hold that his argument counts *against* Deism, but not against Christian theism.[23] Smith correctly follows the classical theologians in distinguishing originating creation (*creatio originans*) from continuing creation (*creatio continuans*). But Deists and Christians alike affirmed both of these. What divided them was a further distinction drawn by the classical theologians concerning God's governance (*gubernatio*) of the world. They distinguished between God's ordinary providence (*providentia ordinaria*) and His exceptional providence (*providentia extraordinaria*). The governance of His ordinary providence roughly coincides with what Smith calls 'rational continuous creation'.[24] But the world also includes events governed by His extraordinary providence, which we would call 'miracles'. Such events need not

---

having exactly similar world histories up to the time of choice. Such a counterfactual cannot therefore be made true by the fact that in all the antecedent-permitting worlds most similar to the actual world up to the point of decision, one chooses the alternative described in the consequent.

[21] Essay VII, Sect. 5.

[22] Essay VIII, Sect. 1.

[23] Essay IX, Sect. 4.

[24] However, providence also involves the intentional aspect that the states of the world are in some way planned or arranged by God. Moreover, nomological conservation would not be interpreted to abrogate creaturely freedom of the will.

be characterized as 'violations of the laws of nature', since natural laws have implicit *ceteris paribus* clauses stipulating that no natural or supernatural causes are intervening.[25] An act of God's exceptional providence is an event which He brings about at time t and location l which could not have been brought about at t,l solely as the effect of the natural causes and agents at t,l. Smith's position is Deistic in that he rejects works of exceptional providence.

But other than simply *labelling* ordinary providence or conservation 'rational' (and thus tacitly relegating exceptional providence to the realm of the 'irrational'), I do not see any new argument on Smith's part for denying the possibility of exceptional providence. According to Aquinas, 'it can be manifested in no better way, that the whole of nature is subject to the divine will, than by the fact that sometimes He does something outside the order of nature. Indeed, this makes it evident that the order of things has proceeded from Him, not by natural necessity, but by free will.'[26] In this respect the God of revelation and the God of the philosophers coincide, what Morris has aptly called 'the God of Abraham, Isaac, and Anselm'.[27] The God Smith describes is not the God of classical theism, but the God of Spinoza's *Tractatus* and Enlightenment rationalism.

As for Smith's argument from efficiency, it will be recalled that I made two points: (1) efficiency is relative to the ends desired, and (2) efficiency is significant only to someone with limited time and/or power. In response to (2), Smith now claims that being efficient is a positive aesthetic value which God must have.[28] This strikes me as an extremely tenuous value judgement on the basis of which to deny the existence of God. But even if we grant this judgement, its importance depends, as Smith says, on 'all else being equal'. Mitigating factors pertinent to one's desired ends easily override the importance of the aesthetic value of efficiency. Would we dare to call an artist wanting in aesthetic value for preferring the creative labour of executing his oil on canvas rather than simply having, if he could, the finished painting? I suggested that the Creator likewise perhaps delights in the work of creation. Smith responds that this

---

[25] For an outstanding treatment of the relation between natural law and miracle see Stephen Bilinskyj, 'God, Nature, and the Concept of Miracle', Ph.D. thesis, University of Notre Dame, 1982.

[26] Thomas Aquinas, *Summa contra Gentiles*, 3.100.10.

[27] Morris, *Anselmian Explorations*, 10.

[28] Essay IX, Sect. 4.

is impossible because it would be inefficient and irrational. This renewed charge of inefficiency closes a vicious circle on Smith's part and condemns artists, chefs, and boys building model airplanes as persons who ' "delight" in doing something inefficient, irrational, or aesthetically disvaluable'.[29] The point is that the delight of engaging in creative activity can itself be a justification for what the rationalist deems inefficient and aesthetically disvaluable activity.

Smith's further charge of irrationality is based on the premiss that a person is irrational if he performs some action which fails to advance his goals rather than an action within his power which would advance his goals.[30] But God's creating the initial singularity *does* serve to advance His goals, for it furnishes Him with the raw material for His creative activity. Moreover, what if His goals include, not merely the having of a created order, but the divine pleasure of fashioning a creation? By focussing too narrowly on the end-product, Smith fails to see the wider purposes which God may have in view. Smith's is the viewpoint of the manufacturer, God's the viewpoint of the artist.

I also suggested that God may have created the world as He did in order to leave a general revelation of Himself in nature. Smith responds that this gets things backwards; the evolution of an animate universe through random chance and improbable occurrences suggests that God does not exist. But this is surely a misreading of the evidence on Smith's part, as is evident from the heated debate surrounding the Anthropic Principle and the new life which this has breathed into the teleological argument.[31] Popularized in novels like Updike's *Roger's Version* or meticulously examined as in John Leslie's *Universes*, the anthropic coincidences are seen by many as so unlikely and finely tuned that they bespeak divine design.[32] Tony Rothman muses, 'It's not a big step from the [Anthropic Principle] to the Argument from Design . . . When confronted with the order and beauty of the universe and the strange coincidences

---

[29] Ibid.

[30] Ibid., Sect. 4.

[31] See William Lane Craig, 'The Teleological Argument and the Anthropic Principle', in William Lane Craig and Mark S. McLeod (eds.), *The Logic of Rational Theism*, Problems in Contemporary Philosophy, xxiv (Lewiston, NY: Edwin Mellen, 1990), 127–53; L. Stafford Betty and Bruce Cordell, 'New Life for the Teleological Argument', *International Philosophical Quarterly*, 27 (1987), 409–35.

[32] John Updike, *Roger's Version* (London: Deutsch, 1986); John Leslie, *Universes* (London: Routledge, 1989).

of nature, it's very tempting to take the leap of faith from science into religion. I am sure many physicists want to. I only wish they would admit it.'[33] P. C. W. Davies is a good example of a physicist who does admit that the anthropic coincidences persuade him of God's existence.[34] The point is that it is unimaginably more probable that the universe should be life-prohibiting rather than life-permitting, and the best explanation for the cosmos as it is may well be intelligent design. Of course, God could have broadcast His existence even more clearly in creation, but if, as John Hick surmises, God wanted to place creation at a certain 'epistemic distance' from Himself so as not to be coercive, then we should expect His revelation to be somewhat subtle and ambiguous, and discernible only to those who have eyes to see.[35]

Finally, in response to my suggestion that God may have reasons for creating as He did which we are unaware of, Smith admits that this blocks a deductive argument against God's existence, but leaves a probabilistic argument intact. Here I think we can learn a lesson from recent work in the philosophy of religion on the problem of evil. There, too, we have a deductive and an inductive (or probabilistic) version of an argument against God's existence, and it is now generally recognized that the deductive version is a failure, since it seems at least possible that God has morally sufficient reasons for permitting evil, even if these remain unbeknown to us. But some non-theists insist that it is none the less highly improbable that God has morally sufficient reasons for permitting the evils in the world. One response to this inductive version of the argument is to point out that there is no probability that we *should* be able to discern all God's reasons for permitting evil, so that our failure to

[33] Tony Rothman, 'A "What you See Is What you Beget" Theory', *Discover* (May 1987), 99.
[34] Paul Davies, *The Mind of God: Science and the Search for Ultimate Meaning* (New York: Simon & Schuster, 1992).
[35] As Pascal wrote, 'It was not then right that He should appear in a manner manifestly divine, and completely capable of convincing all men; but it was also not right that He should come in so hidden a manner that He could not be known by those who should sincerely seek Him. He has willed to make Himself quite recognizable by those; and thus, willing to appear openly to those who seek Him with all their heart, and to be hidden from those who flee from Him with all their heart, He so regulates the knowledge of Himself that He has given signs of Himself, visible to those who seek Him, and not to those who seek Him not. There is enough light for those who only desire to see, and enough obscurity for those who have a contrary disposition.' (Blaise Pascal, *Pensées*, trans. W. F. Trotter (London: J. M. Dent, 1932), no. 430, p. 118.)

do so does not render it improbable that God has such reasons. In a recent development of this response, William Alston exposits six 'cognitive limits' which make it impossible for us to judge that God lacks morally sufficient reasons for permitting evil. One of these limits, particularly relevant to our discussion, is the difficulty of knowing what is metaphysically possible. Alston writes,

We don't have a clue as to what essential natures are in God's creative repertoire and still less do we have a clue as to which combinations of these into total lawful systems are do-able. We are in no position to make a sufficiently informed judgment as to what God could or could not create by way of a natural order that contains the goods of this one without its disadvantages.[36]

Take quantum mechanics, for example. I dare say that we have no idea of whether God could have created a world order comparable in goods to this one while sacrificing quantum physics. This is important because a physical universe governed by quantum-mechanical laws not merely allows for the *possibility* of miracles, but, if God is to be provident and sovereign without recourse to middle knowledge, actually *necessitates* acts of extraordinary providence. For quantum indeterminacy serves to render certain macroscopic systems chaotic, that is, sensitive to small changes in their initial conditions and therefore unpredictable in their outcome. John Barrow gives a striking example from a game of billiards:

What could be more deterministic than the motion of billiard balls on a billiard table? . . . However, cue games like billiards and pool exhibit that extreme sensitivity and instability . . . If we could know the starting state as accurately as the quantum Uncertainty Principle of Heisenberg allows, then this would enable us to reduce our uncertainty as to the starting position of the cue-ball to a distance less than one billion times the size of a single atomic nucleus (this is totally unrealistic in practice of course, but suspend all practicality for one moment). Yet, after the ball is struck, this uncertainty is so amplified by every collision with other balls and with the edges of the table that after only fifteen such encounters our irreducible infinitesimal uncertainty concerning its initial position will have grown as large as the size of the entire table. We can then predict nothing at all about the ensuing motion of the ball on the table using Newton's laws of motion.[37]

[36] William Alston, 'The Inductive Problem of Evil', *Philosophical Perspectives*, 5 (1991), 65.

[37] John D. Barrow, *The World within the World* (Oxford: Clarendon Press, 1988), 277.

Barrow points out that all the important laws of nature are described by equations which exhibit this chaotic sensitivity. What this seems to imply is that if quantum indeterminacy is not merely epistemic, but ontic, then, in the absence of middle knowledge, it is simply impossible for God to direct providentially a world governed by such laws to His previsioned ends without miraculous intervention; in particular, it is impossible for Him to ensure (even with high probability) that an animate universe should evolve from an initially inanimate state. Given the chaotic nature of macro-systems, miracles are not merely necessary, but recurrent, at a very fundamental and probably indiscernible level. Given this exigency, what possible rationale remains for debarring God's interventions prior to the Planck time and at the singularity? What Smith must say is that God could have created a universe of animate creatures described by different laws of nature which are neither indeterministic nor chaotic:

the proponent of the atheological argument may grant that God could not have created an animate universe [governed by the laws of quantum mechanics] without creating a Big Bang singularity, but he will point out that it would be irrational and incompetent on the part of God to create an animate universe; the rational thing to do is to create an animate universe$_1$, or an animate universe$_2$, etc., such that these systems do not require divine interventions for animate states to be ensured.[38]

But this is where Alston's point becomes relevant: we simply have no idea whether God could have created such a world order and even less whether it would have involved the goods which this system does without greater disadvantages. We can imagine such worlds, but that does nothing to prove that they are either possible or feasible. Paraphrasing Alston, I should say that

the judgments required by the inductive argument from [Big Bang cosmology] are of a very special and enormously ambitious type and our cognitive capacities are not equal to this one ... We are simply not in a position to justifiably assert that God would have no sufficient reason for [creating the Big Bang singularity]. And if that is right, then the inductive argument from [Big Bang cosmology] is in no better shape than its late lamented deductive cousin.[39]

[38] Essay IX, Sect. 8. Of course, God could have created a quantum universe without the extrapolated initial state, but then we run into the same problem mentioned in n. 3.

[39] Alston, 'The Inductive Problem of Evil', 65, 61.

It seems to me, therefore, that Smith's argument is based on such multiply mootable premisses that we can repose no confidence in it.

3. ATHEISTIC VERSUS THEISTIC INTERPRETATION OF THE BIG BANG

But what, in any case, is Smith's 'atheistic interpretation' of the Big Bang and what warrant does it enjoy? Although he does not develop this interpretation at any length, it would appear to be that the initial cosmological singularity inexplicably 'exists and emits the four-dimensional spatiotemporal universe'.[40] But at this point one must be very careful. For although Smith uses here tenseless language to describe the origin of the universe, Smith is no B-theorist of time who thinks that the entire spacetime manifold (plus any singular points) exists tenselessly. Rather Smith is an ardent A-theorist who rejects strictly tenseless language and regards even abstract objects as having temporal duration. Hence, in no sense of the term are we to think of the initial cosmological singularity as possessing the property of *permanence*, which has been so effectively analysed elsewhere by Smith.[41] On an A-theory of time, the singularity is neither sempiternal, omnitemporal, everlasting, infinite in the past and future, beginningless and endless in time, endlessly recurrent, eternal, nor merely timeless. In order for any of these predicates to apply to the singularity, one must adopt a B-theory, according to which the singularity does not come to be or pass away, but tenselessly exists. On Smith's A-theoretic view, the first physical state of the universe came to be without any temporally preceding states whatsoever and immediately emitted the spacetime manifold. Moreover, this coming to be is admitted to be unexplained, that is, without cause or reason.[42]

---

[40] Essay IX, Sect. 7.

[41] See Smith's helpful analysis in 'A New Typology of Temporal and Atemporal Permanence', *Noûs*, 23 (1989), 307–30.

[42] In discussing the origin of the universe, one runs the risk of being bamboozled by his own language, for expressions like 'The universe came to be' or 'The universe came into being out of nothing without a cause' or 'God created the universe out of nothing' might lead the uninitiated to infer that one means that there was a state of nothingness temporally prior to the first event from which the universe was created. But as Aquinas recognized, the import of *creatio ex nihilo* is that there was not anything temporally or metaphysically prior to the universe out of which it was

What possible warrant could there be for such an incredible scenario? If it enjoys no independent support or inherent plausibility apart from the alleged inconsistency of the theistic interpretation, then with the failure of Smith's argument, its epistemic warrant shrinks to zero. Smith, however, does offer an argument in favour of his interpretation: *it is simpler than the theistic hypothesis*. Noting that the singularity has zero spatial volume, zero temporal duration, and non-finite values for its density, temperature, and curvature, Smith contends that it is the simplest possible physical object, even as God is the simplest possible person. They are thus on a par with each other. Both God and the initial cosmological singularity exist unexplained and so are also on a par in this respect. But 'It is simpler to suppose that the four-dimensional physical universe began from the simplest instance of the same basic kind as itself, namely, something physical, than it is to suppose that this universe began from the simplest instance of a different kind, namely, something non-physical and personal.'[43]

Smith's argument, however, depends on a parallelism between God and the initial cosmological singularity which seems clearly exaggerated. For the sense in which God is unexplained is radically different from the sense in which the initial cosmological singularity is unexplained. Both can be said to be without cause or reason. But when we say that God is uncaused we imply that He is eternal, that He exists either timelessly or sempiternally. His being uncaused implies that He exists *permanently*. But the singularity is uncaused in the sense that it comes into being without any efficient cause. It is *impermanent*, indeed, vanishingly so. These hypotheses can therefore hardly be said to be on a par with each other. Moreover, God is without a reason for His existence in the sense that His existence is metaphysically necessary. But the singularity's coming to be is

---

made (*Summa contra Gentiles*, 2.16.4; 2.17.2; 2.36.7). It is very difficult to express this idea in a non-misleading way because the mere assertion that the universe or time began to exist can be interpreted by the B-theorist in such a way as to obscure the radicalness of this claim, whereas attempts to capture the A-theoretic sense of the assertion (e.g. 'The universe came into being out of nothing') may sound analogous to the statement 'John came into the house out of the rain', which betrays one's true meaning. What one means is that the universe started to exist without any temporal or causal antecedents and that this is a tensed fact. Fortunately, Smith understands this and nowhere objects to such expressions and occasionally even uses them himself.

[43] Essay IX, Sect. 7.

without a reason in the sense that, despite its contingency, it lacks any reason for happening. Again these hypotheses are fundamentally different. The hypothesis that the universe was brought into being by an eternal, metaphysically necessary being hardly seems on a par with the hypothesis that the singularity inexplicably and causelessly came into being. Thus, Smith's parallelism between God and the singularity evaporates once the alleged parallels are examined.

As for the simplicity argument itself, Smith's case for the superiority of the atheistic interpretation is, in effect, that only on the atheistic hypothesis does the spacetime universe have a material cause, namely, the singularity. But that is a red herring. For the theist could also maintain that the universe emerged from a physical singularity, adding that the latter was created by God. The real issue is rather the origin of the singularity itself. On the theistic hypothesis the spacetime manifold plus its initial singular point was brought into being by God. But on Smith's hypothesis the spacetime manifold plus its initial singularity came to be without any cause or reason. Hence, atheism is not explanatorily simpler than theism after all, since physical reality did not begin from an 'instance of the same basic kind as itself, namely, something physical'.[44] In fact, on Smith's own principle concerning simplicity and difference in kind, theism is arguably a simpler hypothesis, since, as Duns Scotus put it, there is an infinite distance between being and non-being, and theism posits the origin of being by being, whereas atheism posits the origin of being from non-being.

Smith opened his essay with the confession that 'the reason for the apparent embarrassment of non-theists' when faced with the prospect of the beginning of the universe 'is not hard to find': they must believe that 'the universe came from nothing and by nothing'. Like C. D. Broad, I find this notion insupportable, and any worldview taking this thesis on board will be eventually pulled under by its weight. The principle that something cannot come out of absolutely nothing strikes me as a sort of metaphysical first principle, one of the most obvious truths we intuit when we reflect philosophically. Smith, on the other hand, maintains that this principle is neither a necessary a posteriori nor a necessary a priori truth.[45] It

---

[44] Ibid.
[45] Essay VI. Cf. Smith, 'A Big Bang Cosmological Argument for God's Non-existence', 230–3, to which my remarks were originally directed.

cannot be necessary a posteriori because the sentence 'Everything that begins to exist has a cause' cannot express different propositions in different possible worlds while obeying its actual rule of use. Now I personally see no reason at all to think that all necessary a posteriori truths must conform to the analysis Smith lays down. According to Kripke, all of his examples of metaphysically necessary a posteriori truths have a character such that we see that if they are true at all, they are necessarily true, so that any empirical knowledge of their truth is automatically empirical knowledge of their necessity.[46] So why is it implausible that we should see that the proposition 'Everything that begins to exist has a cause' is necessarily true, if true at all, and see on the basis of experience that it is true? I can think of other metaphysically necessary truths that seem analogous; for example, 'No effect precedes its cause' and 'No event precedes itself', which are metaphysically necessary due to the A-theoretical nature of time and becoming,[47] but which perhaps require some experience of time in order to be seen as true.

In any case, can we not construct a scenario meeting Smith's criteria? When I say that 'Everything that begins to exist has a cause', the sentence is significantly tensed and expresses a tensed fact. Everything that did, does, or will begin to exist had, has, or will have a cause. Let us imagine, then, a world *W* exactly like the actual world except that it is devoid of ontological tense. All things in *W* exist tenselessly at their appointed spacetime coordinates and so never really begin (significantly tensed) to exist. Things begin (tenselessly) to exist only in the sense that their world lines have front edges, or are finite in the 'earlier than' direction of time. Persons in *W* believe and utter the tensed sentence 'Everything that begins to exist has a cause', but that sentence does not express the same fact that it does in the actual world, for there are no tensed facts in *W*. Indeed, I should say that the tenseless proposition it expresses, that 'Everything that *begins* to exist *has* a cause', is not necessarily true, since such things never undergo temporal becoming, that is, they never come (significantly tensed) to exist and so do not need a cause of their world lines' being finite in the 'earlier than' direction.

Smith might reject this example because such a tenseless world

[46] Saul A. Kripke, *Naming and Necessity*, rev. edn. (Oxford: Basil Blackwell, 1980), 159.

[47] See discussion in William Lane Craig, *Divine Foreknowledge and Human Freedom*, Studies in Intellectual History, xix (Leiden: E. J. Brill, 1991), 111–13, 150–3.

is metaphysically impossible and, hence, strictly inconceivable. I should agree that it is *inconceivable*; but it is not *unimaginable*. In general it seems to me that Smith confuses conceiving a world with imagining a world. To the same extent that we can imagine a world in which water is not H_2O, we can imagine a world in which tense does not exist or in which things come into being without a cause. But this amounts to no more than our ability to form mental pictures and to give them labels like 'A World in which Water is not H_2O' or 'A World in which Something Begins without a Cause'. Strictly speaking, they are alike inconceivable.

This prompts us to ask why 'Everything that begins to exist has a cause' cannot be a necessary a priori truth which can be immediately known. Smith's response is, in effect, that we can imagine a world in which, say, the universe comes into being without a cause. I agree that we can form such a picture in our imagination. But that does nothing to prove that the proposition is not a necessary a priori truth. Consider Aquinas's point with which we began this paper. A pure potentiality cannot be conceived to actualize itself. Therefore, there must be an actual cause for anything's coming to exist. In the case of creation, there was not anything physically prior to the singularity. Therefore, it is impossible that the potentiality of the existence of the universe lay in itself, since it did not exist. On the theistic view, the potentiality of the universe's existence lay in the power of God to create it. On the atheistic interpretation, on the other hand, there did not even exist any potentiality for the existence of the universe. But then it seems inconceivable that the universe should come to be actual if there did not exist any potentiality for its existence. It seems to me therefore that a little reflection discloses that our mental picture of the universe arising uncaused out of absolutely nothing is just that: pure imagination. Philosophical reflection reveals it to be inconceivable.

Hence, far from being simpler than the theistic hypothesis of creation, the atheistic interpretation is less simple, has zero explanatory power, and in the end degenerates into metaphysical absurdity.

4. CONCLUSION

Enjoying no greater consistency than its theistic rival, with no positive argument to commend it, and unable to escape the charge

276 William Lane Craig

of metaphysical absurdity levelled against it, Smith's atheistic interpretation of the Big Bang appears to be untenable. If the standard model is correct, it does seem to constitute a powerful argument for the existence of a Creator of the universe. Smith leaves it open that the model may be false and some other model not involving an initial cosmological singularity be true. Perhaps, though there are reasons to doubt that an absolute beginning can be avoided through such models; but that is a debate for another day.

3

Theism, Atheism, and Hawking's Quantum Cosmology

XI

'What Place, then, for a Creator?': Hawking on God and Creation

WILLIAM LANE CRAIG

I. INTRODUCTION

Scientists working in the field of cosmology seem to be irresistibly drawn by the lure of philosophy. Now Stephen Hawking has followed the lead of Fred Hoyle, Carl Sagan, Robert Jastrow, and P. C. W. Davies in speculating on what philosophical implications current cosmological models have for the existence of God. Although his recent best-seller *A Brief History of Time* is refreshingly free of the acrimony that characterized the works of some of his predecessors, one still might come away with the impression that Hawking is no more sympathetic to theism than they were. A recent article on Hawking's book in the German tabloid *Stern*, for example, headlined 'Kein Platz für den lieben Gott', and concluded, 'In his system of thought there is no room for a Creator God. Not that God is dead: God never existed.'[1] This impression is no doubt abetted by the fact that the book carries an introduction by Sagan, in which he writes,

This is also a book about God . . . or perhaps about the absence of God. The word God fills these pages. Hawking embarks on a quest to answer Einstein's famous question about whether God had any choice in creating the universe. Hawking is attempting, as he explicitly states, to understand the mind of God. And this makes all the more unexpected the conclusion of the effort, at least so far: a universe with no edge in space, no beginning nor end in time, and nothing for a Creator to do.[2]

First pub. in *British Journal of Philosophical Science*, 41 (1990), 473–91.

[1] *Stern* (undated photocopy), 209. 'In seinem Gedankengebäude ist für einen schöpferischen Gott kein Raum. Gott ist nicht einmal tot, Gott hat nie existiert.'
[2] Carl Sagan, in S. W. Hawking, *A Brief History of Time: from the Big Bang to Black Holes*, intro. (New York: Bantam Books, 1988), p. x.

2. GOD AS SUFFICIENT REASON

But such a characterization of Hawking's position is quite misleading. In point of fact, it is false that there is no place for God in Hawking's system or that God is absent. For while it is true that he rejects God's role as Creator of the universe in the sense of an efficient cause producing an absolutely first temporal effect, nevertheless Hawking appears to retain God's role as the Sufficient Reason for the existence of the universe, the final answer to the question, Why is there something rather than nothing? He distinguishes between the questions *what* the universe is and *why* the universe is, asserting that scientists have been too occupied with the former question to be able to ask the latter, whereas philosophers, whose job it is to ask why-questions, have been unable to keep up with the technical scientific theories concerning the origin of the universe and so have shunned metaphysical questions in favour of linguistic analysis. But Hawking himself is clear that having (to his satisfaction at least) answered the question what the universe is, he is still left with the unanswered why-question:

The usual approach of science of constructing a mathematical model cannot answer the questions of why there should be a universe for the model to describe. Why does the universe go to all the bother of existing? Is the unified theory so compelling that it brings about its own existence? Or does it need a creator, and, if so, does he have any other effect on the universe? And who created him?[3]

Pursuing the question why we and the universe exist is a quest that, in Hawking's view, should occupy people in every walk of life. 'If we find the answer to that, it would be the ultimate triumph of human reason—for then we should know the mind of God.'[4]

At face value, then, God for Hawking serves as the Sufficient Reason for the existence of the universe. Of course, 'the mind of God' might well be a mere *façon de parler*, signifying something like 'the meaning of existence';[5] but, as Sagan noted, Hawking seems very much in earnest about determining the proper role of God as traditionally conceived in the scheme of things. And it is interesting

[3] Hawking, *A Brief History of Time*, 174.
[4] Ibid. 175.
[5] Cf. the remark by Pagels: 'Physicists, regardless of their belief, may invoke God when they feel issues of principle are at stake because the God of the physicists is cosmic order.' (Heinz Pagels, *The Cosmic Code* (London: Michael Joseph, 1982), 83.)

to note that when a reader of an earlier summary draft of Hawking's book in *American Scientist*[6] complained that Hawking seemed afraid to admit the existence of a Supreme Being, Hawking countered that 'I thought I had left the question of the existence of a Supreme Being completely open. . . . It would be perfectly consistent with all we know to say that there was a Being who was responsible for the laws of physics.'[7]

Now it might seem at first somewhat baffling that Hawking senses the need to explain why the universe exists, since, as we shall see, he proposes a model of the universe according to which the universe is 'completely self-contained and not affected by anything outside itself', is 'neither created nor destroyed', but just *is*.[8] On his analysis, the universe is eternal (in the sense that it has neither beginning nor end and exists tenselessly) and therefore has no temporally antecedent cause. But if the cosmos is eternal and uncaused, what sense does it make to ask why it exists?

Leibniz, however, saw the sense of such a question.[9] He held that it is intelligible to ask why it is that an eternal being exists, since the existence of such a being is still logically contingent. Since it is possible that nothing exists, why is it that an eternal cosmos exists rather than nothing? There must still be a Sufficient Reason why there exists something—even an eternal something—rather than nothing. Leibniz concluded that this Sufficient Reason can only be found in a metaphysically necessary being, that is, a being whose nature is such that if it exists, it exists in all possible worlds. Hawking would be interested to learn that analytic philosophy in the past two decades has burst the skins of linguistic analysis and that certain analytic philosophers doing metaphysics have defended Leibniz's conception of God as a metaphysically necessary being.[10] Given the existence of such a being, Hawking need not trouble

[6] S. W. Hawking, 'The Edge of Spacetime', *American Scientist*, 72 (1984), 355–9.
[7] S. W. Hawking, 'Letters to the Editor: Time and the Universe', *American Scientist*, 73 (1985), 12.
[8] Hawking, *A Brief History of Time*, 136.
[9] See G. W. Leibniz, 'On the Radical Origination of Things', 'The Principles of Nature and Grace, Based on Reason', and 'The Monadology', in *Philosophical Papers and Letters*, ed. L. E. Loemker, 2nd edn. (Dordrecht: D. Reidel, 1969), 486–91, 636–53.
[10] See e.g. Robert M. Adams, 'Has it Been Proved that All Real Existence Is Contingent?' *American Philosophical Quarterly*, 8 (1971), 284–91; Alvin Plantinga, *The Nature of Necessity*, Clarendon Library of Logic and Philosophy (Oxford: Clarendon Press, 1974), 197–221; William L. Rowe, *The Cosmological Argument* (Princeton, NJ: Princeton University Press, 1975), 202–21.

himself about who created God, since God, being metaphysically necessary and ultimate, can have no cause or ground of being.[11]

Thus, it seems to me that far from banishing God from reality, Hawking invites us to make Him the basis of reality. Indeed, I think Hawking's book may rightly be read as a discussion of two forms of the cosmological argument: the so-called *kalām* cosmological argument for a temporally First Cause of the universe, which he rejects, and the Leibnizian cosmological argument for a Sufficient Reason of the universe, which he prefers.[12] In this essay, I am not concerned to evaluate the Leibnizian cosmological argument. Like Hawking, I feel the force of Leibniz's reasoning and am inclined to accept it; but unlike Hawking, it seems to me that the *kalām* argument is plausible as well. Accordingly, we need to ask, has Hawking eliminated the need for a Creator?

3. GOD AS METAPHYSICALLY FIRST CAUSE

Now at one level, the answer to that question is an immediate 'No.' For Hawking has a theologically deficient understanding of creation. Traditionally creation was thought to involve two aspects: *creatio originans* and *creatio continuans*. The first concerned God's bringing finite reality into being at a point in time before which no such reality existed, whereas the second involved (among other things) God's preservation of finite reality in being moment by moment. Only the first notion involves the idea of a beginning. *Creatio continuans* could involve a universe existing from everlasting to everlasting, that is to say, a universe temporally infinite in both the past and the future at any point of time. Thus, for example, Thomas Aquinas, confronted on the one hand with Aristotelian and Neoplatonic arguments for the eternity of the world, and, on the other hand, with Arabic *kalām*-style arguments for the finitude of the past, concluded after a lengthy consideration of arguments both

---

[11] On God as the ground of being for other metaphysically necessary entities, see Thomas V. Morris and Christopher Menzel, 'Absolute Creation', *American Philosophical Quarterly*, 23 (1986), 353–62; Christopher Menzel, 'Theism, Platonism, and the Metaphysics of Mathematics', *Faith and Philosophy*, 4 (1987), 365–82. These bold essays should convince Hawking that the great tradition of metaphysics has been fully restored in analytic philosophy!

[12] On these arguments, as well as the Thomist argument, see William Lane Craig, *The Cosmological Argument from Plato to Leibniz*, Library of Philosophy and Religion (London: Macmillan, 1980).

pro and *contra* that it can be proved neither that the universe had a
beginning nor that it did not, but that the question of the temporal
origin of the universe must be decided on the basis of divine
revelation, that is, the teaching of the Scriptures.[13] Given this
position, it appears at first paradoxical that Aquinas also held that
the doctrine of divine *creatio ex nihilo* can be proved.[14] But once we
understand that creation in the sense of *creatio continuans* involves
no notion of a temporal beginning the paradox disappears. To
affirm that God creates the world out of nothing is to affirm that
God is the immediate cause of the world's existence, that there is
no metaphysical intermediary between God and the universe.

Actually, what Hawking has done is fail to distinguish from the
kalām argument yet a third form of the cosmological argument,
which we may call the Thomist cosmological argument, that comes
to expression in Thomas's Third Way[15] and his *De Ente et Essentia*,
3. According to Aquinas, all finite beings, even those like the
heavenly spheres or prime matter which have absolutely no potential
for generation or corruption and are therefore by nature everlasting,
are nevertheless metaphysically contingent in that they are composed
of essence and existence, that is to say, their essential properties
do not entail that such beings exist. If these essences are to be
exemplified, therefore, there must be a being in whom essence and
existence are not distinct and which therefore is uncaused, and it is
this being which is the Creator of all finite beings, which he
produces by instantiating their essences. Hence, *creatio ex nihilo*
does not, in Aquinas's view, entail a temporal beginning of the
universe.

Even if we maintain that a full-blooded doctrine of creation does
entail a temporal beginning of the universe, the point remains that
this doctrine also entails much more than that, so that even if God
did not bring the universe into being at a point of time as in
Hawking's model, it is still the case that there is much for Him to
do, for without His active and continual bestowal of existence to the
universe, the whole of finite reality would be instantly annihilated
and lapse into non-being. Thus, any claim that Hawking has
eliminated the Creator is seen to be theologically frivolous.

[13] Thomas Aquinas, *Summa contra Gentiles*, 2.32–8; cf. Thomas Aquinas, *De Aeternitate Mundi contra Murmurantes.*
[14] Aquinas, *Summa contra Gentiles*, 2.16.
[15] Aquinas, *Summa Theologiae*, 1a.2.3.

4. GOD AS TEMPORALLY FIRST CAUSE

But has Hawking succeeded even in obviating the role of the Creator as temporally First Cause? This seems to me highly dubious, for Hawking's model is founded on philosophical assumptions that are at best unexamined and unjustified and at worst false. To see this, let us recall the fundamental form of the *kalām* cosmological argument, so that the salient points of Hawking's refutation will emerge. Proponents of that argument have presented a simple syllogism:

(1) Whatever begins to exist has a cause.
(2) The universe began to exist.
(3) Therefore, the universe has a cause.

Analysis of the cause of the universe established in (3) further discloses it to be uncaused, changeless, timeless, immaterial, and personal.

4.1. *Hawking's Critique*

Hawking is vaguely aware of the tradition of this argument in Christian, Muslim, and Jewish thought and presents a somewhat muddled version of it in chapter 1.[16] But it is interesting that, unlike Davies, Hawking does not attack premiss (1); on the contrary, he implicitly assents to it. Hawking repeatedly states that on the classical General Theory of Relativity (GTR) Big Bang model of the universe an initial spacetime singularity is unavoidable, and he does not dispute that the origin of the universe must therefore require a supernatural cause. He points out that one could identify the Big Bang as the instant at which God created the universe.[17] He thinks that a number of attempts to avoid the Big Bang were probably motivated by the feeling that a beginning of time 'smacks of divine intervention'.[18] It is not clear what part such a motivation plays in Hawking's own proposal, but he touts his model as preferable because 'There would be no singularities at which the laws of science broke down and no edge of space-time at which one would have to appeal to God or some new law to set the boundary conditions for space-time.'[19] On Hawking's view, then, given the

---

[16] Hawking, *A Brief History of Time*, 7.
[17] Ibid. 9.   [18] Ibid. 46.   [19] Ibid. 136.

classical Big Bang model, the inference to a Creator or temporally First Cause seems natural and unobjectionable.

Hawking's strategy is rather to dispute premiss (2). Typically, proponents of *kalām* supported (2) by arguing against the possibility of an infinite temporal regress of events. This tradition eventually became enshrined in the thesis of Kant's First Antinomy concerning time.[20] Hawking's response to this line of argument is very ingenious. He claims that the argument of the thesis and antithesis 'are both based on his unspoken assumption that time continues back forever, whether or not the universe had existed forever', but that this assumption is false because 'the concept of time has no meaning before the beginning of the universe'.[21] This brief retort is somewhat muddled, but I think the sense of it is the following: In the antithesis Kant assumes that 'Since the beginning is an existence which is preceded by a time in which the thing is not, there must have been a preceding time in which the world was not, i.e. an empty time.'[22] But on some version of a relational view of time, time does not exist apart from change; therefore, the first event marked the inception of time. Thus, there was no empty time prior to the beginning of the universe. In the thesis, on the other hand, Kant states, 'If we assume that the world has no beginning in time, then up to every given moment an eternity has elapsed and there has passed away in the world an infinite series of successive states of things.'[23] To my knowledge, scarcely anyone has ever thought to call into question this apparently innocuous assumption, but it is precisely here that Hawking launches his attack. Unlike other detractors of Kant's argument, Hawking does not dispute the impossibility of forming an actual infinite by successive addition; rather he challenges the more fundamental assumption that a beginningless universe entails an infinite past. The central thrust of Hawking's book and of his proposed cosmological model is to show that a beginningless universe may be temporally finite. Hence, *kalām*-style arguments aimed at proving the finitude of the past need not be disputed, for such arguments do not succeed in estab-

---

[20] For discussion, see William Lane Craig, 'Kant's First Antinomy and the Beginning of the Universe', *Zeitschrift für philosophische Forschung*, 33 (1979), 553–67.
[21] Hawking, *A Brief History of Time*, 8.
[22] Immanuel Kant, *Critique of Pure Reason*, trans. Norman Kemp-Smith (London: Macmillan, 1929), 397 (A427–8/B455–6).
[23] Ibid.

lishing (2), that the universe began to exist. Therefore, the universe need not have a cause, and God's role as Creator is circumscribed to that envisioned in the Thomist and Leibnizian versions of the cosmological argument.

This is a highly original, if not unique, line of attack on the *kalām* cosmological argument, and it will be interesting to see how Hawking essays to put it through.[24] It is Hawking's belief that the introduction of quantum mechanics into the GTR-based Big Bang model will be the key to success. Noting that at the Big Bang the density of the universe and the curvature of spacetime become infinite, Hawking explains that 'there must have been a time in the very early universe when the universe was so small, that one could no longer ignore the small scale effects of . . . quantum mechanics' and that the initial singularity predicted by the GTR 'can disappear once quantum effects are taken into account'.[25] What is needed here is a quantum theory of gravity, and although Hawking admits that no such theory exists, still he insists that we do have a good idea of what some of its central features will be.[26] First, it will incorporate Feynman's sum-over-histories approach to quantum mechanics. According to this approach to quantum theory, an elementary particle does not follow a single path between two spacetime points (that is, have a single history), but it is rather conceived as taking all possible paths connecting those points. In order to calculate the probability of a particle's passing through any given spacetime point, one sums the waves associated with every possible history that passes through that point, histories represented by waves having equal amplitude and opposite phase mutually cancelling so that only the most probable histories remain. But in order to do this without generating intractable infinities, Hawking explains, one must use imaginary numbers for the values of the time coordinate. When this is done, it 'has an interesting effect on

---

[24] One feels a bit diffident about criticizing someone's views as they are expressed in a popular exposition of his thought rather than in his technical papers. But the fact is that it is only in his popular exposition that Hawking feels free to reflect philosophically on the metaphysical implications of his model. For example, imaginary time, which plays so critical a role in his thought, is scarcely even mentioned in his relevant technical paper (J. Hartle and S. Hawking, 'Wave Function of the Universe', *Physical Review*, D28 (1983), 2960–75). In any case, I have in no instance based my criticism on the infelicities inherent in popular exposition of technical subjects.

[25] Hawking, *A Brief History of Time*, 50–1.

[26] Ibid. 133.

space-time: the distinction between time and space disappears completely'.[27] The resulting spacetime is Euclidean.

The second feature which any theory of quantum gravity must possess is that the gravitational field is represented by curved spacetime. When this feature of the theory is combined with the first, the analogue of the history of a particle now becomes a complete curved spacetime that represents the history of the whole universe. Moreover, 'To avoid the technical difficulties in actually performing the sum over histories, these curved space-times must be taken to be Euclidean. That is, time is imaginary and is indistinguishable from directions in space.'[28]

On the basis of these two features, Hawking proposes a model in which spacetime is the four-dimensional analogue to the surface of a sphere. It is finite, but boundless, and so possesses no initial or terminal singularities. Hawking writes,

In the classical theory of gravity, which is based on real space-time, there are only two possible ways the universe can behave: either it has existed for an infinite time, or else it had a beginning at a singularity at some finite time in the past. In the quantum theory of gravity, on the other hand, a third possibility arises. Because one is using Euclidean space-times, in which the time direction is on the same footing as directions in space, it is possible for space-time to be finite in extent and yet to have no singularities that formed a boundary or edge. . . . There would be no singularities at which the laws of science broke down and no edge of space-time at which one would have to appeal to God or some new law to set the boundary conditions for space-time. . . . The universe would be completely self-contained and not affected by anything outside itself. It would be neither created nor destroyed. It would just BE.[29]

Hawking emphasizes that his model is merely a proposal, and so far as he describes it, it makes no unique successful predictions, which would be necessary to transform it from a metaphysical theory to a plausible scientific theory. Still Hawking believes that

The idea that space and time may form a closed surface without boundary . . . has profound implications for the role of God in the affairs of the universe. . . . So long as the universe had a beginning, we could suppose it had a creator. But if the universe is really completely self-contained, having no boundary or edge, it would have neither beginning nor end. What place, then, for a creator?[30]

---

[27] Ibid. 134.   [28] Ibid. 135.   [29] Ibid. 135–6.   [30] Ibid. 140–1.

4.2. *Assessment*

Unfortunately, Hawking's model is rife with controversial philosophical assumptions, to which he gives no attention. Since Hawking is trying to explain how the universe could exist without the necessity of God's bringing it into being at a point of time, it is evident that he construes his theory to be, not merely an engaging mathematical model, but a realistic description of the universe. On a non-realist interpretation of science, there would be no contradiction between his model and temporal *creatio ex nihilo*. Hence, the central question that needs to be addressed in assessing his model as an alternative to divine creation is whether it represents a realistic picture of the world.

Now to me at least it seems painfully obvious that Hawking faces severe difficulties here. Both quantum theory and relativity theory inspire acute philosophical questions concerning the extent to which they picture reality. To begin with quantum theory, most philosophers and reflective physicists would not disagree with the remarks of Hawking's erstwhile collaborator Roger Penrose:

> I should begin by expressing my general attitude to present-day quantum theory, by which I mean standard, non-relativistic quantum mechanics. The theory has, indeed, two powerful bodies of fact in its favour, and only one thing against it. First, in its favour are all the marvellous agreements that the theory has had with every experimental result to date. Second, and to me almost as important, it is a theory of astonishing and profound mathematical beauty. The one thing that can be said against it is that it makes absolutely no sense![31]

Does Hawking believe, for example, that Feynman's sum-over-histories approach describes what really happens, that an elementary particle really does follow all possible spacetime paths until its wave function is collapsed by measurement? I think most people would find this fantastic. If he does interpret this approach realistically, then what justification is there for such an interpretation? Why not a Copenhagen Interpretation which eschews realism altogether with regard to the quantum world? Or an alternative version of the Copenhagen Interpretation which holds that no quantum reality exists until it is measured? Why not hold that the uncollapsed wave

---

[31] Roger Penrose, 'Gravity and State Vector Reduction', in Roger Penrose and C. J. Isham (eds.), *Quantum Concepts in Space and Time* (Oxford: Clarendon Press, 1986), 129.

function is, in Bohr's words, 'only an abstract quantum mechanical description' rather than a description of how nature is? A disavowal of realism on the quantum level does not imply a rejection of a critical realism on the macroscopic level. Or why not interpret quantum mechanics as a statistical theory about ensembles of particles rather than about the behaviour of any individual particle? On this interpretation, the wave function describes the collective behaviour of particles in identical systems, and we could quit worrying about the measurement problem. Or again, what about a neo-realist interpretation along the lines of the de Broglie–Bohm pilot wave? A non-local hidden variables theory, in which a particle follows a definite spacetime trajectory, is compatible with all the experiment and evidence for quantum theory, is mathematically rigorous and complete, and yet avoids the philosophical difficulties occasioned by the typical wave functional analysis. Obviously, it is not my intention to endorse any one of these views, but merely to point out that a realistic interpretation of Feynman's sum-over-histories approach on Hawking's part would be gratuitous.

In general, I think we should do well to reflect on de Broglie's attitude to the mathematical formalism of quantum theory. As Georges Lochak notes, 'He does not consider that mathematical models have any ontological value, especially geometrical representations in abstract spaces; he sees them as practical mathematical instruments among others and only uses them as such.'[32] The principle of the superposition of wave functions is a case in point. Simply because a mathematical model is operationally successful, we are not entitled to construe its representations physically. Feynman himself gave this sharp advice: 'I think it is safe to say that no one understands quantum mechanics. Do not keep saying to yourself, if you possibly can avoid it, "But how can it be like that?" because you will go "down the drain" into a blind alley from which nobody has yet escaped. Nobody can know how it can be like that.'[33] One can use the equations without taking them as literal representations of reality.

[32] Georges Lochak, 'The Evolution of the Ideas of Louis de Broglie on the Interpretation of Wave Mechanics', in A. O. Barut, A. V. D. Merwe, and J.-P. Vigier (eds.), *Quantum, Space, and Time*, Cambridge Monographs on Physics (Cambridge: Cambridge University Press, 1984), 20.
[33] Cited in N. Herbert, *Quantum Reality: Beyond the New Physics* (Garden City, NY: Doubleday Anchor Books, 1985), p. xiii.

Now it might be said that Hawking's use of Feynman's sum-over-histories approach may be merely instrumental and that no commitment to a physical description is implied. But it is not evident that such a response will work for Hawking. For his model, based on the application of quantum theory to classical geo-metrodynamics, must posit the existence of a superspace which is ontologically prior to the approximations of classical spacetime that are slices of this superspace. This superspace is no *ens fictum*, but the primary reality. The various 3-geometries surrounding the classical spacetime slice in superspace are fluctuations of the classical slice. By 'summing the histories' of these 3-geometries one can construct a leaf of history in superspace which can be mapped on to a spacetime manifold. Since, as we have seen, Hawking takes the wave function of a particle to be the analogue of a physical spacetime that represents the history of the universe, an instrumentalist inter-pretation of the sum-over-histories approach leads to an equally instrumentalist, non-realist view of spacetime, which betrays Hawking's whole intent.

In short, Hawking's wave functional analysis of the universe requires the Many Worlds Interpretation of quantum physics, and in another place Hawking admits as much.[34] But why should we adopt this interpretation of quantum physics with its bloated ontology and miraculous splitting of the universe? John Barrow has recently remarked that the Many Worlds Interpretation is 'essential' to quantum cosmology because without it one is left, on the standard Copenhagen Interpretation, with the question, Who or what collapses the wave function of the universe?—some Ultimate Observer out-side of space and time?[35] This answer has obvious theistic implica-tions. Indeed, although 'the theologians have not been very eager to ascribe to God the role of Ultimate Observer who brings the entire quantum Universe into being', still Barrow admits that 'such a picture is logically consistent with the mathematics. To escape this step cosmologists have been forced to invoke Everett's "Many Worlds" interpretation of quantum theory in order to make any sense of quantum cosmology.'[36] 'It is no coincidence', he says, 'that

---

[34] S. W. Hawking, 'Quantum Cosmology', repr. in L. Z. Fang and R. Ruffini (eds.), *Quantum Cosmology*, Advanced Series in Astrophysics and Cosmology, iii (Singapore: World Scientific, 1987), 192–3.

[35] John Barrow, *The World within the World* (Oxford: Clarendon Press, 1988), 156.

[36] Ibid. 232.

all the main supporters of the Many Worlds interpretation of quantum reality are involved in quantum cosmology.'[37] But if we, like most physicists, find the Many Worlds Interpretation outlandish, then quantum cosmology, far from obviating the place of a Creator, might be seen to create for Him a dramatic new role. Again, my intention is not to endorse this view, but simply to underscore the fact that a realist construal of Hawking's account involves extravagant and dubious metaphysical commitments, such that his model can hardly be said to have eliminated the place of a Creator.

The impression that Hawking's model is thoroughly non-realist is heightened by his use of imaginary time in summing the waves for particle histories and, hence, in his final model of spacetime. But does anyone seriously believe that one has thereby done anything more than perform a mathematical operation on paper, that one has thereby altered the nature of time itself? Hawking asserts, 'Imaginary time may sound like science fiction but it is in fact a well-defined mathematical concept.'[38] But that is not the issue; the question is whether that mathematical concept has any counterpart in physical reality. Already in 1920, Eddington suggested that his readers who found it difficult to think in terms of the unfamiliar non-Euclidean geometry of relativistic spacetime might evade that difficulty by means of the 'dodge' of using imaginary numbers for the time coordinate, but he thought it 'not very profitable' to speculate on the implications of this, for 'it can scarcely be regarded as more than an analytical device'.[39] Imaginary time was merely an illustrative tool which 'certainly do[es] not correspond to any physical reality'.[40] Even Hawking himself maintains, 'In any case, as far as everyday quantum mechanics is concerned, we may regard our use of imaginary time and Euclidean spacetime as merely a mathematical device (or trick) to calculate answers about real space-time.'[41] But now in his model this imaginary time and Euclidean spacetime are suddenly supposed to be, not merely conceptual devices, but actual representations (however unimaginable) of physical reality in the very early history of the universe. This 'ontologizing' of mathematical opera-

[37] Ibid. 156.
[38] Hawking, *A Brief History of Time*, 134.
[39] Arthur Eddington, *Space, Time and Gravitation*, Cambridge Science Classics (first pub. 1920; repr. Cambridge: Cambridge University Press, 1987), 48.
[40] Ibid. 181.
[41] Hawking, *A Brief History of Time*, 134–5.

tions is not only neither explained nor justified, but is, to my mind, metaphysically absurd. For what possible physical meaning can we give to imaginary time? Having the opposite sign of ordinary 'real' time, would imaginary time be a sort of negative time? But what intelligible sense can be given, for example, to a physical object's enduring for, say, two negative moments, or an event's having occurred two negative moments ago or going to occur in two negative moments? If we are A-theorists and take temporal becoming as objective and real, what does it mean to speak of the lapse of imaginary time or the becoming of events in imaginary time? Since imaginary time is on Hawking's view merely another spatial dimension, he admits that there is no direction to time, even though the ordinary time with which we are acquainted is asymmetric.[42] Could anything be more obvious than that imaginary time is a mathematical fiction?[43]

Hawking recognizes that the history of the universe in real (=ordinary) time would look very different than its history in imaginary time. In real time, the universe expands from a singularity and collapses back again into a singularity. 'Only if we could picture the universe in terms of imaginary time would there be no singularities. . . . When one goes back to the real time in which we live, however, there will still appear to be singularities.'[44] This might lead one to conclude that Hawking's model is a mere mathematical construct without ontological import. Instead, Hawking draws the astounding conclusion,

This might suggest that the so-called imaginary time is really the real time, and that what we call real time is just a figment of our imaginations. In real time, the universe has a beginning and an end at singularities that form a boundary to space-time and at which the laws of science break down. But in imaginary time, there are no singularities or boundaries. So maybe what we call imaginary time is really more basic, and what we call real is just an idea that we invent to help us describe what we think the universe is like.[45]

I can think of no more egregious example of self-deception than this. One employs mathematical devices (tricks) such as sum-over-

---

[42] Ibid. 144.
[43] As Mary Cleugh nicely puts it, 'What is the wildest absurdity of dreams is merely altering the sign to the physicist.' (Mary Cleugh, *Time and its Importance in Modern Thought* (London: Methuen, 1937), 46.)
[44] Hawking, *A Brief History of Time*, 138–9.
[45] Ibid. 139.

histories and changing the sign of the time coordinate in order to construct a model spacetime, a model which is physically unintelligible, and then one invests that model with reality and declares that the time in which we live is in fact unreal.

Hawking defends his position by arguing that 'a scientific theory is just a mathematical model we make to describe our observations: it exists only in our minds. So it is meaningless to ask: Which is real, "real" or "imaginary" time? It is simply a matter of which is the more useful description.'[46] But this reasoning is fallacious and relapses into an instrumentalist view of science which contradicts Hawking's realist expressions and intentions. One may adopt a sort of nominalist view of the ontological status of theories themselves, but this says absolutely nothing about whether those theories are meant to describe, in approximate limits, physical reality or are merely pragmatic instruments for making new discoveries and advancing technology. I should like to know on what theory of meaning Hawking dismisses the question concerning physical time as meaningless. We seem to see here the vestige of a defunct positivism, which surfaces elsewhere in Hawking's book.[47] But a verificationist theory of meaning is today widely recognized as being simply indefensible.[48] The question Hawking brushes aside is not only obviously meaningful, but crucial for the purposes of his book, for only if he can prove that imaginary time is ontologically real and real time fictitious has he succeeded in obviating the need for a Creator. Which brings us again to his scientific realism: it seems clear that for Hawking the ontological status of time is not just a matter of the more useful description. He believes that 'The eventual goal of science is to provide a single theory that describes

---

[46] Ibid.

[47] Ibid. 55, 126.

[48] Healey describes the contemporary attitude toward positivism: 'Positivists attempted to impose restrictions on the content of scientific theories in order to ensure that they were empirically meaningful. An effect of these restrictions was to limit both the claims to truth of theoretical sentences only distantly related to observation, and the claims to existence of unobservable theoretical entities. More recently positivism has come under such sustained attack that opposition to it has become almost orthodoxy in the philosophy of science.' (Richard Healey (ed.), *Reduction, Time and Reality* (Cambridge: Cambridge University Press, 1981), p. vii.) For a disinterested and devastating critique of positivism, see Frederick Suppe, 'The Search for Philosophic Understanding of Scientific Theories,' in F. Suppe (ed.), *The Structure of Scientific Theories*, 2nd edn. (Urbana, Ill.: University of Illinois Press, 1977), 62–118.

the whole universe' and that this goal should be pursued even though the theory 'may not even affect our lifestyle'.[49] Hawking yearns to understand 'the underlying order of the world'.[50] Knowing the mind of God is for him not just a matter of pragmatic utility. Thus, he both needs and believes in scientific realism.

To address as meaningful the question posed above, then, it is evident that imaginary time is not ontological time. This is apparent not only from its physically unintelligible nature, but also from the fact that it transforms time into a spatial dimension, thus confounding the distinction between space and time. According to Hawking, the use of imaginary numbers 'has an interesting effect on space-time: the distinction between time and space disappears completely . . . there is no difference between the time direction and directions in space . . . time is imaginary and is indistinguishable from directions in space'.[51] This decisively disqualifies Hawking's model as a representation of reality, since in fact time is not ontologically a spatial dimension. Contemporary expositors of the Special Theory of Relativity (STR) have been exercised to dissociate themselves from the frequent statements of early proponents of the theory to the effect that Einstein's theory had made time the fourth dimension of space.[52] B-theorists of time have been especially sensitive to the allegation by A-theorists that they have been guilty of 'spatializing' time and have pointed to the opposite sign of the time coordinate as evidence that the temporal dimension is in fact not a mere fourth dimension of space. By changing the sign, Hawking conflates the temporal dimension with the spatial ones. Hawking apparently feels justified in this move because he, like certain early interpreters

---

[49] Hawking, *A Brief History of Time*, 10, 13; cf. his remarks in S. W. Hawking, 'The Boundary Conditions of the Universe', in H. A. Bruck, G. V. Coyne, and M. S. Longair (eds.), *Astrophysical Cosmology*, Pontificiae Academiae Scientiarum Scripta Varia, xlviii (Vatican City: Pontificia Academia Scientiarum, 1982), 563.

[50] Hawking, *A Brief History of Time*, 13.

[51] Ibid. 134–5.

[52] See the interesting citations in E. Meyerson, 'On Various Interpretations of Relativistic Time', repr. in M. Capek (ed.), *The Concepts of Space and Time*, Boston Studies in the Philosophy of Science, ii (Dordrecht: D. Reidel, 1976), 354–5. In his comments on Meyerson's book, Einstein repudiated the 'extravagances of the popularizers and even many scientists' who construed STR to teach that time is a spatial dimension: 'Time and space are fused into one and the same *continuum*, but this continuum is not isotropic. The element of spatial distance and the element of duration remain distinct in nature . . .' (A. Einstein, 'Comment on Meyerson's "La Déduction relativiste"', repr. in ibid. 367.)

of STR, believes that STR itself treats time as a spatial dimension. He writes, 'In relativity, there is no real distinction between the space and time coordinates, just as there is no real difference between any two space coordinates.'[53] He justifies this statement by pointing out that one could construct a new time coordinate by combining the old time coordinate with one of the spatial coordinates. In spatializing time, Hawking implicitly rejects an A-theory and identifies himself as a B-theorist. His statement concerning the universe as he models it that 'It would just BE' is an expression of the tenseless character of its existence. Unfortunately, he provides no justification whatsoever for adopting a B-theory of time. Perhaps he thinks that STR entails a B-theory; but A-theorists have argued repeatedly that the Special Theory is neutral with regard to the issue of temporal becoming, and the most sophisticated B-theorists do not appeal to it as proof of their view.[54] The debate between the A-theory and the B-theory is controversial. But in the absence of some overwhelming proof of the B-theory, I see no reason to abandon our experience of temporal becoming as objective. D. H. Mellor, himself a B-theorist, agrees, commenting, 'Tense is so striking an aspect of reality that only the most compelling argument justifies denying it: namely, that the tensed view of time is self-contradictory and so cannot be true.'[55] Mellor accordingly tries to rehabilitate McTaggart's proof against the objectivity of the A-series, but, to my thinking, to no avail.[56] Moreover, it seems to me (although space does not permit me to argue it here) that no B-

---

[53] Hawking, *A Brief History of Time*, 24.

[54] For A-theoretic approaches to STR, see M. Capek, 'Time in Relativity Theory: Arguments for a Philosophy of Becoming', in J. T. Fraser (ed.), *Voices of Time* (New York: Braziller, 1966), 434–54; H. Stein, 'On Einstein–Minkowski Space-Time', *Journal of Philosophy*, 65 (1968), 5–23; K. Denbigh, 'Past, Present, and Future', in J. T. Fraser (ed.), *The Study of Time*, iii (Berlin: Springer Verlag, 1978), 301–29; G. J. Whitrow, *The Natural Philosophy of Time*, 2nd edn. (Oxford: Clarendon Press, 1980), 283–307, 371; and D. Dieks, 'Special Relativity and the Flow of Time', *Philosophy of Science*, 55 (1988), 456–60. A Grünbaum, 'The Status of Temporal Becoming', in R. M. Gale (ed.), *The Philosophy of Time* (London: Macmillan, 1968), 322–54, makes no appeal to STR to defend a B-theory.

[55] D. H. Mellor, *Real Time* (Cambridge: Cambridge University Press, 1981), 5.

[56] See the relevant chapters in my *God, Time, and Eternity* (forthcoming). McTaggart's fundamental error was his mistaken assumption that the A-theory combines a B-theoretical ontology with temporal becoming. Mellor makes the same assumption, but also errs decisively in making sentence tokens rather than propositions his truth-bearers, which leads to his misconceived token-reflexive truth conditions of tensed sentences.

theorist has successfully defended that theory against the incoherence that if external becoming is mind-dependent, still the subjective experience of becoming is objective, that is, there is an objective succession of contents of consciousness, so that becoming in the mental realm is real. If an A-theory of time is correct, then Hawking's model is clearly a mere mathematical abstraction.

Whether the opposite sign of the time coordinate in the relativity equations is sufficient to establish a 'real difference' between time and space dimensions in the Special Theory need not be adjudicated here. If it is not sufficient, that only goes to show that the mathematical formalism of the theory is insufficient to capture the ontology of time and space, but is a useful mathematical abstraction from reality.[57] That time and space are ontologically distinct is evident from the fact that a series of mental events alone is sufficient to set up a temporal series of events even in the absence of spatial events.[58] Imagine, for example, that God led up to creation by counting, '1, 2, 3, . . . *fiat lux!*' In that case, time begins with the first mental event of counting, though the physical universe does not appear until later.[59] Clearly, then, time and space are ontologically distinct.

But what, then, of the oft-repeated claim of Minkowski that, 'Henceforth, space by itself, and time by itself, are doomed to fade away into mere shadows, and only a kind of union of the two will preserve an independent reality'?[60] This claim is based on one of the most widespread and persistent errors concerning the interpretation of the Special Theory that exists, namely, the failure to

[57] See helpful discussions in Cleugh, *Time and its Importance in Modern Thought*, 46–9, and Peter Kroes, *Time: Its Structure and Role in Physical Theories*, Synthese Library, clxxix (Dordrecht: D. Reidel, 1985), 60–96.

[58] On Minkowski spacetime, Wenzl cautions, 'From the standpoint of the physicist, this is a thoroughly consistent solution. But the physicist will [doubtless] understand the objection, raised by philosophy, that time is by no means merely a physical matter. Time is, as Kant put it, the form not merely of our outer but also of our inner sense. . . . Should our experiences of successiveness and of memory be mere illusion . . . ?' (A. Wenzl, 'Einstein's Theory of Relativity, Viewed from the Standpoint of Critical Realism, and its Significance for Philosophy', in P. A. Schilpp (ed.), *Albert Einstein: Philosopher-Scientist*, Library of Living Philosophers, vii [La Salle, Ill.: Open Court, 1949], 587–8).

[59] For more on God's relationship to time see William Lane Craig, 'God and Real Time', *Religious Studies*, 26 (1990), 335–47; William Lane Craig, 'Theories of Divine Eternity and the Special Theory of Relativity', *Faith and Philosophy*, 11 (1994), forthcoming.

[60] H. Minkowski, 'Space and Time', repr. in *The Principle of Relativity*, trans. W. Perrett and G. B. Jeffery (New York: Dover, 1952), 75.

distinguish between what we may call measured or empirical time and ontological or real time. According to Hawking, 'the theory of relativity put an end to the idea of absolute time. . . . The theory of relativity does force us to change fundamentally our ideas of space and time. We must accept that time is not completely separate from and independent of space, but is combined with it to form an object called space-time.'[61] Nothing could be farther from the truth. Einstein did not eliminate absolute simultaneity: he merely redefined simultaneity. In the absence of a detectable ether, Einstein, under the influence of Ernst Mach's positivism,[62] believed that it was quite literally meaningless to speak of events' occurring absolutely simultaneously because there was no empirical means of determining that simultaneity. By proposing to redefine simultaneity in terms of the light signal method of synchronization, Einstein was able to give meaning to the notion of simultaneity, only now the simultaneity was relative due to the invariant velocity of light and the absence of the ether frame. In so doing, Einstein established a sort of empirical time, which would be subject to dilation and in which the occurrence of identical events could be variously measured. But it is evident that he did nothing to 'put an end' to absolute time or absolute simultaneity.[63] To say that those notions are meaningless

---

[61] Hawking, *A Brief History of Time*, 21, 23.

[62] The positivistic foundations of Einstein's STR are widely recognized by historians of science, but are surprisingly rarely discussed by philosophers exploring the philosophical foundations of that theory. For discussion, see G. Holton, 'Mach, Einstein, and the Search for Reality', in *Ernst Mach: Physicist and Philosopher*, Boston Studies in the Philosophy of Science, vi (Dordrecht: D. Reidel, 1970), 165–99; P. Frank, 'Einstein, Mach, and Logical Positivism', 271–86, H. Reichenbach, 'The Philosophical Significance of the Theory of Relativity', 289–311, P. Bridgman, 'Einstein's Theories and the Operational Point of View', 335–54, and V. Lenzen, 'Einstein's Theory of Knowledge', 357–84, all in Schilpp (ed.), *Albert Einstein*. According to Sklar, 'Certainly the original arguments in favor of the relativistic viewpoint are rife with verificationist presuppositions about meaning, etc. And despite Einstein's later disavowal of the verificationist point of view, no one to my knowledge has provided an adequate account of the foundations of relativity which isn't verificationist in essence.' (L. Sklar, 'Time, Reality, and Relativity', in Healey (ed.), *Reduction, Time, and Reality*, 141.) 'I can see no way of rejecting the old aether-compensatory theories . . . without invoking a verificationist critique of some kind or other.' (Ibid. 132.)

[63] Cleugh hits the essential point: 'It cannot be too often emphasized that physics is concerned with the measurement of time, rather than with the essentially metaphysical question as to its nature'; 'however useful "*t*" may be for physics, its *complete* identification with Time is fallacious.' (Cleugh, *Time and its Importance in Modern Thought*, 51, 30.)

is to revert to the dead dogmas of positivism and the verificationist theory of meaning. J. S. Bell asserts that apart from matters of style, it is primarily this philosophical positivism which serves to differentiate the received interpretation from the Lorentz–Larmor interpretation, which distinguishes between empirical, local time and ontological, real time. Bell writes,

The difference of philosophy is this. Since it is experimentally impossible to say which of two uniformly moving systems is *really* at rest, Einstein declares the notions 'really resting' and 'really moving' as meaningless. For him, only the relative motion of two or more uniformly moving objects is real. Lorentz, on the other hand, preferred the view that there is indeed a state of *real* rest, defined by the 'aether,' even though the laws of physics conspire to prevent us identifying it experimentally. The facts of physics do not oblige us to accept one philosophy rather than the other.[64]

Since verificationism is hopelessly flawed as a theory of meaning, it is idle to talk about STR's 'forcing' us to change our fundamental ideas of space and time. Lawrence Sklar concludes, 'One thing is certain. Acceptance of relativity cannot force one into the acceptance or rejection of any of the traditional metaphysical views about the reality of past and future.'[65]

Of course, Hawking might retort that ontological time is scientifically useless and may therefore be left to the metaphysician. Granted, but then the point is surely this: *Hawking is doing metaphysics.* When he begins to speculate about the nature of space and time and to claim that he has eliminated the need for a Creator, then he has, as I said, entered the realm of the philosopher, and here he must be prepared to do battle with philosophical weapons on a broader conceptual field or else retreat within the walls of a limited scientific domain.

What is ironic is that even within that restricted domain there may now be empirical evidence for rejecting the received interpretation of STR. For the experimental results of the Aspect experiments on the inequalities predicted by Bell's Theorem have apparently established that widely separated elementary particles are in some way correlated such that measurements on one result instantly in the collapse of the wave function of the other, so that

---

[64] John S. Bell, 'How to Teach Special Relativity', in *Speakable and Unspeakable in Quantum Mechanics* (Cambridge: Cambridge University Press, 1987), 77.
[65] Sklar, 'Time, Reality, and Relativity', 140.

locality is violated. Even a hidden variables interpretation of the fabled EPR experiment must be a non-local theory. Nor is the violation of locality dependent upon the validity of quantum theory; it can be demonstrated on the macro-level, so that even if quantum theory should be superseded, any new theory will apparently have to include non-locality. But these data contradict the received interpretation of STR, not because non-locality posits superluminal signals, but rather because it goes to establish empirically relations of absolute simultaneity. Hence, disclaimers that STR is not violated because no signal or information is sent from one particle to another are beside the point. Rather the salient point is that the collapse of the wave function in both correlated particles occurs *simultaneously*, wholly apart from considerations of synchronization by light signals. Karl Popper thus regards the Aspect experiments as the first crucial test between Lorentz's and Einstein's interpretation of STR, commenting,

The reason for this assertion is that the mere existence of an infinite velocity entails that of an absolute simultaneity and thereby of an absolute space. Whether or not an infinite velocity can be attained *in the transmission of signals* is irrelevant for this argument: the one inertial system for which Einsteinian simultaneity coincides with absolute simultaneity . . . would be the system at absolute rest—whether or not this system at absolute rest can be experimentally identified.[66]

The establishment of non-local correlations in spacetime could thus vindicate even within the scientific domain the validity of Lorentz's distinction between local time and true time in opposition to the positivistic conflation of the two in the received view.

What this lengthy excursus goes to show is that it is metaphysically misguided to identify ontological time as a dimension of space. Since Hawking reduces empirical time in the very early history of the universe to a spatial dimension and conflates empirical time with ontological time, his model requires a tenselessly existing spacetime which he wishes to pass off as reality. Add to these errors the fact that the time involved is imaginary in its early stages, and the metaphysical absurdity of Hawking's vision of the world seems starkly apparent.

---

[66] Karl Popper, 'A Critical Note on the Greatest Days of Quantum Theory', in Barut *et al.*, *Quantum, Space, and Time*, 54.

5. CONCLUSION

There are many other things which one should like to say about
Hawking's view (for example, his misuse of the Anthropic Principle),
but I think enough has been said to answer his fundamental question,
'What place, then, for a Creator?' We have seen that contrary to
popular impression, God plays for Hawking an important role as a
sort of Leibnizian Sufficient Reason for the universe. With regard
to God's role as Creator, we saw that Hawking failed to distinguish
between *creatio originans* and *creatio continuans*, so that even if God
failed to play the former role, He may still carry out the latter as a
sort of Thomistic ground of being. But finally we have seen that
Hawking's critique of God's assuming the office of temporally First
Cause as demonstrated by the *kalām* cosmological argument is rife
with unexamined and unjustified philosophical assumptions, assumptions that, when examined, degenerate to metaphysical absurdity.
The success of Hawking's model appears to depend on a realist
application of Feynman's sum-over-histories approach to the
derivation of spacetime from an ontologically prior superspace, a
construal which is implausible and in any case unjustified. Essential
to Hawking's scheme is the identification of imaginary time with
physical time in the early history of the universe, a construal which
is again never justified and is in any case physically unintelligible.
Hawking's model depends, moreover, on certain questionable philosophical assumptions about relativity theory as well, for example,
the identification of time as a dimension of space, a move which is
extremely dubious, since time can exist without space. Hawking's
appeal to the Special Theory to justify this move rests on an
interpretation of that theory which fails to distinguish empirical
time from ontological time, an interpretation essentially dependent
on a defunct positivistic theory of meaning and now perhaps called
into question by empirical facts as well. Any attempt to interpret
the temporal dimension as a tenselessly existing spatial dimension
betrays the true nature of time.

The postulate of metaphysical superspace, the metamorphosis of
real to imaginary time, the conflation of time and space: all these
seem extravagant lengths to which to go in order to avoid classical
theism's doctrine of *creatio ex nihilo*—which forces us and Hawking
to confront squarely a different question: What price, then, for no
Creator?

XII

The Wave Function of a Godless Universe
QUENTIN SMITH

I. NOTHINGNESS AND THE INTERPRETATION OF QUANTUM COSMOLOGY

Classical Big Bang cosmology represents the universe as beginning at a lawless singularity about 15 billion years ago. One main difference between classical and quantum cosmology is that the latter represents the universe as beginning about 15 billion years ago in accordance with a physical law. The first to develop such a theory was Tryon,[1] but the most developed version of quantum cosmology is worked out by Hawking and his collaborators.[2] In this version, the universe is described as beginning from nothing in accordance with some law. Here 'nothing' does not mean the quantum-mechanical vacuum (which it often means in quantum cosmologies, such as Tryon's), but *literally* nothing, i.e. the absence of all concrete objects (mass, energy, spacetime). In this respect, Hawking follows the early Vilenkin,[3] rather than Tryon and others, since both Hawking and the early Vilenkin regard the universe as beginning from literally nothing rather than a pre-existent spacetime.

It is worth emphasizing that Vilenkin's early theory (which influenced Hawking more than Vilenkin's later theories) stated that the universe begins from *literally* nothing, since this is frequently misunderstood by philosophers. Vilenkin's early theory (of 1982) is widely misconstrued as theory about the origination of the universe

Not previously published.

[1] E. Tryon, 'Is the Universe a Vacuum Fluctuation?' *Nature*, 246 (1973), 396–7.
[2] Many of the relevant articles are reprinted in Li Zhi Fang and Remo Ruffini (eds.), *Quantum Cosmology* (Singapore: World Scientific, 1987).
[3] A. Vilenkin, 'Creation of Universes from Nothing', *Physical Letters* 117B (1982), 25–8.

from 'nothing' in the sense of a quantum-mechanical vacuum; Craig, Leslie, Munitz, Carroll,[4] and other philosophers interpret Vilenkin in this way. This interpretation is based on a confusion of Vilenkin's 1982 theory with his later theories. In the later theories, Vilenkin holds that the universe begins from 'nothing' in the sense of a 'realm of unrestrained quantum gravity',[5] but in his early theory he believes it begins 'from literally *nothing*',[6] i.e. the absence of all particulars. This interpretation of Vilenkin's early theory is confirmed by Vilenkin's own remarks in a conversation with Heinz Pagels. Pagels writes:

All the models of the origin of the universe I have discussed so far assume the preexistence of some kind of empty space—the vacuum whence it all began. . . . Alex Vilenkin, a theoretical physicist at Tufts University, was not satisfied with any of these notions of 'nothing'. 'Space is still something,' Alex once remarked to me, 'and I think the universe should really begin as nothing. No space, no time—nothing'. When Alex first mentioned this possibility to me, I said, 'What do you mean by nothing?' He just shrugged his shoulders and declared emphatically, 'Nothing is nothing!'[7]

It is nothing in this literal sense that is implied in Hawking's theory that the universe began from nothing about 15 billion years ago in accordance with the physical law encoded in his wave function of the universe.

In this article I shall examine Hawking's theory with three ends in mind. First, I shall show that it is physically intelligible. Secondly, I shall argue that it is inconsistent with theism and therefore that Hawking's wave function of the universe is literally a 'wave function of a godless universe'. (However, I shall show that Hawking misunderstood the respects in which his theory 'has no place for a Creator'.) Thirdly, I shall argue that Hawking's theory has superior explanatory value to theism.

The task of showing Hawking's quantum cosmology is physically intelligible corresponds to one of Craig's main arguments in Essay

[4] This claim about Vilenkin's theory is made by Munitz in *Cosmic Understanding* (Princeton, NJ: Princeton University Press, 1986), 130–7, esp. 136 n. 18, W. E. Carroll, 'Big Bang Cosmology, Quantum Tunnelling from Nothing, and Creation', *Laval Théologique et Philosophique*, 44 (1988), 59–75, John Leslie, *Universes* (New York: Routledge, 1989), 80, Craig, Essay V, and other philosophers.
[5] A. Vilenkin, 'Birth of Inflationary Universes', *Physical Review*, D27 (1983), 2848–55, esp. 2851.
[6] Vilenkin, 'Creation of Universes from Nothing', 26.
[7] H. Pagels, *Perfect Symmetry* (New York: Simon & Schuster, 1985), 343.

XI. Craig subjects Hawking's theory to a penetrating criticism and charges it with metaphysical absurdity. Craig concludes his article with the comment:

> The postulation of metaphysical superspace, the metamorphosis of real to imaginary time, the conflation of time and space: all these seem extravagant lengths to which to go in order to avoid classical theism's doctrine of *creatio ex nihilo*—which forces us and Hawking to confront squarely a different question: What price, then, for no Creator?[8]

I recognize the force of Craig's criticism and will endeavour to meet it by constructing an interpretation of Hawking's theory that does not require the posits of superspace, the metamorphosis of real to imaginary time, and the conflation of time and space. I will argue that Hawking's quantum cosmology commits us to little more than the familiar Friedman expanding universe and an early inflationary phase, but without a Big Bang singularity. According to this interpretation, we do not have to pay the price of metaphysical absurdity (imaginary time, etc.) for no Creator.

Craig's article poses a fundamental challenge for any defender of Hawking's theory; it presents the defender with the following dilemma: either we adopt a realist interpretation of Hawking's theory, in which case we are faced with metaphysical absurdities, or we adopt an instrumentalist interpretation, in which case we cannot take it as a representation of reality and thus cannot take it as an alternative world-picture to theism. Instrumentalism gives us no picture of reality but merely a device for calculating observations. Craig correctly points out that in Hawking's popular book *A Brief History of Time* Hawking inconsistently slides back and forth between an instrumentalist and realist interpretation of his cosmological model. However, I shall not follow Hawking's unsatisfactory presentation in his popular book but shall concentrate on his technical articles. The main aim of the first half of this essay is to show that there is a mid-point between a realist and instrumentalist interpretation of Hawking's theory—what I shall call a *quasi-instrumentalist interpretation*—and on this interpretation Hawking's theory gives a picture of reality that is an alternative to theism and does not involve the absurdities or excesses of imaginary time, splitting universes, infinite dimensional superspace, etc. I shall

[8] Essay XI, Sect. 5.

show, moreover, that this quasi-instrumentalist interpretation is implicit in Hawking's technical articles (but not in his *Brief History of Time* or his other popular writings).

After I develop this quasi-instrumentalist interpretation of Hawking's theory, I shall argue it is inconsistent with theism and that it is more reasonable to believe that the universe exists *because* of Hawking's wave function of the universe than that it exists *because* God created it. The first 'because' is an acausal explanatory 'because' and the second a causal explanatory 'because'. I shall maintain that Hawking's quantum cosmology is superior on explanatory grounds to theism since Hawking's cosmology but not theism is consistent with the strongest viable version of the Principle of Sufficient Reason.

2. HAWKING'S QUASI-INSTRUMENTALIST INTERPRETATION OF QUANTUM MECHANICS

The key to the physical intelligibility of Hawking's quantum cosmology is the instrumentalist/realist distinction in the interpretation of quantum mechanics. I shall argue that it is reasonable to interpret Hawking's theory as instrumentalist *except* in so far as it approximates classical General Relativity, in which case it is to be interpreted realistically. (This is what I mean to convey by 'quasi-instrumentalist'. I could also have said equivalently 'quasi-realist'.) Such a view, moreover, is implicit in Hawking's own writings, although he does not develop it in a clear or consistent fashion, as we shall see.

Hawking accepts the Everett or Many Worlds Interpretation (MWI) of quantum mechanics. But this does not commit him to splitting universes. Craig writes of this realist strain in Hawking's theory that 'Hawking's wave-functional analysis of the universe requires the Many Worlds Interpretation of quantum physics, and in another place Hawking admits as much. But why should we adopt this interpretation of quantum physics with its bloated ontology and miraculous splitting of the universe?'[9] Willem B. Drees also attributes this bloated ontology to Hawking; according to Drees, 'Hawking accepts the Many Worlds Interpretation of quantum

---

[9] Ibid., Sect. 4.2.

physics, and therefore there is no choice left once a probability distribution is given: all possibilities are actual.'[10] But this mistakenly supposes that the MWI can only be accepted on a full-blown realist construal. Hawking accepts it on basically instrumentalist grounds:

This is usually called the 'Many Worlds' Interpretation. I think it is a rather misleading name because it suggests that the universe is continually branching or dividing and people object that they do not feel themselves branching. In fact I think that the Many Worlds Interpretation simply involves the use of conditional probabilities, that is, the probability that A will occur given B. These conditional probabilities or correlations can be calculated by the standard rules of quantum mechanics.[11]

Hawking accepts the MWI merely as a calculation device for predicting states of our universe. Regarding the realist interpretation of MWI, he writes: 'In my opinion there is no problem in applying quantum mechanics to the whole universe and my attitude to those who argue about the interpretation of quantum mechanics is reflected in a paraphrase of Goering's remark: "When I hear of Schrödinger's cat, I reach for my gun." '[12] Hawking uses the MWI as a mere calculation device to derive the probability that state A of our universe will occur given the state B. However, as we shall see later, Hawking's wave function has the unique feature that it gives the unconditional probability that our universe will begin to exist with a certain state B.

Hawking's version of MWI does not imply that he accepts as real only sense data or middle-sized physical objects (trees, tables, etc.), with concepts of electrons and superclusters of galaxies lacking objective reference. He is not an instrumentalist in the traditional sense, e.g. of Stephen Toulmin or John Dewey. He is quasi-instrumentalist since he denies only that the concept of splitting universes lacks objective reference; the only real universe is the expanding Friedman universe to which we belong. Hawking's theory implies a certain agreement with J. S. Bell's 'one-world version' of the Everett theory of quantum mechanics; Bell writes of the Everett theory:

[10] Willem B. Drees, *Beyond the Big Bang* (La Salle, Ill.: Open Court, 1990), 45.
[11] S. W. Hawking, 'Quantum Cosmology', in B. S. DeWitt and R. Stora (eds.), *Relativity, Groups and Topology*, ii (Amsterdam: Elsevier Science Publishers, 1984), 336.
[12] Ibid. 336–7.

Now it seems to me that this multiplication of universes is extravagant, and serves no real purpose in the theory, and can simply be dropped without repercussions. So I see no reason to insist on this particular difference between the Everett theory and the pilot-wave theory—where, although the *wave* is never reduced, only *one* set of values of the variables x is realized at any instant.[13]

However, the apparent agreement is only partial, since Hawking nowhere commits himself to Bell's thesis that there is no unique past associated with the present, a thesis that has been effectively criticized by Richard Healey.[14] More to the point, Bell's one-world version of MWI has been superseded in the late 1980s and early 1990s by another one-world version of MWI, the 'consistent histories' approach of Griffiths, Omnes, Hartel, and Gell-Mann,[15] and this is the version normally associated with recent quantum cosmologies such as Hawking's. Some of the common principles accepted by most contemporary quantum cosmologists (Barrow and Tipler[16] being two exceptions) have been summarized by Chris Isham:

Many of those working in this area [quantum cosmology] agree that any discussion of quantum states of the entire universe must be within an interpretative scheme in which: (i) there are no references to measurements of the entire system being made by an external observer; (ii) there is no concept of collapse of the state vector induced by such measurements; (iii) an interpretation of probability is used that avoids the notion of 'ensembles' of universes; (iv) it is possible to reproduce the usual statistical results when applied to a sufficiently small sub-system of the physical universe.[17]

[13] J. S. Bell, *Speakable and Unspeakable in Quantum Mechanics* (Cambridge: Cambridge University Press, 1987), 133–4.

[14] Richard A. Healey, 'How Many Worlds', *Noûs*, 18 (1984), 591–616.

[15] R. B. Griffiths, 'Consistent Histories and the Interpretation of Quantum Mechanics', *Journal of Statistical Physics*, 36 (1984), 216–72; R. Omnes, 'Logical Reformulation of Quantum Mechanics', *Journal of Statistical Physics*, 53 (1988), 893–975; J. Hartle, 'The Quantum Mechanics of Cosmology', in S. Coleman, J. Hartle, T. Piran, and S. Weinberg (eds.), *Quantum Cosmology and Baby Universes* (Singapore: World Scientific, 1991).

[16] John Barrow and Frank Tipler, *The Anthropic Cosmological Principle* (New York: Oxford University Press, 1986), 472 ff.

[17] Chris J. Isham, 'Creation of the Universe as a Quantum Process', in R. J. Russell, W. R. Stoeger, and G. V. Coyne (eds.), *Physics, Philosophy, and Theology* (Notre Dame, Ind.: University Press, 1988), 394. Also see C. Isham, 'Quantum Theories of the Creation of the Universe', typescript (1992). Isham advocates that we should accept the view that the only statements that have any direct physical meaning are those that affirm something with probability 1. This is an issue too complex to discuss in the present paper, but I shall assume in this paper that statements affirming a probability less than 1 also have physical meaning.

The same sort of quasi-instrumentalism can be applied to Wheeler's superspace, which Hawking also uses in his theory construction. Craig says of the realist interpretation of Hawking's model that 'his model, based on the application of quantum theory to classical geometrodynamics, must posit the existence of a superspace which is ontologically prior to the approximations of classical space-time that are slices of this super-space'.[18] However, we need only posit an ontological reality corresponding to some of the points in superspace that are mapped out by a relevant solution to the Wheeler–DeWitt equation.[19] These points will describe our universe. Each point in superspace is a three-dimensional space (a 3-geometry) with a certain matter field configuration. A trajectory through superspace constitutes a 4-geometry, which is a history of the evolution of the geometry of space. On the quasi-instrumentalist interpretation I am suggesting, 'superspace', '3-geometry', 'trajectory through superspace' as used in the present context do not refer to physical realities but to abstract mathematical representations. Here 'superspace' means a *configuration space* (an abstract mathematical space) and a 3-geometry is a part of this abstract space. A physical space will be a four-dimensional general relativistic spacetime (with small quantum corrections) that corresponds to some of the points (3-geometries) in this configuration space.

Superspace can be explained more exactly as follows. There is a distinction between the configuration space and the state space. The *configuration space* is the set of all curved three-dimensional spaces (specified by a metric h_{ij}), each with its own matter field, ϕ. The *state space* is the set of all complex-valued wave functions ψ defined on the configuration space. Each wave function ψ associates a complex number $\psi(n)$ with each point (each 3-geometry) in the configuration space. The probability of the 3-geometry is proportional to the real number $|\psi(n)|^2$.

In order to determine the possible paths of systems in the con-

---

[18] Essay XI, Sect. 4.2.

[19] The Wheeler–DeWitt equation is

$$[-G_{ijkl}\frac{\delta^2}{\delta h_{ij}\delta h_{kl}} - {}^3R(h)h^{1/2} + 2\lambda h^{1/2}]\,\psi[h_{ij}] = 0.$$

G_{ijkl} is the metric on superspace:

$$G_{ijkl} = \tfrac{1}{2}h^{-1/2}(h_{ik}h_{jl} + h_{il}h_{jk} - h_{ij}h_{kl}).$$

${}^3R$ is the scalar curvature of the intrinsic geometry of the 3-surface. λ is the cosmological constant. $\psi[h_{ij}]$ is the amplitude for the 3-geometry h_{ij}.

figuration space, we turn to the Wheeler–DeWitt equation. This equation yields possible paths in the configuration space after a choice of internal time has been made. This equation does not provide an external or independent time variable t, but if time is defined internally in terms of one of the physical variables (e.g. the curvature of the 3-spaces), then solutions to the equation may be taken as describing a path through the configuration space. The Wheeler–DeWitt equation describes how the dependence of the wave function ψ on the metric h_{ij} and matter field ϕ is related to the dependence of the wave function ψ on the internal time (e.g. the time defined in terms of the curvature of the three-dimensional spaces). The equation gives us the probability of getting from one curved 3-space with a certain matter field to a later curved 3-space with matter field. Thus, the paths corresponding to solutions to this equation are not evolutions of the physical values of the quantum gravity system but of the probabilities of the system to have certain physical values. As the quantum gravity system 'moves' through the configuration space, the probability of finding it in a small region corresponding to the area around n in the configuration space is proportional to the real number $|\psi(n)|^2$.

The above ideas suggest how the Feynman path integral method or sum-over-histories method (which is also a part of Hawking's theory) can be given a quasi-instrumentalist interpretation. The time development of the wave function ψ can be described by Feynman's path integral method. This is a method for obtaining the wave function. The sum-over-histories is done by associating with each history two numbers, one representing the amplitude (the size of the wave) and the other its phase (its position in the cycle). The wave function ψ is defined as a sum of terms, one for each possible path in the configuration space that join together some initial point p_1 (at the internal time t_1) and some final point p_2 (at time t_2). Most of the paths cancel each other, leaving the probability of the system being at p_2 equivalent to a solution of classical equations of motion (with small quantum corrections).

The question of physical interpretation arises only for the non-cancelled paths through the configuration space. We need not adopt the full-blown realism that Craig suggests a realist interpretation of Hawking's theory implies. Craig asks: 'Does Hawking believe, for example, that Feynman's sum-over-histories approach describes what really happens, that an elementary particle really does follow all possible spacetime paths until its wave function is collapsed by measurement?'[20] First, there is no collapse of the wave function on the MWI of quantum mechanics, which Hawking accepts. Secondly, as Craig himself notes, almost all of the paths through the configuration space cancel each other (destructive interference) and the remaining paths (which satisfy the condition of constructive interference) describe the most probable paths of the system. These remaining paths are the only candidates for physical interpretation.

Hawking's solution to the Wheeler–DeWitt equation corresponds to many non-cancelled paths through the configuration space (and these paths represent many possible four-dimensional spacetimes) and a further argument is needed to show that only one of these possible spacetimes is actual (which would be our universe). In this regard, we may invoke *decoherence* to argue that only one of the possible four-dimensional spacetimes is actual. This method of argument is developed at length in J. J. Halliwell's 'Decoherence in Quantum Cosmology',[21] where a procedure is introduced to select the actual spacetime from the many possible ones permitted by the solution to the Wheeler–DeWitt equation. If quasi-instrumentalism about Hawking's wave function implies a commitment only to our universe, then some method such as decoherence needs to be a part of the quasi-instrumentalist interpretation of Hawking's wave function.

An early version of Hawking's quantum cosmology, the Hartle–Hawking theory of 1983, uses minisuperspace rather than superspace for its configuration space, since the Wheeler–DeWitt equation is too complex to solve for superspace. Hartle and Hawking solved the equation for a finite dimensional submanifold of superspace, a minisuperspace. This does not mean it is false that all 3-spaces are included. The difference between super- and minisuperspace is that the infinite number of degrees of freedom of the gravitational and matter fields in superspace are restricted to a finite number. In both cases, all possible 3-spaces are summed over in the path integral method.

In summary, a quasi-instrumentalist interpretation of superspace (or minisuperspace) and the Feynman path integral method does

---

[20] Essay XI, Sect. 4.2.
[21] J. J. Halliwell, 'Decoherence in Quantum Cosmology', *Physical Review*, D39 (1989), 2912–23.

not give us a physically real infinite dimensional superspace (or finite dimensional minisuperspace) and systems that 'follows all paths' to their destination. Rather, it allows us to take as physically real only a single and approximately classical (i.e. general relativistic) history of the geometry and matter field of a three-dimensional spatial system. By this means, a quasi-instrumentalist interpretation of Hawking's quantum cosmology gives us our universe alone, with its three spatial dimensions and matter field and its one temporal dimension. As is the case with Hawking's version of the MWI of quantum mechanics, we are not saddled with the metaphysical absurdities or extravagances that Craig suggests are implied by a non-instrumentalist interpretation of Hawking's theory.

But what of Hawking's infamous 'imaginary time'?

3. IMAGINARY TIME

The most intriguing aspect of Hawking's quantum cosmology is its use of the concept of imaginary time. Craig believes Hawking's use of this concept counts as a decisive refutation of his cosmology:

> But now in his model this imaginary time and Euclidean spacetime are suddenly supposed to be, not merely conceptual devices, but actual representations (however unimaginable) of physical reality in the very early history of the universe. This 'ontologizing' of mathematical operations is not only neither explained nor justified, but is, to my mind, metaphysically absurd. For what possible physical meaning can we give to imaginary time? Having the opposite sign of ordinary 'real' time, would imaginary time be a sort of negative time? But what intelligible sense can be given, for example, to a physical object's enduring for, say, two negative moments, or an event's having occurred two negative moments ago or going to occur in two negative moments?[22]

The notion of imaginary time requires some explanation before we can see whether Craig's remarks count against the viability of Hawking's cosmology.

A novel aspect of Hawking's cosmology is that superspace is a configuration space defined in terms of Euclidean metrics. A Euclidean metric belongs to a Euclidean spacetime. Euclidean spacetime may be regarded as a four-dimensional space, without a temporal dimension. The reason the 'time' in Euclidean spacetime may

---

[22] Ibid., Sect. 4.2.

be regarded as a fourth dimension of space is that the real number time variable (in our familiar Lorentzian spacetime) is replaced by the imaginary number time variable (in Euclidean spacetime), leaving no distinction between directions in time and space. The time sign is positive like the three spatial signs, so the signature is Euclidean, i.e. positive definite $(+++++)$ rather than Lorentzian $(-+++)$.

But since we are still talking about a configuration space and not about anything physical that corresponds to (any part of) it, we are not obligated to define or describe some kind of physical reality that corresponds to the positive time sign. In effect, we are only talking about mathematical entities and their interrelations.

Hawking's wave function is obtained by summing over all compact four-dimensional spaces (Euclidean spacetimes) with a three-dimensional space S as the boundary. Each Euclidean 4-space has a certain metric g_{uv} and matter field ϕ, such that the 3-space S that constitutes its boundary has a given metric h_{ij} and matter field ϕ. Again, '4-spaces', '3-space S', 'matter field ϕ', etc. should be understood as referring to mathematical entities and interrelations in a configuration space. A 'picture' of the configuration space relevant to Hawking's wave function would look like Fig. XII.1. The fact that the configuration space includes the 3-space S as the boundary of the 4-spaces enables Hawking to calculate the *unconditional* probability of the 3-space S, without needing to presuppose some earlier 3-space S' as the 'initial condition' from which the probability of the 'final condition' S needs to be calculated. This means in effect that the wave function $\psi[h_{ij}]$ gives us the amplitude for a universe that includes as one of its 3-spaces the 3-space S with the metric h_{ij} to arise from nothing. (The other 3-spaces S', S'', etc. that belong to the history of the universe can be calculated from the 3-space S.) Since the amplitude squared gives the probability, $\psi|[h_{ij}]|^2$ gives us the probability that the universe will arise from nothing with the 3-geometry h_{ij}.

In the 1983 Hartle–Hawking paper, this idea is explained as follows: 'One can interpret the functional integral over all compact four-geometries bounded by a given three-geometry as giving the amplitude for that three-geometry to arise from a zero three-geometry, i.e. a single point. In other words, the ground state is the amplitude for the Universe to appear from nothing.'[23] It is

---

[23] J. Hartle and S. W. Hawking, 'Wave Function of the Universe', *Physical Review*, D28 (1983), 2960–75: 2961.

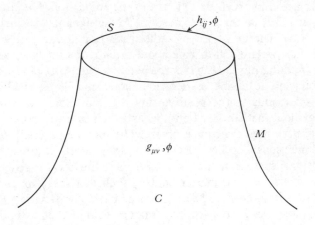

FIG. XII.1. Here S is the three-dimensional space that has the metric h_{ij} and the matter field ϕ. The class C is the class of 4-spaces $g_{\mu\nu}$ and matter fields ϕ that have S as their boundary. The wave function $[h_{ij}, \phi]$ is given by integrating over all the 4-spaces and matter fields that belong to the class C of the manifold M.

important to note here that this 'nothing' is not literally nothing, but a single zero-dimensional spatial point. W. B. Drees correctly notes that 'a weird structure, like "a zero three-geometry, i.e. a single point", as the citation of Hartle and Hawking has it, is not nothing'.[24] John Leslie remarks about the Hartle–Hawking theory,

But even assuming it made clear sense to speak of tunnelling from a zero volume, could this truly mean that all creation's puzzles had at last been solved? Presumably not; for a zero volume *with three dimensional geometry*

---

[24] Willem B. Drees, 'Interpretation of "the Wave Function of the Universe"', *Journal for Theoretical Physics* (1987), 940.

and sufficiently subject to the laws of quantum physics to allow for talk of 'tunnelling' from it can look interestingly different from pure nothingness.[25]

Leslie is basically correct, but the phrase 'a zero three-geometry' should not be interpreted to mean that the zero volume has a three-dimensional geometry (which would be a logical contradiction). This phrase means only what Hartle and Hawking says it means, 'i.e. a single point'.

However, there is no reason to regard this single point as essential to Hawking's cosmology and it may reasonably be regarded as a *façon de parler*, with no physical significance. The phrase 'the amplitude for that three-geometry to arise from a zero three-geometry, i.e. a single point' can be omitted from the Hartle–Hawking paper without affecting the theory. The crucial phrase in the passage I quoted is: 'the ground state is the amplitude for the Universe to appear from nothing'. Hawking's later formulations will not include this single point but will represent the first state of the universe as the first state of a four-dimensional general relativistic spacetime manifold, as we shall see shortly. In these formulations, we can say that the wave function gives the amplitude for the universe to arise from *literally* nothing.

The issue of the reality of the 'single point' in Hawking's quantum cosmology is different from the issue of the reality of the 'single point Big Bang singularity' in classical General Relativity, since the point is predicted by the classical Friedman equations for a finite universe but is not predicted by Hawking's equations. His equations predict instead that the universe begins with a three-dimensional space, as I shall indicate.

The Hartle–Hawking equation gives the amplitude for a certain state of the universe, namely, the ground state ψ_0, which is the state of minimum excitation. Besides calculating the metric h_{ij} of the ground state they also calculate its matter field, represented in the equations by a conformally invariant scalar field ϕ. Accordingly, the wave function ψ of their model is $\psi[h_{ij},\phi]$ and the probability that a universe with this ground state metric and matter field appears from nothing is $\psi_0|[h_{ij},\phi]|^2$. The wave function equation ((3.1) in the Hartle–Hawking paper) is

(3.1) $\quad \psi_0[h_{ij},\phi] = N \int \sigma g \, \sigma\phi \, \exp(-I[g,\phi])$.

---

[25] Leslie, *Universes*, 81.

Here N is the normalization constant, I is the Euclidean Einstein action. It is derived by replacing the time t with $(-i\tau)$ in the usual Einstein action and adjusting the sign so that I is positive. The integral is a path integral over all positive definite $(++++)$ 4-geometries and matter fields which have the 3-geometry h_{ij} and matter field ϕ as a boundary.

The wave function equation (3.1) is a solution of the Wheeler–DeWitt equation and this is picked out from other solutions as a candidate for physical interpretation. This reflects the quasi-instrumentalist interpretation we are adopting. A full-blown realist interpretation would regard all possible solutions to the Wheeler–DeWitt equations as describing a physical reality, but a quasi-instrumentalist takes as a candidate for physical interpretation only the solution (3.1).

But is (3.1) consistent with observations of our universe?

The equation (3.1) is not in fact consistent with observations of our universe since (3.1) represents only zero rest mass fields, whereas our universe has a massive scalar field. Hartle and Hawking mention this problem at the end of their article:

> The ground-state wave function in the simple minisuperspace model that we have considered with a conformally invariant field does not correspond to the quantum state of the Universe that we live in because the matter wave function does not oscillate. However, it seems that this may be a consequence of using only zero rest mass fields and that the ground-state wave function for a universe with a massive scalar field would be much more complicated and might provide a model of [the] quantum state of the observed Universe.[26]

Hawking later noted that in our universe the expansion is strongly coupled to the matter content. However, with the *conformally* invariant scalar field of the Hartle–Hawking model, there is no such coupling. Accordingly, 'such a model cannot describe the observed universe in which the expansion is strongly coupled to the matter content. In order to obtain a model which corresponds to what we observe, it is necessary to introduce non-conformally invariant scalar fields.'[27] This can be done by including a spatially constant massive scalar field.

---

[26] Hartle and Hawking, 'Wave Function of the Universe', esp. 2975.
[27] S. W. Hawking, 'The Quantum State of the Universe', *Nuclear Physics*, B239 (1984), 257–76, esp. 267.

It is this new model (subsequently refined in numerous articles) which Hawking non-technically discusses in his book *A Brief History of Time*, and thus it is not entirely accurate for Craig to suggest that the 'relevant technical paper'[28] for Hawking's popular book is the original Hartle–Hawking paper. His popular book is based on an equation in which the matter wave function oscillates, and the wave function equation in the Hartle–Hawking paper is rejected 'because the matter wave function does not oscillate'.[29]

This new quantum cosmology, with its idea of a massive scalar field, was originally suggested to Hawking by Linde at the Shelter Island conference in June 1983. This cosmology is based on the semi-classical approximation to the path integral, which is used to estimate the form of the solution to the wave equation on lines of a constant large massive scalar field ϕ. To use the semi-classical approximation, we need to solve the classical Euclidean field equations. The relevant metric can be expressed as follows (cf. equation (4.10) in Hawking's 1984 article 'The Quantum State of the Universe' and equation (8.25) in his 1983 article 'Quantum Cosmology'[30]):

$$(4.10) \quad \mathrm{d}s^2 = \sigma^2 \, (\mathrm{d}\tau^2 + a^2(\tau)\mathrm{d}\Omega_3^2).$$

Here a is the radius of the 3-sphere spacelike surfaces, τ the time coordinate, and $\mathrm{d}\Omega_3^2$ the metric on a unit 3-sphere. $\sigma^2 = \frac{2}{3}\pi m_{\mathrm{p}}^2$. If we have a massive scalar field ϕ, then we will be able to divide the minisuperspace into two regions, the Euclidean region and the Lorentzian region. The latter region will be reached only by the oscillatory part of the wave function ψ. This division of minisuperspace depends on the values of the massive scalar field ϕ and the radius a of the 3-sphere spacelike surface. Hawking notes that 'if both $\phi > 1$ and $a > 1/m\phi$ . . . the solution which determines the semi-classical approximation to the wave function will be complex and so the wave function will be oscillatory'.[31] If $a < 1/m\phi$, the wave function will be exponential, which relates to the Euclidean region of minisuperspace. Hawking suggests that only the oscillating part should be given a physical interpretation: 'the oscillating component in the wave function should be interpreted as corresponding to a lorentzian geometry and exponentially growing component

[28] Essay XI, n. 24.
[29] Hartle and Hawking, 'Wave Function of the Universe', 2975.
[30] Hawking, 'Quantum Cosmology', 372, and 'The Quantum State of the Universe', 268.
[31] Hawking, 'Quantum Cosmology', 374.

should be interpreted as corresponding to a euclidean geometry. We live in a lorentzian geometry and therefore we are interested really only in the oscillating part of the wave function.'[32]

This correspondence to Lorentzian geometries is possible since the rapid oscillation of the wave function enables one to use the WKB method. In the WKB approximation, the Wheeler–DeWitt equation for the oscillating component of the wave function can be reduced to the Hamilton–Jacobi equation, which describes a Lorentzian universe. One can write the wave function in the form (equation (4.20) in Hawking's 'The Quantum State of the Universe'[33]):

$$(4.20) \quad \psi(a,\phi) = C(a,\phi) \cos[S(a,\phi)].$$

Here $S(a,\phi)$ is a rapidly varying phase and $C(a,\phi)$ a slowly varying amplitude. The trajectories of ∇S in the (a,ϕ) plane 'correspond to classical lorentzian solutions $a(t)$, $\phi(t)$ of the field equations'.[34] Hawking continues:

> The semi-classical approximation for large ϕ correspond to a classical solution which starts with $\phi(t) = \phi_1$, $d\phi/dt = 0$, at a minimum radius $a(t) = (1/m\phi_1)$ at $t = 0$ and then expands in a de Sitter-like manner. . . . For large ϕ_1 this gives a long inflationary period. . . . Eventually . . . the universe goes over to a matter-dominated phase with $a(t) \propto t^{\frac{2}{3}}$. The universe will expand to a maximum radius of order $(1/m\phi_1)\exp(6\phi_1^2)$ and then recollapse. In general, it will recollapse to a singularity.[35]

This provides us with the key to the physical interpretation of Hawking's quantum cosmology, namely, *that only the oscillating component in the wave functions should be given a physical interpretation.* We have none of the metaphysical absurdities and extravagances that Craig suggested were implied by a full-blown realist interpretation of Hawking's theory. On the present quasi-instrumentalist interpretation, we have a classical system as our only physical reality. We have a plausible and indeed familiar universe, namely, a universe that begins at a minimum three-dimensional radius, $a(t) = (1/m\phi_1)$, inflates, and then goes over to a normal Friedman expansion until it reaches a maximum radius, at which time it recollapses to a singularity. (As I have already implied, since the oscillating

---

[32] Hawking, 'The Quantum State of the Universe', 272.
[33] Ibid.
[34] Hawking, 'Quantum Cosmology', 374.
[35] Ibid. 374–5.

component describes many possible classical universes, we need to invoke *decoherence* to argue that only one is actual.[36]) The plausibility of this quasi-instrumentalist interpretation of Hawking's cosmology becomes apparent if we try to give a physical interpretation to the exponential part of the wave function. Hawking himself tried this in his popular writings, with disastrous results, as we shall now see.

4. HAWKING'S ABSURD PHYSICAL INTERPRETATIONS OF THE EXPONENTIAL PART OF THE WAVE FUNCTION

It is distressing to see what results if we attempt to interpret physically the Euclidean region of minisuperspace, corresponding to the exponentially growing component of the wave function. Although in his technical papers Hawking implicitly suggests that the exponential component is not to be interpreted physically, he explicitly suggests otherwise in his popular writings. In his 'The Edge of SpaceTime' (1984) he writes:

In the very early universe, when space was very compressed, the smearing effect of the uncertainty principle can change this basic distinction between space and time. It is possible for the square of the time separation to become positive under some circumstances. When this is the case, space and time lose their remaining distinction—we might say that time becomes fully spatialized—and it is then more accurate to talk, not of spacetime, but of a four-dimensional space. Calculations suggest that this state of affairs cannot be avoided when one considers the geometry of the universe during the first minute fraction of a second. The question then arises as to the geometry of the four-dimensional space which has to somehow smoothly join onto the more familiar spacetime once the quantum smearing effects subside. One possibility is that this four-dimensional space curves around to form a closed surface, without any edge or boundary, in much the same way as the surface of a ball or the Einsteinian universe, but this time in *four* [spatial] dimensions.[37]

Clearly Hawking is offering a physical interpretation of four-dimensional space, i.e. Euclidean spacetime. But such an interpretation is implicitly logically self-contradictory. The problem appears in the statement that the four-dimensional space joins on to

[36] Halliwell, 'Decoherence in Quantum Cosmology'.

[37] S. W. Hawking, 'The Edge of SpaceTime', in Paul Davies (ed.), *The New Physics* (Cambridge: Cambridge University Press, 1989), 68.

the real (Lorentzian) spacetime 'once' (i.e. after) the quantum smearing effects subside: 'The question then arises as to the geometry of the four-dimensional space which has to somehow smoothly join onto the more familiar spacetime *once* the quantum smearing effects subside' (my italics). If the four-dimensional space does not possess a real time value, how can it stand in relation to the four-dimensional spacetime of being earlier than it? If the four-dimensional space is not in real (Lorentzian) time, then it is not really earlier than, later than, or simultaneous with the four-dimensional spacetime manifold. Accordingly, it is false that the 4-sphere joins on to the familiar spacetime *once* (i.e. after in real time) the quantum effects dissipate. Nor can this 'once' refer to imaginary time, which would imply that the real spacetime is later in imaginary time than the 4-sphere, which it is not. There is no imaginary time in real spacetime and real spacetime is not located in imaginary time. (To be so, its temporal dimension would have to be a fourth spatial dimension, which it is not.) Indeed, 'later' is not an appropriate expression to use of imaginary time, since the earlier–later direction is a feature only of real time. (Imaginary time is instead like a spatial dimension, in which there is no direction.) It is false, then, that 'in the very *early* universe' (my italics) there is a quantum phase in which the universe is a 4-sphere. 'The very early universe' can refer only to the very earliest *times*, i.e. the lowest values of the real time coordinate. It is rather the case that the earliest time (t_0) is the time at which the four-dimensional spacetime manifold has the radius $a(t)$ $= (1/m\phi_1)$.

It might be suggested that the four-dimensional space exists timelessly and is spatially but not temporally joined on to the universe. But how is that possible? Which three-dimensional spatial hypersurface is the four-dimensional space connected to? Presumably the earliest hypersurface, where $a(t) = (1/m\phi_1)$. This hypersurface would then be connected at both ends, one to the four-dimensional geometry and one to the three-dimensional hypersurface that is the beginning of the de Sitter inflationary phase. But 'ends' here means 'ends of the temporal interval occupied by the hypersurface where $a(t) = (1/m\phi_1)$'. But this implies, of logical necessity, that the four-dimensional geometry is connected to the *earlier end* of this hypersurface and thus has a real time value, contradicting the original supposition.

In his book *A Brief History of Time* Hawking puts forth a different but equally preposterous account of the Euclidean region of mini-superspace. He says the history of the universe in imaginary time can be pictured as the earth, with 'the distance from the North Pole representing imaginary time and the size of a circle of constant distance from the North Pole representing the spatial size of the universe'.[38] The universe starts at the North Pole. As one moves south, the circles of latitude at constant distance from the North Pole get bigger, corresponding to the universe expanding in imaginary time. The universe reaches a maximum radius at the equator and then recollapses to a singularity at the South Pole. 'The history of the universe in real time, however, would look very different. At about ten or twenty thousand million years ago, it would have a minimum size, which was equal to the maximum radius of the history in imaginary time. At later real times, the universe would expand like the chaotic inflationary model'[39] and eventually recollapse to a singularity.

Now in this book, Hawking does not say that the imaginary time phase is physically real and earlier than the real time phase, and thus avoids the logical contradiction in his 'The Edge of SpaceTime'. Nor, however, does he take the sensible line suggested in his technical articles that only the real time phase is physically meaningful. Rather, he adopts an absurd form of instrumentalism:

This might suggest that the so-called imaginary time is really the real time, and that what we call real time is just a figment of our imaginations. In real time, the universe has a beginning and an end at singularities that form a boundary to space-time and at which the laws of science break down. But in imaginary time, there are no singularities or boundaries. So maybe what we call imaginary time is really more basic, and what we call real is just an idea we invent to help us describe what we think the universe is like. But according to the approach I described in Chapter I, a scientific theory is just a mathematical model we make to describe our observations: it exists only in our minds. So it is meaningless to ask: Which is real, 'real' or 'imaginary' time? It is simply a matter of which is the more useful description.[40]

This is absurd, at least observationally, since it is perfectly obvious that the universe in which we exist lapses in real rather than imaginary time, that it is a Lorentzian rather than Euclidean space-

---

[38] S. W. Hawking, *A Brief History of Time* (New York: Bantam, 1989), 137.
[39] Ibid. 138.   [40] Ibid. 139.

time. To quote from one of Hawking's technical articles, 'we live in a Lorentzian universe, not a Euclidean one'.[41] Perhaps we can interpret the above paragraph charitably as allowing of the interpretation that 'the more useful description is obviously that the universe exists in real time', but this does not seem to be the tenor of Hawking's remarks, which seem designed to baffle and intrigue his readers. One can sympathize with Craig's exasperated remarks about this and related passages in Hawking's book.

However, the picture of the earth (on page 138) in Hawking's popular book can be given a sensible interpretation. This picture has a bearing on physical reality in that the minimum size of the Lorentzian (real) spacetime at its very beginning is equal to the size represented by the earth's equator, i.e. it is 'equal to the maximum radius of the history in imaginary time'. The picture of the earth is a picture, not of the history of the universe, but (in a Pickwickian sense) of the part of the equation that is used to derive the minimum size of the universe at its beginning. We know from the oscillating part of the wave function that the minimum radius is $a(t) = (1/m\phi_1)$. This radius corresponds to the maximum radius of the history of the universe in imaginary time, such that the Lorentzian universe begins at the radius represented by the equator of the earth. Hawking writes in his 1987 article 'Quantum Cosmology': 'The Lorentzian solutions will be the analytic continuation of the Euclidean solutions. They will start in a smooth and non-singular state at a minimum radius equal to the radius of the 4-sphere and will expand and become more irregular.'[42] The 4-sphere (Euclidean spacetime) is what is depicted by the picture of the earth.

Robin Le Poidevin's article[43] on Hawking's theory is mistaken in implying Hawking's theory represents time as closed. Rather there is an earliest time t_0 where $a(t) = (1/m\phi_1)$ and a latest time $t > t_0$, where the universe ends with a big crunch singularity; in Hawking's theory, time is open, like a finite line, rather than closed like a circle. The big crunch singularity does not cause the phase where $a(t) = (1/m\phi_1)$. Rather, the big crunch occurs at the last moment of

[41] Hawking, 'Quantum Cosmology', in S. W. Hawking and W. Israel (eds.), *Three Hundred Years of Gravitation* (Cambridge: Cambridge University Press, 1987), 639. This is a different article from his 1983 article of the same title.
[42] Ibid. 650.
[43] Robin Le Poidevin, 'Creation in a Closed Universe; or, Have Physicists Disproved the Existence of God?' *Religious Studies*, 27 (1991), 39–48.

time and causes nothing and the era where $a(t) = (1/m\phi_1)$ occurs at the first moment of time and is caused by nothing.

The foregoing remarks suggest that we should ignore Hawking's popular writings and rely on his technical articles for our understanding of the physical meaning of his theory. If we do this, we shall find that there is a plausible interpretation of his theory, namely, that it describes a Lorentzian universe that begins at a minimum and non-zero radius, inflates, and then expands normally to a maximum and recontracts to a singularity. (Note that Hawking's technical articles contradict the quoted statement from his *Brief History of Time* that in 'real time, the universe has a beginning and end at singularities that form a boundary to space-time and at which the laws of science break down'. His articles allow that the universe *ends* at a singularity but not that it *begins* at a singularity.)

We may agree with Craig that 'the central question that needs to be addressed in assessing his [Hawking's] model as an alternative to divine creation is whether it represents a realistic picture of the world'.[44] I believe I have shown it does represent a realistic picture (we are not committed to splitting universes, superspace, a particle taking all possible paths, or imaginary time) and thus that it is an alternative to divine creation. I shall now argue that it is not merely an alternative to divine creation but that it is inconsistent with and more reasonable than the theistic theory.

5. NO PLACE FOR A CREATOR

Hawking misidentifies the respect in which there is 'no place for a Creator' in his theory. His 'atheistic argument' is that the universe has no beginning in real time and therefore that there is no place for a Creator. Hawking writes:

Because one is using Euclidean space-times, in which the time direction is on the same footing as directions in space, it is possible for space-time to be finite in extent and yet have no singularities that formed a boundary or edge . . . at which one would have to appeal to God or some new law to set the boundary conditions for space-time. . . . So long as the universe had a beginning, we could suppose it had a creator. But if the universe is really completely self-contained, having no boundary or edge, it would have

[44] Essay XI, Sect. 4.2.

neither beginning nor end; it would simply be. What place, then, for a creator?[45]

Hawking's atheistic argument amounts to this howler:

(1) If Euclidean spacetime is physically existent and Lorentzian spacetime a theoretical fiction, then there is no real time at which the universe began to exist.

(2) If there is no real time at which the universe began to exist, then there is no real beginning that God created.

(3) Euclidean spacetime physically exists and Lorentzian spacetime does not.

Therefore,

(4) There is no real beginning created by God.

Therefore,

(5) God does not exist.

This is probably the worst atheistic argument in the history of Western thought and I shall not waste the reader's time by refuting it. In any case, most of the problems with it have already been exposed by Craig in Essay XI. I shall concentrate instead on showing that there are two respects, unmentioned by Hawking, which render his theory (as interpreted quasi-instrumentally) inconsistent with theism.

The first respect concerns the fact that his wave function gives the amplitude that a universe comes into existence with a certain structure and this amplitude, when squared, gives a probability estimate that a universe with this structure will come into existence. Hawking's calculations show that it is highly probable that there is a universe that begins with a radius of $a(t) = (1/m\phi_1)$, has early density fluctuations that will lead to galaxies, stars, and planets, expands in an inflationary phase, changes over to a normally expanding homogeneous and isotropic Friedman universe, is almost flat (i.e. is near to the critical density $\Omega = 1$), and recollapses to a big crunch singularity. Let us call such a universe (which, incidentally, is consistent with all current observations) a Hawking-universe. And let us say for the sake of simple numerical illustration that a Hawking-universe has an 85 per cent probability of arising from nothingness. Then Hawking's wave function gives us the proposition:

---

[45] Hawking, *A Brief History of Time*, 135, 136, 140–1.

(1) The probability that there is a Hawking-universe is 85 per cent.

Now consider the hypothesis that God ordains that Hawking's wave function law obtains. God could create the sort of universe that is most probable or one that is second most probable, etc. Suppose in fact he decides to create the most probable one, as specified by the law he chose to make obtain. Then this universe has a sufficient condition of its existence, not a probabilistic one. Note that

(2) God wills that the Hawking-universe exist

entails

(3) The probability that there is a Hawking-universe is 100 per cent.

Clearly, '*C* occurs and *C* is a sufficient condition of *E*' entails 'The probability that *E* occurs is 1 (i.e. 100 per cent).

However, (3) is inconsistent with (1). Since (1) is true if Hawking's quantum cosmology is true and (2) entails (3), it follows that (2) is false if Hawking's quantum cosmology is true. Put another way, it is implicitly self-contradictory that

(4) God wills that Hawking's wave function law obtains and that a Hawking-universe exists,

since if the law obtains the probability is 85 per cent and if God wills that a Hawking-universe exists the probability is 100 per cent.

Suppose the theist argues that (1) really means 'Without supernatural causation, the probability that there is a Hawking-universe is 85 per cent'. But this conflicts with the various necessary truths to which theists such as Craig subscribe, e.g. that it is necessarily true that everything that begins to exist has a cause. Given that this is a necessary truth, it is instead true that 'Without supernatural causation, the probability that there is a Hawking-universe is 0 per cent', since there is no metaphysically possible world in which a universe begins to exist uncaused. If the wave function is 'the amplitude for the universe to appear [naturally uncaused] from nothing'[46] and the amplitude squared gives the probability, then the probability *contra* Hawking is 0 if it is true that the universe cannot begin naturally uncaused.

Thus, the theist who accepts Craig's causal principle faces a

---

[46] Hartle and Hawking, 'Wave Function of the Universe', 2961.

dilemma if she tries to show that Hawking's theory is consistent with theism. *Either* the probability estimate Hawking derives for a Hawking-universe to begin to exist is unconditional, i.e. takes into account all relevant factors, in which case it conflicts with the theistic principle that the unconditional probability is 100 per cent (which takes into account the factor that God wills a Hawking-universe), *or* it is conditional only upon the set of all naturalistic considerations (including the absence of natural causes), in which case it conflicts with Craig's causal principle which implies the probability is then 0 per cent.

A second respect in which Hawking's theory is incompatible with theism is that it adopts the Many Worlds Interpretation of quantum mechanics and rejects the Copenhagen Interpretation. Craig appreciates part of the significance of this fact and quotes John Barrow's remarks on this issue:

> John Barrow has recently remarked that the Many Worlds Interpretation is 'essential' to quantum cosmology because without it one is left, on the standard Copenhagen Interpretation, with the question, 'Who or what collapses the wave function of the universe?'—some Ultimate Observer outside of space and time? This answer has obvious theistic implications. Indeed, although 'the theologians have not been very eager to ascribe to God the role of Ultimate Observer who brings the entire quantum Universe into being', still Barrow admits that 'such a picture is logically consistent with the mathematics. To escape this step cosmologists have been forced to invoke Everett's "Many Worlds" interpretation of quantum theory in order to make any sense of quantum cosmology' . . . But if we, like most physicists, find the Many Worlds interpretation of quantum reality outlandish, then quantum cosmology, far from obviating the place of a Creator, might be seen to create for Him a dramatic new role. Again, my intention is not to endorse this view, but simply to underscore the fact that a realist construal of Hawking's account involves extravagant and dubious metaphysical commitments, such that his model can hardly be said to have eliminated the place of a Creator.[47]

Now we have already seen that Hawking adopts a quasi-instrumentalist interpretation of MWI, so we need not have the 'extravagant and dubious metaphysical commitments' to which Craig alludes. However, the point is that theism is consistent with the Copenhagen Interpretation but not with the MWI and Hawking's theory incorporates the latter. According to the Copen-

---

[47] Essay XI, Sect. 4.2.

hagen Interpretation, the wave function needs to be collapsed by something outside of the system being measured. The wave function of the universe, accordingly, needs to be collapsed by something outside of the universe. Now most versions of the Copenhagen Interpretation regard the observer (often explicitly identified with consciousness) as what collapses the wave function. In this respect, the cosmological application of the Copenhagen Interpretation may reasonably be thought to posit God (or a disembodied person that has superhuman attributes) outside of the universe. Indeed, this seems to be the best scientific argument for God that is present in twentieth-century science. I agree with Barrow that it is surprising that this argument is not taken up by theists.

But the issue at hand is whether theism is consistent with Hawking's quantum cosmology. Hawking rejects the Copenhagen Interpretation; for Hawking, the wave function never collapses and therefore there is no role for an Ultimate Observer to collapse it. On the Copenhagen Interpretation of quantum cosmology, the state of the universe at some time is insufficient to produce a measurable state at a later time. In order for there to be a subsequent measurable state, a supernatural agency is required to collapse the wave function. But on the MWI, the state of the universe at one time is sufficient to produce some subsequent measurable state (except in the case of the big crunch that ends the universe). No outside agency is needed.

This suggests that the theory of 'continuous creation' (as I defined it elsewhere[48]) is not logically compatible with the MWI of quantum mechanics. The supposition that God continuously creates the universe implies that the prior state is insufficient by itself to produce a subsequent state. It supposes that, left to itself, the state of the universe would be followed by nothingness. A divine agency is required for the next state. As Craig says, 'without his active and continual bestowal of existence to the universe, the whole of finite reality would be instantly annihilated and lapse into non-being'.[49] But this hypothesis contradicts the principles of the MWI. The MWI entails that the prior state *is* sufficient to produce some subsequent state. (Note that this is not a unique implication of MWI but also is true of classical General Relativity, which supposes

---

[48] Essay IX.   [49] Essay XI, Sect. 3

that a prior state is sufficient by itself for the occurrence of a subsequent state. It is an interesting question whether this is also true of Newtonian cosmology *as originally formulated by Newton*, since Newton explicitly introduced God into his theory.)

In order for the theory of continuous creation to be consistent with a given cosmological theory, the theory must imply that a natural state is insufficient to produce a subsequent natural state and requires a supernatural agency for the subsequent state to be brought about. The cosmological application of the Copenhagen Interpretation meets this condition but the MWI does not. Thus, if we accept Hawking's quantum cosmology we are left with 'no place for a continuous creator'.

I shall now address the question of whether it is more reasonable to believe Hawking's cosmology or theism.

6. QUANTUM COSMOLOGY, THEISM, AND MAXIMAL EXPLANATIONS

There are several ways to argue that Hawking's quantum cosmology is rationally preferable to theism. For example, it could be argued that his cosmology has greater predictive power since it predicts that it is highly probable there is a universe with the properties we observe, namely, the early inflationary era, the density fluctuations in the background radiation (discovered by the 1992 COBE satellite) that explain the origins of galaxies, the large-scale homogeneity and isotropy of the universe, and the era of normal Friedman expansion of the galactic superclusters, and that the universe is near the critical density. Obviously, these observations are not predicted by the hypothesis that God exists or creates a universe. If they were predicted, they could have been derived by Augustine or Aquinas on the basis of theological considerations alone. In this sense, theism is relatively 'empty of empirical content' in comparison with Hawking's quantum cosmology. It also could be argued that theism predicts that there is no gratuitous natural evil, whereas quantum cosmology does not predict this, and that in this respect theism but not quantum cosmology is disconfirmed.[50]

---

[50] See Quentin Smith, 'An Atheological Argument from Evil Natural Laws', *International Journal for the Philosophy of Religion*, 29 (1991), 159–74.

However, I shall not develop the two above-mentioned lines of argument since they will take us down well-worn paths in the debate about atheism versus theism. Instead, I shall explore a new path. I shall argue that Hawking's quantum cosmology has a superior explanatory value to theism in a rather surprising respect. If Hawking's cosmology is true it is epistemically (and perhaps metaphysically) possible that the existence and nature of the universe has a *maximal explanation*, whereas if theism is true it is epistemically and metaphysically impossible that the universe has such an explanation. In order to define 'maximal explanation' we need to distinguish several versions of the Principle of Sufficient Reason. In order of increasing strength (with III being the strongest possible version) some of the different versions are

(I) Everything that begins to exist has a reason why it begins to exist.

(II) Everything that exists has a reason why it exists.

(III) Everything that exists has a reason why it exists and why it possesses the properties it in fact possesses.

'Everything' here quantifies over all concrete and abstract objects. We can now define a maximal explanation of the universe: *The universe has a maximal explanation if and only if the universe exists and the strongest possible version of the Principle of Sufficient Reason, version III, is true.*

I know of no contemporary philosopher who believes III is true. I shall argue, however, that the best arguments against III are unsound and that it is epistemically possible that III is true. (By '*p* is epistemically possible' I shall here mean 'We are not justified in believing not-*p*'.)

Perhaps the best recent argument against III is William Rowe's in *The Cosmological Argument*. He argues that a relevantly analogous principle is false:

(PSR₁) Every actual state of affairs has a reason either within itself or in some other state of affairs.

We may suppose with Rowe that 'the actual state of affairs S has a reason within itself' is logically equivalent to 'it is metaphysically necessary that S obtain'. Rowe's argument against PSR₁ is that there are positive, contingent states of affairs and that if there are such states of affairs, then there is some state of affairs for which there is no reason. He writes:

Let us introduce the idea of *a positive, contingent state of affairs* as follows: X is a positive, contingent state of affairs if and only if from the fact that X obtains it follows that at least one contingent being exists. That there are elephants, for example, is a positive, contingent state of affairs.[51]

Rowe then asks us to consider the general state of affairs t:

(t) that there are positive, contingent states of affairs.

t is a contingent state of affairs, Rowe argues, since there is a possible world in which the only beings that exist are necessary. In such a world, there would exist numbers, properties, propositions, and other abstract objects (and God, if God is a necessary being), but nothing contingent. Rowe argues that t has no explanation and therefore that PSR_1 is false. Let q be a state of affairs that allegedly explains t. Rowe continues:

Suppose that q is the state of affairs that explains t and that 'q explains t' is made true by the fact that the actual state of affairs q stands in a certain relation R to t. The actual state of affairs qRt must entail the state of affairs t, otherwise the fact that qRt would not make it true that q explains t. . . . Now the actual state of affairs qRt is either necessary or contingent. It cannot be necessary, for t would then be necessary . . . This means that the actual state of affairs qRt is *a positive, contingent state of affairs*. This being so, it is clear that qRt cannot make it true that q explains t. For to explain t, q must explain why there are positive, contingent states of affairs—and clearly q cannot serve this explanatory role by virtue of standing in relation R to t, if the fact that q stands in relation R to t is itself a positive, contingent state of affairs.[52]

It follows (according to Rowe) that t has no explanation and that PSR_1 is false. It also follows that III is false, for then there would be something, the state of affairs t, that possessed a property (namely, the property of obtaining) for no reason.

Rowe's argument is sound only if 'entails' means 'wholly entails' rather than 'wholly or partly entails'. 'qRt partly entails t' may be defined as follows: qRt partly entails t if and only if qRt makes the probability of t greater than $\frac{1}{2}$ and less than 1. qRt wholly entails t if and only if qRt makes t probable to degree 1. If qRt partly entails t, then qRt could be necessary without t being necessary. Suppose

[51] William Rowe, *The Cosmological Argument* (Princeton, NJ: Princeton University Press, 1975), 103.
[52] Ibid. 105.

q = a wave function law,
t = there are contingent state of affairs,
R = makes probable to degree 0.9,
qRt = the wave function law makes probable to degree 0.9 that there are contingent states of affairs.

Now q and qRt could be necessary, consistently with t being contingent. The wave function law could be true in all possible worlds and it could be true in all possible worlds that the wave function makes probable to degree 0.9 that there be contingent states of affairs. But this very fact implies that t does not obtain in all possible worlds; it implies that t obtains only in 90 per cent of all possible worlds. Since t does not obtain in all possible worlds, t is contingent.

At the risk of belabouring this point, we may make more explicit the sort of explanation qRt provides. Suppose qRt obtains necessarily. It follows that there are some possible worlds (namely, 10 per cent of them) in which qRt obtains but t does not. In 90 per cent of all worlds, t obtains. In these worlds, t is explained by q, the wave function law. It is not explained deductively but probabilistically; that is, given q, t does not follow, but it does follow that t is 90 per cent probable. t is explained, not by virtue of being wholly entailed by its explanans, but by virtue of being partly entailed by its explanans. Thus, Rowe's refutation of PSR_1 and of III is unsound. His mistake is to suppose all explanations are deductive, whereas some are probabilistic. I believe other arguments against III (e.g. Ross's[53]) can also be refuted in this manner and thus that it is epistemically possible that III is true. III is epistemically possible in the sense that we are not justified in believing the negation of III. We are not justified in believing the negation of III if we know of no argument that refutes III. This does not show, of course, that III is true, but merely that we are not justified in believing it to be false.

If it is objected to my argument against Rowe that PSR_1 and III were called 'principles of sufficient reason' and therefore cannot be satisfied by probabilist explanations, I would respond that the title of this principle is intended to convey that the reasons mentioned

---

[53] James Ross, *Philosophical Theology* (Indianapolis: Bobbs-Merrill, 1969), 295–304.

are 'sufficient' in the sense that they are adequate or complete by themselves (i.e. 'nothing more is needed') to explain the explanandum. The title of this principle is not intended to convey (or *should not* be intended to convey, in case certain historical figures have in fact used this title in this way) that the reasons are sufficient in the sense of being deterministic causes or deductive reasons that wholly entail the explanandum. Clearly, probabilistic reasons are sufficient/adequate to explain their explananada, as has become common lore in the philosophy of science since at least the 1960s.

The establishment of the epistemic possibility of the strongest possible version of the Principle of Sufficient Reason is all I need to show that Hawking's quantum cosmology is explanatorily superior in a certain respect to theism. I shall show that it is epistemically and metaphysically impossible that theism maximally explains the universe, but epistemically possible (and perhaps metaphysically possible) that quantum cosmology maximally explains it.

In order to make this argument precise, we need some more definitions. A theory T *explains* our universe U if and only if U exists and T wholly or partly entails that there exists some universe in a set of very similar possible universes that includes among its members U. We shall say that a universe U' is *very similar to* our universe U if and only if U' is an expanding Friedman universe that has an early inflationary phase, density fluctuations in the background radiation that lead to galactic formations, a large-scale isotropy and homogeneity, and a critical density near to one, and contains planets some of which are inhabited by intelligent organisms. A theory T *maximally explains* our universe U if and only if T explains U and the proposition, *that T explains U*, entails the Principle of Sufficient Reason III.

If theism is true, then it is epistemically and metaphysically impossible that our universe U has a maximal explanation. Theism is the theory that God exists, i.e. that there exists a disembodied person that is omnipotent, omnibenevolent, omniscient, and the creator of whatever concrete objects contingently exist. Some theists, such as Swinburne, hold that God contingently exists, but we shall assume for the sake of argument Craig's brand of theism, i.e. that God necessarily exists; this assumption is the relevant one since only with this assumption is there any apparent hope that theism can provide a maximal explanation of our universe. Let us suppose, then, that it is true that God necessarily exists, i.e. that God exists in all metaphysically possible worlds. Theism as characterized above may thus be summarized by the sentence 'God exists necessarily'. This sentence does not wholly or partly entail 'There exists some universe in a set of very similar possible universes that includes among its members U'. There are many possible worlds (a world is not a universe[54]) in which God creates nothing at all and many other possible worlds in which God creates a universe very different from ours. Furthermore, there are many possible worlds in which God creates no universe but only angels, i.e. disembodied finite persons. It is false that (and no part of classical theistic theory entails that) in most of the worlds in which God exists, he creates our universe or some other expanding Friedman universe that has an early inflationary phase, density fluctuations in the background radiation that lead to galactic formations, a large-scale isotropy and homogeneity, and a critical density near to one, and contains planets some of which are inhabited by intelligent organisms. No theist has even attempted to argue this. Accordingly, we cannot say that the hypothesis that God necessarily exists *maximally explains* the universe U since this hypothesis does not *explain* U.

But suppose that 'God necessarily exists' does partly entail 'There exists some universe in a set of very similar possible universes that includes among its members U'. It may at least be supposed that this is epistemically possible. Would it then be epistemically possible that theism maximally explains our universe?

The answer must be negative, since if theism explains our universe then there is something (namely, God) that possesses some property (namely, a property God acquires by virtue of creating our universe) for no reason. Since this violates the Principle of Sufficient Reason III, it follows that theism does not maximally explain U. (Recall that a theory T *maximally explains* our universe U if and only if T explains U and the proposition *that* T *explains* U entails the Principle of Sufficient Reason III.)

This is proven as follows. If theism explains our universe, then

---

[54] A possible world is an abstract object (a proposition or state of affairs) and thereby differs from the universe, which is a concrete object, roughly a maximal spatial–temporal system. In terms of propositions, a possible world W is a maximal proposition P, such that for every proposition p, P entails p or $-p$. See Quentin Smith, 'Tensed States of Affairs and Possible Worlds', *Grazer Philosophische Studien*, 31 (1988), 225–35.

there is a certain action, God's creation of U, that is free. This action is free in the libertarian sense, i.e. no event causes it. Now some defenders of libertarianism, such as Chisholm and Taylor,[55] hold that free actions are caused, but not by events; rather they are caused by the agents (persons) who perform them. Let us call God's-standing-in-the-relation-of-causation-to-the-act-of-creating-U the state S. Is there a reason why S exists? This is distinct from the question of whether God's free action has a reason, which we have already answered in the affirmative; God's free action has a causal reason, namely, God, who causes the free action in the sense of agent causality. The question we are asking is about the reason for the state consisting of God's causing his free act of creating the universe. Is there a reason why God possesses the relational property of *standing in the relation of causation to the act of creating U*? Is there a necessitating cause, probabilistic cause, or sufficient condition for God's exemplification of this property? I think God's freedom precludes such reasons. For example, one cannot say that some desire that God possesses, such as the desire to create U, causes God to exemplify this relational property, since this would conflict with his freedom. There is no reason for S and this entails that the Principle of Sufficient Reason III is false.

However, there is an alternative version of libertarianism that may seem to avoid this objection. This version, most thoroughly developed by Stewart Goetz,[56] is that free actions are not caused by the agent but that they none the less have reasons, namely, teleological reasons. These reasons are the ends the person intends to accomplish by means of his actions. The reason for God's free action of *creating U* would be the end he intends to accomplish by this action, an end that may be generally characterized as *the end of realizing goodness*. Thus, this theory also gives us a reason for the free action of creating U, but in this case the reason is a teleological rather than causal one. However, the same problem arises as before. Let us call God's-freely-creating-U-in-order-to-realize-goodness the state S'. God's freely creating U has a reason, but is there a reason for S'? Is there a necessitating cause, probabilistic cause, sufficient

---

[55] R. Chisholm, 'Freedom and Action', in Keith Lehrer (ed.), *Freedom and Determinism* (New York: Random House, 1966), 11–12, and R. Taylor, *Action and Purpose* (Englewood Cliffs, NJ: Prentice Hall, 1966), 88–9. Chisholm and Taylor have since withdrawn their commitment to agent causation.

[56] Stewart Goetz, 'A Noncausal Theory of Agency', *Philosophy and Phenomenological Research*, 44 (1988), 303–16.

condition, teleological reason, or any other sort of reason for the existence of S'? Clearly not. Since S' exists contingently, it also does not contain its reason within itself. (Recall that we are following Rowe in using 'contains its reason within itself' to be logically equivalent to 'is metaphysically necessary'.) Thus, S' exists without reason, which again entails that the Principle of Sufficient Reason III is false.

It follows that it is epistemically and metaphysically impossible that theism maximally explains our universe. Theism fails to explain maximally our universe U since the proposition *that theism explains U* entails the negation of the Principle of Sufficient Reason III. If theism is true, the existence of the universe is absurd in the sense that the terminus of explanation of the universe's existence is a contingent state that is itself unexplained, namely, the state S or S'. But if theism is false, the universe (and by implication, human life) is not absurd, if the following holds true of Hawking's quantum cosmology.

It is epistemically possible that Hawking's quantum cosmology maximally explains the universe only if it is epistemically possible that his wave function law is necessarily true. Interestingly, the grounds for thinking this to be possible are approximately the same grounds for thinking it to be possible that his law is necessarily false. Hawking notes that 'we do not yet have a complete and consistent theory that combines quantum mechanics and gravity. However, we are fairly certain of some features that such a unified theory should have,'[57] such as Feynman's sum-over-histories method and the concept of Euclidean spacetime. This implies that Hawking's quantum cosmology is not a complete theory but merely a working model that attempts to approximate in outline form a part of such a theory. Since we do not yet have such a theory, it is epistemically possible that Hawking's specific ideas are not merely false but necessarily false because inconsistent. Chris Isham makes an even stronger sceptical claim about the 1983 Hartle–Hawking theory, that 'it is most unlikely that the theory is mathematically consistent in the form presented above. Quantum theories of gravity tend to be plagued with ill-defined expressions which are singularly difficult to remove.'[58] Isham elaborates that the 'mathematical inconsistency is the proven perturbative non-renormalisability of

[57] Hawking, *A Brief History of Time*, 133.
[58] Isham, 'Creation of the Universe as a Quantum Process', 402.

quantum gravity. True, this does not rule out a non-perturbative construction of the functional integral, but it makes it extremely unlikely; hence the great interest in superstring theory.'[59] I would add to this that Hawking has argued in 1990 that even superstring theory is non-renormalizable, which puts a complete and consistent unified theory even further out of reach. Hawking concludes that the theory 'will appear to be non-renormalizable, with an infinite number of undetermined coupling constants. This rather seems to undercut the main reason for studying strings: to get a theory of quantum gravity that didn't have an infinite number of coupling constants.'[60]

Related to the idea that extant theories in their present form are probably inconsistent is the epistemic possibility that there is only one mathematically consistent theory of a maximal physical system or universe, or, more specifically, of the gravitational-cum-quantum-mechanical structure of reality. If we know so little about the complete unified theory that we do not even know if any of the extant working models for this theory are mathematically consistent, then we do not know enough to rule out the possibility that there is only one complete unified theory that is mathematically consistent. If there is more than one complete theory that is consistent, then observational considerations will decide among them and none will be logically true. But if there is only one theory that is consistent, then it will be true of logical necessity. In this case, Wheeler's idea of a 'physics as manifestation of logic'[61] will have some meaning.

Does this show that it is epistemically possible that Hawking's wave function law is necessarily true? Perhaps a more plausible way to phrase this is to ask if it is epistemically possible that Hawking's wave function approximates in outline form a complete unified theory of the universe that is true of logical necessity and that leads to the same predictions as Hawking's cosmology, namely, that it is probable that a Hawking-universe exists. Call such a theory T. I think the above considerations about our lack of knowledge of a complete and consistent unified theory give a reason to think that it is epistemically possible that T is necessarily true.

[59] C. Isham, private communication, 5 Sept. 1992.
[60] S. W. Hawking, 'Baby Universes', *Modern Physics Letters*, A52 (1990), 145–55, esp. 154.
[61] Cf. C. Misner, K. Thorne, and J. Wheeler, *Gravitation* (New York: W. H. Freeman, 1973), 1212.

If T is necessarily true, would it maximally explain our universe *U*? Suppose T entails

(P) There is an 85 per cent probability that there exists some universe in a set of very similar possible universes that includes among its members *U*.

If T is true, *U* would then exist for a reason, namely, the probabilistic reason (P). The proposition (P)'s property of being true would then have an analogous role to God's freely-creating-*U*-in-order-to-realize-goodness, namely, that it serves to *explain U*. However, since T is necessarily true, (P) is necessarily true. Consequently, (P)'s property of being true, in contradistinction from God's property of freely-creating-*U*-in-order-to-realize-goodness, would be necessarily exemplified and thus (P)'s being true would have a reason, namely, it would contain its reason within itself (i.e. would be metaphysically necessary). In this case, the terminus of explanation of the universe's existence would not be a contingent and unexplained state, but a state that contains its reason within itself (this state being (P)'s exemplification of truth). Therefore, we would have the further difference from the theistic case that the proposition *that P explains U* would not entail the negation of the Principle of Sufficient Reason III.

However, more must be said to show that the Principle of Sufficient Reason III would indeed be true in this case. For example, it would have to be the case that the *compatibilist* rather than libertarian theory of free will is the true one, since if the latter were true then the free decisions of humans would ultimately be without reason and therefore there would be something, namely, some human, that possessed some property, namely, freely acting in order to achieve an end *E*, for no reason. Given compatibilism, every human action has a causal reason. Given the current state of controversy between defenders of compatibilism and defenders of libertarianism, it seems reasonable to think that it is epistemically possible that compatibilism is true.

Furthermore, we would have to reject Hawking's conceptualism, i.e. that scientific theories exist only in the human mind, for in that case the theory T would be dependent for its existence upon the contingent existence of human minds. Hawking writes that 'a scientific theory . . . exists only in our minds and does not have any other reality (whatever that might mean)'.[62] In this case, 'T is necessarily true' would be equivalent to an assertion such as 'T is true if and only if T exists and T exists if and only if there exists humans who are conceiving T'. Given this, the proposition *that T exists* would be contingently true and thus we would be left with a contingent fact as the terminus of explanation. But if we assume a realist (Platonist) theory of propositions, then we may regard T's existence as necessary. Given the lack of any definitive knowledge or consensus about conceptualism versus Platonism, I think it is safe to assume that it is epistemically possible that Platonism is true.

Given the specifics of my refutation of Rowe's argument against the strongest possible version of the Principle of Sufficient Reason III, as well as the plausible theses that it is epistemically possible that there is only one mathematically consistent cosmological theory, that compatibilism is true and that Platonism is true, it follows that there is reason to believe that it is epistemically possible that a quantum cosmology of the sort that Hawking's working model approximates provides a maximal explanation of our universe. By implication, it is epistemically possible that human life is not ultimately absurd (without reason). Since we have seen that it is epistemically impossible that theism maximally explains our universe, it follows that in this respect quantum cosmology has superior explanatory value to theism.

Now Craig is likely to respond that this argument is unsound since it is epistemically impossible for the universe to begin uncaused and thus epistemically impossible that a quantum cosmology of Hawking's sort is true. The wave function of the universe is purported to be an acausal explanation of the universe, but acausal explanations of the universe (Craig might argue) are epistemically impossible explanations. The only possibility is a causal explanation, specifically, the theistic explanation. This in effect reverts to the question of whether the causal principle

(C) Necessarily, everything that begins to exist has a cause

can be conceived to be false. Craig argues (C) is intuitively obvious (and thus epistemically necessary) but I have argued (in Essay VI) it is neither self-evident nor justifiable by any other means. Since the arguments of Craig and myself have already been laid out, the

---

[62] Hawking, *A Brief History of Time*, 9.

reader will have to decide for himself or herself which argument is more plausible and thus whether theism has superior or inferior explanatory value to quantum cosmology.

I should like to close this discussion by noting a similar propositional attitude that underlies the different beliefs defended by Craig and myself in this book. There is an underlying agreement in attitude that motivates Craig's and my various efforts to fathom the universe's existence, namely, a wonder or awe that there is not nothingness and that the universe (to borrow a phrase from Hawking[63]) goes 'to all the bother of existing'. I will end with a quote from Craig about the emotion that we agree underlies the enterprise of this book: 'I want to underline the fact that I in no wise denigrate Smith's profound astonishment, which he poetically expresses, that the universe exists at all—on the contrary, I feel it, too.'[64] Craig adds that this 'astonishment should not end in a mute stupefaction, but lead us . . . to the intelligible explanation of the universe'. The considerations adduced in Part 3 suggest that we may agree on this point as well, with the difference between us coming down to the question: Is the intelligible explanation of the universe causal or acausal?

[63] Ibid. 174.   [64] Essay VIII, Sect. 2.

Index of Selected Names

Index of Subjects